# Evolution Education Re-considered

Ute Harms · Michael J. Reiss
Editors

# Evolution Education Re-considered

## Understanding What Works

Springer

*Editors*
Ute Harms
IPN - Leibniz Institute for Science and Mathematics Education at Kiel University
Kiel, Germany

Michael J. Reiss
Institute of Education
University College London
London, UK

ISBN 978-3-030-14697-9 ISBN 978-3-030-14698-6 (eBook)
https://doi.org/10.1007/978-3-030-14698-6

Library of Congress Control Number: 2019934529



This Springer imprint is published by the registered company Springer Nature Switzerland AG.
The registered company address is: Gewerbestrasse 11, 6330 Cham, Switzerland

*To Uwe Bertsch and David Ogden*

# Foreword

"Another book on evolution? Aren't there many out there already?" you may ask. Indeed, there are. Nevertheless, the public understanding and acceptance of evolutionary theory are relatively low, at least compared to other, equally important and well-established scientific theories. Therefore, there must be more we need to do to improve the situation. Evolution is perhaps a special case because teachers have to face two different kinds of obstacles: conceptual obstacles, which exist for all scientific theories, and emotional obstacles, which do not usually exist for other scientific theories. Whereas learning about gravity or atomic theory does not usually raise any personal concerns, learning about evolution touches upon important aspects of our identity: Who we are, where we come from, and where we go. Thus, there are emotional obstacles that may make an evolutionary theory to seem to be in conflict with one's worldviews. Furthermore, evolutionary theory is also counterintuitive: The inherent unpredictability and contingency of evolutionary outcomes seem to conflict with our design stance that makes us look for plan and purpose in nature.

So, what do we need? We certainly need teachers who have understood evolution and who feel confident in teaching it. But we also need to find effective ways of teaching evolution, ones that will promote our students' conceptual understanding without making them feel that their worldviews are threatened. We need to find ways to motivate our students to understand this counterintuitive theory that provides the best explanation for both the unity and the diversity of life. In principle, this might seem simple because there is already a lot of research on students' preconceptions about evolution. One might thus expect that given this research, effective teaching sequences that would address them could be easily designed. Well, design is one thing, and implementation is another; sometimes, they can lie far apart from each other. Therefore, the time has come to: (1) investigate under controlled conditions what factors might cause students' misunderstandings about evolution (e.g., threshold concepts like randomness and probability); (2) design teaching sequences that explicitly address well-known preconceptions; and (3) investigate the conditions for successful implementation of such teaching sequences.

This is what makes the contributions to this book very important, and the book itself is extremely useful. The editors have produced a fine compilation of empirical studies on teaching and learning evolution for learners of different ages and from different countries in the world. No matter what strategies we have devised in order to address learners' conceptions and promote conceptual change in evolution, it is only when these strategies are applied in classrooms, museums or elsewhere that one can really see if they are effective or not. Empirical research in such places is the cornerstone for improving the teaching of evolution. If a strategy works well, we can use it; but even if it fails terribly, there are still useful lessons to learn when we realize what went wrong and why.

Of special importance are those studies that focus on the teaching of evolution, or of evolution-related concepts, at the elementary/primary school. Like others, I have been arguing for years that perhaps it is too late to address students' preconceptions about evolution at the secondary school. Rather, what we need is to start challenging the design stance and gradually building an evolutionary habit of mind from very young ages. This is far from simple and straightforward, and this is why the related empirical research is valuable. In addition to such studies, the present book also includes topics that have not been investigated in depth such as students' understanding of probabilities, a skill that generally is not taught at schools and the lack of which deprives students from understanding several domains of contemporary science. And there is a lot more in the present book that readers will appreciate.

I will never get bored in quoting Theodosius Dobzhansky who wrote that without evolution biology is a collection of sundry facts that make no meaningful picture as a whole. Nothing in biology makes sense except in light of evolution he noted, and he was right. The present book contains valuable empirical studies on teaching and learning about evolution, the central unifying theory of biology. I expect it should be a very valuable resource *for* years to come.

Geneva, Switzerland

Kostas Kampourakis
University of Geneva
Editor-in-Chief *Science & Education*
Editor-in-Chief *Science: Philosophy, History and Education*

# Contents

# Editors and Contributors

## About the Editors

**Ute Harms** is Director at the IPN - Leibniz Institute for science and mathematics education, Full Professor for Biology Education at the University of Kiel (Germany) since 2007, and Fellow of the Royal Society of Biology (Great Britain). She has a Ph.D. in cell biology and has worked as a high school teacher for several years. In 2000, she got her first Professorship for Biology Education at the Ludwig-Maximilians-University in Munich (Germany). From 2006 to 2007, she held a chair in Biology Education at the University of Bremen. Her main research interests are conceptual learning in biology and in science, focusing on evolution and energy, biology teacher education, biology-related competitions and transfer of contemporary topics in the life sciences to the public.

**Michael J. Reiss** is Professor of Science Education at UCL Institute of Education, University College London, Visiting Professor at the universities of York and Kiel and the Royal Veterinary College, Honorary Fellow of the British Science Association, Docent at the University of Helsinki and Fellow of the Academy of Social Sciences. After undertaking a Ph.D. and postdoctoral research in evolutionary biology and population genetics, he trained to be Science Teacher and taught in schools for five years before returning to higher education. The former Director of Education at the Royal Society, his academic interests are in science education, bioethics and sex education and he has published widely on issues to do with creationism in schools.

## Contributors

**Manzoorul Abedin** Faculty of Education, Canterbury Christ Church University, Canterbury, UK

**Heidemarie Amon** Universität Wien ZLB – Didaktik der Naturwissenschaften Österreichisches Kompetenzzentrum für Didaktik der Biologie – AECC-Bio, Vienna, Austria

**Trevor R. Anderson** Purdue University, West Lafayette, USA

**Julia Arnold** School of Education, University of Applied Sciences and Arts Northwestern Switzerland (FHNW), Basel, Switzerland

**Helene Bauer** Universität Wien ZLB – Didaktik der Naturwissenschaften Österreichisches Kompetenzzentrum für Didaktik der Biologie – AECC-Bio, Vienna, Austria

**Paul M. Beardsley** Department of Biological Sciences, Cal Poly Pomona, Pomona, CA, USA

**Constance M. Bertka** Science and Society Resources, LLC, Potomac, MD, USA

**Berry Billingsley** Faculty of Education, Canterbury Christ Church University, Canterbury, UK

**Franz X. Bogner** Department of Educational Biology, Centre of Math and Science Education, University Bayreuth, Z-MNU, Bayreuth, Germany

**Loredana L. Buchan** Department of Biology and Biochemistry, The Milner Centre for Evolution, University of Bath, Bath, UK

**Alexandra Buck** Department of Educational Biology, Centre of Math and Science Education, University Bayreuth, Z-MNU, Bayreuth, Germany

**Keith Chappell** Faculty of Education, Canterbury Christ Church University, Canterbury, UK

**Paul Davies** Queen's College, London, England, UK

**Daniela Fiedler** Department of Biology Education, IPN - Leibniz Institute for Science and Mathematics Education, Kiel, Germany

**Alejandra García Franco** Universidad Autónoma Metropolitana—Cuajimalpa, Mexico City, Mexico

**A. A. Gómez Galindo** Unidad Monterrey, Cinvestav, Monterrey, Mexico

**Leonardo Gonzáles Galli** Consejo Nacional de Investigaciones Científicas y Técnicas, Buenos Aires, Argentina

**Timothy A. Goodale** Elizabeth City State University, Elizabeth City, NC, USA

**Jorge Groß** Science Education, University of Bamberg, Bamberg, Germany

**Ute Harms** Department of Biology Education, IPN - Leibniz Institute for Science and Mathematics Education, Kiel, Germany

**Chris Hatcher** Loughborough University, Loughborough, UK

**Momna V. Hejmadi** Department of Biology and Biochemistry, The Milner Centre for Evolution, University of Bath, Bath, UK

**Laurence D. Hurst** Department of Biology and Biochemistry, The Milner Centre for Evolution, University of Bath, Bath, UK

**Lisa O. Kenyon** Departments of Biological Sciences and Teacher Education, Wright State University, Dayton, OH, USA

**Yi Kong** College of Education, Fujian Normal University, Fuzhou, Fujian, China

**Kerstin Kremer** Leibniz University of Hanover, Hanover, Germany

**Kathy L. Malone** Nazarbayev University, Astana, Kazakhstan

**Linda McGuigan** University of Liverpool, Liverpool, England, UK

**Jaimie L. Miller-Friedmann** Department of Education, University of Oxford, Oxford, UK

**Jo Nicholl** UCL Institute of Education, London, England, UK

**Jeffrey T. Olimpo** The University of Texas at El Paso, El Paso, USA

**David C. Owens** Georgia Southern University, Savannah, GA, USA

**Nancy Pelaez** Purdue University, West Lafayette, USA

**Briana Pobiner** Human Origins Program, National Museum of Natural History, Smithsonian Institution, Washington, DC, USA

**Michael J. Reiss** UCL Institute of Education, University College London, London, UK

**William L. Romine** Departments of Biological Sciences and Teacher Education, Wright State University, Dayton, OH, USA

**Terry Russell** University of Liverpool, Liverpool, England, UK

**Zakee Sabree** The Ohio State University, Columbus, OH, USA

**Philip M. Sadler** Harvard University, Cambridge, MA, USA

**Jaqueline Scheibstock** Universität Wien ZLB – Didaktik der Naturwissenschaften Österreichisches Kompetenzzentrum für Didaktik der Biologie – AECC-Bio, Vienna, Austria

**Martin Scheuch** Universität Wien ZLB – Didaktik der Naturwissenschaften Österreichisches Kompetenzzentrum für Didaktik der Biologie – AECC-Bio, Vienna, Austria

**Anita M. Schuchardt** University of Minnesota—Twin Cities, Minneapolis, MN, USA

**Sofoklis Sotiriou** Ellinogermaniki Agogi, R&D, Pallini, Athens, Greece

**Susan E. Sunbury** Smithsonian Astrophysical Observatory, Cambridge, MA, USA

**José de la Cruz Torres Frías** UdeG, Guadalajara, Mexico

**Emily M. Walter** Science and Mathematics Education Center, California State University, Fresno, CA, USA

**William A. Watson** Learning Research Across the Divide, Haddon Heights, NJ, USA

# The Present Status of Evolution Education

Ute Harms and Michael J. Reiss

## 1 Evolution—The Core Line of Biology

Evolution through natural selection is a central, unifying and overarching theme in biology. Evolutionary theory is the integrative framework of modern biology and provides explanations for similarities and adaptive differences among organisms, biological diversity, and many features and processes of the physical world. It is also applied in numerous other fields, both biological (e.g. agriculture and medicine) and, increasingly, non-biological (e.g. economics and computer science), though its use in these other fields is contentious and is not considered further here.

The essential tenets of evolutionary theory have long been regarded as key parts of the foundations of science education (e.g. Beardsley, 2004; Bishop & Anderson, 1990; Nehm and Reilly, 2007; Speth et al., 2014). Accordingly, the American Association for the Advancement of Science (AAAS, 2006), the Next Generation Science Standards (NGSS, 2013), the National Education Standards of Germany (Secretariat of the Standing Conference of the Ministers of Education and Cultural Affairs of the Länder in the Federal Republic of Germany [KMK], 2005), the National Curriculum for England (DfE, 2014), as well as the official documentation of many other countries, all describe evolution as an organising principle for biological science and include the topic as a learning goal.

Although evolutionary processes may occur (and be applied) in numerous kinds of systems, unless specified otherwise, evolution generally refers to changes in populations or taxa of organisms due to the generation of variation and natural selection (Gregory, 2009). There is a massive empirical body of work on natural selection, myr-

U. Harms (✉)
IPN - Leibniz Institute for Science and Mathematics Education, Biology Education, Olshausenstr. 62, 24118 Kiel, Germany
e-mail: harms@ipn.uni-kiel.de

M. J. Reiss
UCL Institute of Education, University College London, London, UK
e-mail: m.reiss@ucl.ac.uk



U. Harms and M. J. Reiss (eds.), *Evolution Education Re-considered*,
https://doi.org/10.1007/978-3-030-14698-6_1

iads of processes involved have been elucidated, and extensive terminology has been developed (e.g. Rector, Nehm, & Pearl, 2013). Nevertheless, biologists generally agree that three principles are necessary and sufficient for explaining evolutionary change by natural selection: (1) the generation of variation, (2) heritability of variation and (3) differential survival and/or reproduction of individuals with differing heritable traits (e.g. Gregory, 2009). However, evolutionary change is still poorly understood by students throughout their time in education (Nehm & Reilly, 2007; Shtulman, 2006; Spindler & Doherty, 2009), science teachers (e.g. Nehm et al., 2009), and the general public (Evans et al., 2010). This poor understanding has been attributed to diverse cognitive, epistemological, religious and emotional factors (for an overview see Rosengren, Brem, Evans, & Sinatra, 2012) that evidently evolution education is generally not successfully coping with. Against this background, this chapter will provide an overview of the status of evolution education considering the three central aspects of education: (1) the students, (2) the teachers and (3) the teaching (including the curriculum).

## 2 Students' Understanding of Evolution—What Do We Know?

For decades, scholars in biology education all over the world have investigated students' understandings of various evolutionary concepts (e.g. selection, adaptation). The result today is a substantial body of literature and knowledge of respective misconceptions. This knowledge is an invaluable treasure for further research on evolution education as it reveals obstacles that affect or specifically hinder students' learning of evolution and also elucidates links for fostering evolution understanding. Therefore, initially, we will highlight the main findings on students' conceptions of evolution, addressing the main categories of misconceptions (according to Gregory 2009, complemented by Neubrand and Harms 2017).

### *2.1 Selection and Adaptation*

Frequent misconceptions of students are apparent in Lamarckian, teleological (finalistic) and anthropomorphic explanations of the mechanism of evolution. In the Lamarckian understanding, features of an individual that it acquires during its lifetime are passed onto its offspring (Kampourakis, 2014). These conceptions are similar to the widespread teleological conceptions that describe changes as being goal- or purpose-oriented. New features develop because they are advantageous. The process is directed by a creator or by the organism itself and this process has a natural end rather than being permanently ongoing, as it is when evolution is understood scientifically. Thus, the significance of randomness and probability that trigger mutation and selection, respectively, are not fully appreciated.

Anthropomorphic conceptions can be seen as a particular version of teleological conceptions. Here also, evolutionary change is seen as being steered by the organism itself and, simultaneously, human characteristics are ascribed to the organism no matter whether it is a non-human animal, a plant or a prokaryote. The way that evolution is described in textbooks, both in terms of the wording and formulations, often even supports the anthropomorphic way of explaining evolutionary processes (cf. Alters & Nelson, 2002; Bishop & Anderson, 1990; Nehm, Rector, & Ha, 2010). The large number of studies (Table 1) that have elucidated such misconceptions show the failure to appreciate the clear-cut distinction between the development of features that appeared by chance (e.g. mutation) and prevail because they fitted better to the environment than others (selection), on the one hand (scientific explanation), and a deliberate development of such features by the organism(s) themselves—because they serve a particular function better—on the other hand (misconception), severely hinders the understanding of evolution.

Another difficulty for understanding evolution is the distinction between the individual and the population level. The mechanism of selection affects the individual and its interdependency with the environment. Genetic variability leads to different phenotypes and the individuals of one population often show small differences in morphology, physiology and behaviour from those in other populations of the same species. Often the significance of this kind of variation is not appreciated (Bishop & Anderson, 1990; Brumby, 1979, 1981, 1984; Evans et al., 2005; Gelman, 2004; Shtulman, 2006; Strevens, 2000). But the variability among the individuals in a population is one essential precondition for the process of natural selection. Without variation, the chances for survival and reproduction are the same (as far as natural selection is concerned) for all individuals. Genetic variation as the result of recombination and mutation leads, over the generations, to individuals in a population that are better adapted to the environment than others. Of course, evolution takes place at the population not the individual level (Kutschera, 2006). Only when considering a series of generations, can evolution be observed as changes in the frequency of variants (Campbell & Reece, 2006, p. 513). Very often, students do not realise the meaning of the population in this context. This leads to the misconception that adaptation occurs at the individual level (Brumby, 1979, 1981, 1984; Jimenez-Aleixandre & Fernández-Pérez, 1987).

Another resistant misconception is when learners regard adaptation as a final status or an event that, having started, then comes to an end (Baalman et al., 2004; Brumby, 1979, 1981, 1984; Ferarri & Chi, 1998; Sinatra et al., 2008). Such learners do not understand the process character of adaptation. This notion of the completion of adaptation is already implicit in teleological, anthropomorphic and essentialist conceptions. Whilst misconceptions about adaptation and selection in populations have been investigated thoroughly over the years, the origin of new taxonomic groups as the result of cumulative changes over huge periods of time, i.e. macroevolution, is less researched. However, so far as is known, the misconceptions about macroevolution primarily address the processes of speciation and aspects of phylogeny.

**Table 1** Overview on students' misconceptions of evolution (according to Gregory, 2009, Neubrand and Harms, 2017)

| Concept(s) | Misconception | Description |
|---|---|---|
| Selection and adaptation | | |
| Inheritance | Lamarckian conceptions | Living beings change by active adaptation. These changes are passed on to their progeny |
| | Andrews, Kalinowski, and Leonard (2011), Baalmann, Frerichs, Weitzel, Gropengießer, and Kattmann (2004), Bizzo (1994), Brumby (1979, 1981, 1984), Deadman and Kelly (1978), Demastes, Settlage and Good (1995), Ferrari and Chi (1998); Graf and Soran (2011), Kampourakis and Zogza (2008, 2009), Lammert (2012), Nehm et al. (2009), Nehm and Reilly (2007), Nehm and Schonfeld (2007, 2008), Prinou, Halkia and Skordoulis (2008), Settlage (1994) | |
| Intentionality | Teleological conceptions | Changes arise that are purpose- and goal-directed |
| | Andrews et al. (2011), Baalmann et al. (2004), Beardsley (2004), Bishop and Anderson (1990), Brumby (1979, 1981, 1984), Deadman and Kelly (1978), Engel Clough and Wood-Robinson (1985), Evans, Szymanowski, Smith, and Rosengren (2005), Flanagan and Roseman (2011), Greene (1990), Jensen and Finley (1995, 1996), Jimenez-Aleixandre (1992), Jimenez-Aleixandre and Fernández-Pérez (1987), Johannsen and Krüger (2005), Kampourakis and Zogza (2008, 2009), Kampourakis, Pavlidi, Papadopoulou, and Palaiokrassa (2012), Lammert (2012), MacFadden et al. (2007), Nehm et al. (2009), Nehm and Reilly (2007), Nehm and Schonfeld (2007, 2008), Pedersen and Hallden (1994), Prinou et al. (2008), Settlage (1994), Sinatra, Brem, and Evans (2008), Southerland, Abrams, Cummins, and Anzlmo (2001), Tamir and Zohar (1991), van Dijk and Kattmann (2010), Weitzel and Gropengießer (2009) | |
| | Anthropomorphic conceptions | Transfer of human features to non-human animals and plants. Changes are the result of purposeful and goal-directed action provoked by maladaptation |

(continued)

**Table 1** (continued)

| Concept(s) | Misconception | Description |
|---|---|---|
| | Baalmann et al. (2004), Demastes et al. (1995), Engel Clough and Wood-Robinson (1985), Jimenez-Aleixandre and Fernández-Pérez (1987), Johannsen and Krüger (2005), Jungwirth (1975), Tamir and Zohar (1991) | |
| Individual versus population | | |
| | Essentialistic conceptions | The 'type' and the commonalities of individuals are crucial for evolutionary processes |
| | Alters (2005), Andersson and Wallin (2006), Andrews et al. (2011), Bardapurkar (2008), Brumby (1979, 1981, 1984), Evans et al. (2005), Gelman (2004), Greene (1990), Halldén (1988), Jimenez-Aleixandre (1992), Shtulman (2006), Spindler and Doherty (2009), Strevens (2000) | |
| | Individualisation | Adaptation happens at the individual, not the population level |
| | Brumby (1979, 1981, 1984), Halldén (1988), Jimenez-Aleixandre and Fernández-Pérez (1987) | |
| Insularity | State/event instead of process | Adaptation is not a dynamic process |
| | Baalman et al. (2004); Brumby (1979, 1981, 1984), Chi, Kristensen and Roscoe (2012), Ferarri and Chi (1998), Sinatra et al. (2008) | |
| Speciation | | |
| | Spontaneous speciation | The origin of species is not a dynamic process |
| | Evans (2000), Samarapungavan and Wiers (1997) | |
| | Creationism | All living beings have been created simultaneously and separately by God (creationism) |
| | Berti, Toneatti and Rosati (2010), Evans (2000), Großschedl, Konnemann and Basel (2014); Illner (2000) | |
| Phylogeny | | |
| Deep time | | Deep time is not understood |
| | Graf and Hamdorf (2011), van Dijk and Kattmann (2009) | |
| Taxonomy | | Relatedness and its representation do not depict the principle of the last common ancestor |
| | Baum, DeWitt, and Donovan (2005), Catley, Phillips, and Novick (2013), Gregory (2008), Meikle and Scott (2010), Meir, Perry, Herron, and Kingsolver (2007), Novick and Catley (2006), Novick, Schreiber, and Catley (2014), Phillips, Novick, Catley, and Funk (2012) | |

## 2.2 Speciation

The basic mechanism of macroevolution is the process of speciation. When reproductive barriers arise as a consequence of genetic divergence, new species emerge. Typically, speciation occurs from the accumulation of adaptation and selection processes over many generations (though certain events, e.g. chromosome mutations such as polyploidy, can cause such reproductive barriers to be set in motion very rapidly, even in a single generation). Thus, misconceptions about speciation can result from students' explanations about adaptation and selection. Additionally, creationist conceptions that ascribe speciation to a higher entity are important for significant numbers of students and in more countries than is sometimes realised (Reiss, 2011).

## 2.3 Phylogeny

One further obstacle to grasping the history of life is an adequate understanding of huge extents of time, i.e. some four billion years (Graf & Hamdorf, 2011, p. 32; McVaugh et al., 2011). Understanding this so-called deep time comes up against the limitations of human imagination (Gould, 1992, p. 15). This is mirrored by misconceptions of students. They typically show severe problems in ordering evolutionary events in time (Catley & Novick, 2009; Trend, 2001). The comprehension of deep time affects the understanding of the cumulative development of living beings and consequently of the dynamic of the processes of adaptation (cf. van Dijk & Kattmann, 2010). Another aspect of the concept of phylogeny is the classification of species in taxonomic groups. Cladograms visualise family trees of organisms. Students tend to misinterpret these (Baum et al., 2005; Catley et al., 2013; Gregory, 2008; Meir et al., 2007; Novick & Catley, 2006; Novick et al., 2014; Phillips et al., 2012). In particular, they do not understand the meaning of the last common ancestor (Meikle & Scott, 2010). From this stems, for example, the widespread misconception that humans come from one of the species of apes that is found today.

## 2.4 Genetics, Randomness and Probability, Dimensionality

The described patterns of explanation (Lamarckian, teleological and anthropomorphic) at the phenotype level also appear at the molecular level. Students not infrequently argue that genes become dominant because they are useful to the individual, that genetic information can intentionally be changed for the purpose of adaptation and that this change is carried over to the next generation (Baalmann et al., 2004; Brumby, 1979, 1981, 1984). Accordingly, students do not consider mutation and recombination as random processes (Fiedler, Tröbst, & Harms, 2017; Johannsen & Krüger, 2005; Nehm & Schonfeld, 2007; Robson & Burns, 2011). Furthermore, they expect randomness and processes that rely on probability to be, in the main, inefficient and pointless (Garvin-Doxas & Klymkowsky, 2008).

To understand evolution requires consideration of concepts and principles at different levels of organisation (micro, meso and macro), and this has been shown to be very difficult for students (Ferrari & Chi, 1998; Niebert & Gropengießer, 2015). The processes that make up evolution take place over time periods from the order of seconds (or even more briefly) at the molecular level (e.g. mutation) up to millions of years, regarding the origin of new taxonomical groups at the level of species and above. To understand scales of time and space and be able to apply this knowledge to evolution appropriately are important preconditions for comprehending the theory of evolution, a comprehension that many students don't achieve.

Students' conceptions constitute the starting point for teaching. However, successful education first requires by teachers an adequate understanding of the relevant scientific concepts and information. Therefore, in the next section, we discuss findings on (pre-service) teachers' knowledge of evolution and also their acceptance of the theory of evolution, as we know that knowledge and acceptance of evolution affect each other mutually. Sound subject matter knowledge generally comes with a high acceptance of evolution theory (Akyol et al., 2012; Athanasiou and Papadopoulou, 2012; Deniz et al., 2008; Ha et al., 2012), though some students with sound subject matter knowledge actively reject evolution theory, typically on religious grounds, and also with the willingness of teachers to integrate this topic extensively in their teaching (e.g. Großschedl et al., 2014; Nehm and Schonfeld, 2007).

## 3 Teachers' and Pre-service Teachers' Knowledge and Acceptance of Evolution

There is empirical evidence that university students—even those majoring in science—have problems understanding evolution-related topics. They show comparable Lamarckian, teleological and anthropomorphic misconceptions to those shown by school students (e.g., Gregory, 2009). In a study with 552 high school biology teachers, Rutledge and Warden (2000) found only little knowledge of basic evolutionary concepts. Yates and Marek (2014) tested biology teachers in Oklahoma and showed that the lack of subject matter knowledge by the teachers was a reason for the development of students' misconceptions about evolution; they found that some students even showed poorer knowledge about evolution after the teaching than before. It became clear that the teachers' subject-related competence, independent of their personal university degree, was higher when evolution had played a central role in their study programme. Also, the particular biological content seems to trigger the difficulty to solve problems in evolution. Nehm and Ha (2011), Opfer et al. (2012) showed that college students have fewer problems answering questions on the acquisition than on the loss of features during evolution (e.g. the evolution of webbed feet in ducks, and the loss of the ability to fly in the evolution of penguins, respectively).

The acceptance of the theory of evolution plays an important role for (pre-service) teachers' abilities to teach evolution. Rutledge and Warden (2000) describe acceptance of evolution as 'perceptions of evolutionary theory's scientific validity, ability to explain phenomena, and acceptance within the scientific community' (pp. 13–14). Religious and epistemological beliefs, reflecting the capacity or willingness to consider opposing arguments, seem to affect the acceptance of evolution theory (Deniz et al., 2008). According to the Model of Conceptual Ecology (cf. Deniz et al., 2008), three factor categories can be distinguished: cognitive, affective and contextual ones. For an overview on the literature concerning the factors influencing the acceptance of the theory of evolution, see Großschedl et al. (2014). The most important cognitive factors are the understanding of the theory itself (Akyol et al., 2012; Athanasiou & Papadopoulou, 2012; Deniz et al., 2008; Ha et al., 2012), the understanding of the nature of science (Athanasiou & Papadopoulou, 2012) and knowledge of genetics (Miller et al., 2006). The most relevant affective factors are religious beliefs and personal attitude towards science (Athanasiou & Papadopoulou, 2012; Graf & Soran, 2011; Losh & Nzekwe, 2011; Miller et al., 2006). In addition, gender and academic degree may be the predictive factors for the acceptance of evolutionary theory (Losh & Nzekwe, 2011).

Taking into account the findings from the empirical studies sketched here, it seems clear that to enable future and in service biology teachers to teach evolution well, teacher education should address cognitive, affective and contextual aspects. However, how to do this is still frequently a question for science education research to elaborate.

## 4 Teaching Evolution

One challenge for evolution teaching is its deceptive appearance: the central statements of the theory of evolution can be described in a few sentences and this can give a false impression that the theory is easy to understand. Only on closer examination does its complexity come to the fore. This contrast between superficial facility and masked difficulty can lead to an illusion of understanding (Monod, 1997, p. 390, cited in Graf and Hamdorf, 2011, p. 28) that is uncovered by the various misconceptions sketched above. These misconceptions are often very resistant against instruction (Beardsley, 2004; Ferrari & Chi, 1998; Gregory, 2009; Jensen & Finley, 1995; Kampourakis & Zogza, 2008; Nehm & Reilly, 2007; Spindler & Doherty, 2009). Thus, for teaching evolutionary concepts, educational approaches seem to be reasonable that consider noted misconceptions, making these explicit for students and offering tools to alter these towards a plausible conception (cf. 'conceptual change' according to Posner et al., 1982; 'model of educational reconstruction' according to Kattmann et al., 1997). Several authors recommend this procedure (e.g. Abraham et al., 2009; Grant, 2009; Kalinowski et al., 2010; Kattmann, 2005; Meikle & Scott, 2010; Robbins & Roy, 2007).

Besides this general approach to misconceptions in education, some authors recommend particular approaches for addressing misconceptions about evolution. These approaches include consideration of structural requirements as well as of content-related goal settings. It seems to be widely agreed that evolution education pictures the integrating character of evolution biology. In other words, evolution should not be taught as a distinct topic—like cell biology or physiology can be—but as a/the core principle throughout biology education (Kattmann, 1995; Harms et al., 2004; Nehm et al., 2009; van Dijk & Kattmann, 2010). In this context, Kalinowski et al. (2010) stress the necessity to interrelate genetics and evolution (on higher levels) as a deeper understanding of genetics requires evolution knowledge and vice versa. The authors assume that in this way, many difficulties in teaching and learning evolution could be prevented. To teach evolution as a core principle throughout biology education in the course of schooling is dependent on structural regulations like school curricula. In many countries, evolution is described in the biology curriculum as one topic amongs many others. However, in Germany, currently some *Länder* (e.g. Schleswig-Holstein and Lower-Saxony) define evolution as a core principle throughout biology education at school secondary level. At the moment, it is still an open empirical question whether this approach will foster a better scientific understanding of evolution. Another aspect to be considered is when to begin evolution education in schooling. Campos and Sá-Pinto (2014) call for an early beginning. There is empirical evidence that even very young children (elementary level) are able to grasp correct conceptions about evolution (Catley, Lehrer & Reiser, 2005; Nadelson et al., 2009). However, in many countries, evolution education does not start before middle or upper secondary level. Regarding time for teaching, the duration of time needed for learners to develop a correct understanding—especially when misconceptions already exist—is an open question (e.g. Beardsley, 2004; Demastes et al., 1995).

With respect to the content of evolution, many scholars stress the macroevolutionary aspects in evolution teaching (Novick et al., 2014; Phillips et al., 2012; van Dijk & Kattmann, 2009), focus on random mechanisms like genetic drift (Beggrow & Nehm, 2012), and apply a relative time concept rather than teaching absolute time frames and exact dates (Trend, 2001; van Dijk & Kattmann, 2010). One focal point of recommendations addresses the concept of natural selection, a key concept that is fundamental for the understanding of evolution (Gregory, 2009). Several authors propose different key concepts that should structure evolution understanding. Mayr (1982) describes seven key concepts; Anderson et al. (2002) differentiate the theory of evolution into ten basic ideas, though more recent authors reduce these to three basic principles: variation, inheritance and selection (McVaugh et al., 2011; Nehm et al., 2012; Nehm & Schonfeld, 2010; Tibell & Harms, 2017). Another perspective on evolution education arises from the discussion on so-called threshold concepts (Meyer & Land, 2005). Meyer and Land (2006) proposed a further approach to explain the learning of complex concepts like natural selection and evolution that are abstract, rather than concrete in nature. They defined these threshold concepts metaphorically as portals that, once passed though by a learner, open up new, previously inaccessible, ways to develop knowledge.

Conceptual change theory and the threshold concept model jointly imply that knowledge of core abstract concepts, the 'thresholds', could be essential for the conceptual change required to gain conceptual knowledge of a particular content. In this respect, evolutionary theory can be regarded as resting on a conglomerate of several threshold concepts, including randomness, probability, temporal scales and spatial scales (Ross et al., 2010), that must be understood in order to understand evolution generally and natural selection specifically. For the learner, this opens up new ways of thinking that were not previously possible, and enables new extended understandings of subject matter. Whether evolution understanding will improve when considering these threshold concepts in teaching is still an open question (cf. Fiedler et al., 2017).

In summary, to characterise the present situation of evolution education, there is surprisingly little empirical evidence on how to foster evolution understanding across the phases of education. We know that students, teachers and the public hold a wide range of resistant misconceptions on evolution but we have little knowledge on educational approaches that can successfully change this situation. To acquire such knowledge, intervention studies are needed that give evidence for educational methods and procedures that support a scientifically correct understanding of evolution.

## 5 The Studies in This Book

Against this background, this book presents a collection of studies that investigate a variety of tools to foster students' understanding of evolution. We begin with several studies undertaken in primary (elementary) classrooms. Such work is of particular significance given that some countries have now made evolution a part of the primary curriculum. First, Loredana Buchan, Momna Hejmadi and Laurence Hurst in Chap. 2 look at whether a four-lesson scheme of work (variation, natural selection, geological time lines and homology/common ancestry) can lead to increased understanding in primary and middle school students of all abilities. Then, in Chap. 3, Berry Billingsley, Manzoorul Abedin, Keith Chappell and Chris Hatcher examine pre-service teachers' perceptions of the advantages and disadvantages of a cross-curricular session in their course and also their attitudes to using cross-curricular teaching with their primary students. This cross-curricular session was designed in the light of the fact that evolution is widely seen by teachers and pre-service teachers as an area of science that is challenging to teach, with one of the reasons often given being a concern that the science may conflict with some children's religious beliefs. In Chap. 4, Terry Russell and Linda McGuigan review their research into the teaching and learning of evolution across the 5–14 age range. Their original focus was on the mandatory curricular requirements for 'Evolution and inheritance', newly introduced in England for ages 9–11. Closer engagement with teachers and primary students clarified the challenge and opportunity to take a broader, more universal, view of progression in this curricular domain. The need they perceived was to link disconnected fragments into a coherent experience of progression, reflecting the underpinning breadth, depth and interconnectedness of evolutionary theory.

Martin Scheuch, Jaqueline Scheibstock, Heidemarie Amon and Helene Bauer in Chap. 5 situate their work in the context of the Austrian school curriculum where evolution is only mentioned in grades 7 and 12. They therefore set out to develop a learning progression including grades 8, 9 and 10 to fill the gap and enable year-by-year learning of evolution. To assess the students' learning within this learning progression, a longitudinal interview study was undertaken which revealed students' conceptions of teleological thinking and goal-oriented adaptation. In Chap. 6, Jaimie Miller-Friedmann, Susan Sunbury and Philip Sadler assessed US middle and high school student understanding of national science standards—National Science Educational Standards (NSES) for middle school students and Next Generation Science Standards (NGSS) for high school students—for evolution with a nationally representative sample in diverse settings. They were particularly interested to determine whether students from a wide range of school types, socio-economic status and regions in the United States are being taught and are learning evolution equally.

Yi Kong, Nancy Pelaez, Trevor Anderson and Jeffrey Olimpo in Chap. 7 start from the established finding that a lack of tree-thinking abilities is a factor that hampers deep understanding of evolution. They therefore compared an innovative curriculum intended to develop tree-thinking abilities to that of a traditional tree-thinking curriculum with regard to how these curricula were implemented by graduate teaching assistants in an introductory undergraduate biology classroom. In Chap. 8, Timothy Goodale reports on the effects on beginning science teachers in the USA of using instructional units involving the teaching and learning of genetics and evolution through context-based methods surrounding food security issues in Africa.

Alexandra Buck, Sofoklis Sotiriou and Franz Bogner in Chap. 9 look at the consequences of an inquiry-based, hands-on approach with multimedia workstations focusing on the *Archaeopteryx* fossil for understanding evolution. They argue that this approach is an example of shifting from STEM to STEAM (science, technology, engineering, arts and mathematics) subjects. In Chap. 10, David Owens reports on the results of a gameful, inquiry-based learning intervention with the intention of enhancing motivation among undergraduates to learn in the context of plant evolutionary life history.

Briana Pobiner, William Watson, Paul Beardsley and Constance Bertka in Chap. 11 examine the impact of implementing constructivist, guided-inquiry 'mini-units' that focus on examples of natural selection in humans on advanced US high school students' understanding of key concepts and the frequency of cognitive biases and misconceptions. They also describe the effect of supplementing this instruction with lessons that help teachers negotiate student resistance to learning about evolution due to religious or cultural beliefs. In Chap. 12, Kathy Malone, Anita Schuchardt and Zakee Sabree start by noting that the use of models and modelling in science education has been demonstrated to achieve cognitive gains in several science disciplines. However, there is a dearth of quasi-experimental studies in secondary classrooms that examine *how* the use of models and modelling can affect the cognitive gains of learners in biology in general and evolution in particular. Accordingly, they report on a study of an evolution unit grounded in the use of modelling and its effects on learning in evolution and attitudes towards science in general.

In Chap. 13, Alma Gómez Galindo, Alejandra García Franco, Leonardo Gonzáles Galli and José Torres Frías point out that evolution education has not sufficiently explored the cultural and contextual aspects related to learning. They therefore discuss the possibility of teaching evolution using an intercultural dialogic approach in which they worked with indigenous students in the Mayan Highlands in Mexico, exploring their knowledge about domestication of maize and reflecting on how knowledge about domestication of maize could be relevant for learning evolution. Lisa Kenyon, Emily Walter and William Romine in Chap. 14 transformed a college introductory biology course to more practice-based learning environment, in which students constructed knowledge about evolution through explanation and argumentation, and examined the consequences for conceptual change around natural selection, mechanistic reasoning related to natural selection and engagement in argumentation around data.

In Chap. 15, Ute Harms and Daniela Fiedler report on two studies to test the hypothesis that one central problem of understanding evolution is comprehension of the abstract concepts of randomness and probability. In the first study, they analysed the relationships of students' understanding of randomness and probability with their understanding of evolution; in the second study, three interventions were applied to improve students' understandings of randomness: an animation, a text on randomness and mathematical tasks. Jorge Groß, Kerstin Kremer and Julia Arnold in Chap. 16 present two case studies that research the interplay between creationist conceptions and evolution understanding in informal learning environments. Case study one deals with the topic of the emergence of humankind in an exhibition presented to visitors in an IKEA store; case study two deals with a guided tour about the evolution of life throughout geological eras in a natural history museum. In Chap. 17, Jo Nicholl and Paul Davies discuss the findings of a study in a small Natural History Museum to look at how the use of objects supports pre-service science teachers in both their subject knowledge and their pedagogic knowledge of biological evolution.

Finally, in Chap. 18, the two of us as editors present some overall conclusions for the various studies reported in this book and suggest future avenues for research depending on the characteristics of learners (e.g. age and religious affiliations) and the nature of the learning environment (e.g. in school versus out of school, mediated by teachers versus not mediated by teachers).

## References

Abraham, J. K., Meir, E., Perry, J., Herron, J. C., Maruca, S., & Stal, D. (2009). Addressing Undergraduate Student Misconceptions about Natural Selection with an Interactive Simulated Laboratory. *Evolution: Education & Outreach, 2*, 393–404. https://doi.org/10.1007/s12052-009-0142-3.

Akyol, G., Tekkaya, C., Sungur, S., & Traynor, A. (2012). Modeling the interrelationships among pre-service science teachers' understanding and acceptance of evolution, their views on nature

of science and self-efficacy beliefs regarding teaching evolution. *Journal of Science Teacher Education, 23*(8), 937–957.
Alters, B. J. (2005). Teaching Biological Evolution in Higher Education. M*ethodological, Religious, and Nonreligious Issues*. Sudbury, MA: Jones and Barlett Publishers.
Alters, B. J., & Nelson C. E. (2002). Perspective: Teaching evolution in higher education. Evolution: *International Journal of Organic Evolution, 56*(10), 1891–1901.
American Association for the Advancement of Science (AAAS) (2006). E*volution on the front line: An abbreviated guide for teaching evolution* (Project 2061). Received from http://www.project2061.org/publications/guides/evolution.pdf.
Anderson, D. L., Fisher, K. M., & Norman, G. J. (2002). Development and evaluation of the conceptual inventory of natural selection. *Journal of Research in Science Teaching, 39*(10), 952–978.
Andersson, B., & Wallin, A. (2006). On developing content-oriented theories taking biological evolution as an example. *International Journal of Science Education, 28*(6), 673–695. https://doi.org/10.1080/09500690500498385.
Andrews, T. M., Kalinowski, S. T., & Leonard, M. J. (2011). "Are humans evolving?" A classroom discussion to change student misconceptions regarding natural selection. *Evolution: Education & Outreach, 4*, 456–466. https://doi.org/10.1007/s12052-011-0343-4.
Athanasiou, K., & Papadopoulou, P. (2012). Conceptual ecology of the evolution acceptance among Greek education students: Knowledge, religious practices and social influences. *International Journal of Science Education, 34*(6), 903–924. https://doi.org/10.1080/09500693.2011.586072.
Baalmann, W., Frerichs, V., Weitzel, H., Gropengießer, H., & Kattmann, U. (2004). Schülervorstellungen zu Prozessen der Anpassung: Ergebnisse einer Interviewstudie im Rahmen der Didaktischen Rekonstruktion. *Zeitschrift für Didaktik der Naturwissenschaften, 10*(1), 7–28.
Bardapurkar, A. (2008). Do students see the "selection" in organic evolution? A critical review of the causal structure of student explanations. *Evolution: Education and Outreach, 1*(3), 299–305. https://doi.org/10.1007/s12052-008-0048-5.
Baum, D. A., DeWitt-Smith, S., & Donovan, S. S. S. (2005). The tree-thinking challenge. *Science, 310*(5750), 979–980. https://doi.org/10.1126/science.1117727.
Beardsley, P. M. (2004). Middle school student learning in evolution: Are current standards achievable? *The American Biology Teacher, 66*(9), 604–612. https://doi.org/10.2307/4451757.
Beggrow, E. P., & Nehm, R. H. (2012). Students' mental models of evolutionary causation: Natural selection and genetic drift. *Evolution: Education & Outreach, 5*, 429–444. https://doi.org/10.1007/s12052-012-0432-z.
Berti, A. E., Toneatti, L., & Rosati, V. (2010). Children's conceptions about the origin of species: A study of Italian children's conceptions with and without instruction. *Journal of the Learning Sciences, 19*(4), 506–538.
Bishop, B. A., & Anderson, C. W. (1990). Student conceptions of natural selection and its role in evolution. *Journal of Research in Science Teaching, 27*(5), 415–427.
Bizzo, N. M. V. (1994). From down house landlord to Brazilian high school students: what has happened to evolutionary knowledge on the way? *Journal of Research in Science Teaching, 31*(5), 537–556. https://doi.org/10.1002/tea.3660310508.
Brumby, M. N. (1979). Problems in learning the concept of natural selection. *Journal of Biological Education, 13*(2), 119–122.
Brumby, M. N. (1981). The use of problem-solving in meaningful learning in biology. *Research in Science Education, 11*(1), 103–110. https://doi.org/10.1007/BF02356773.
Brumby, M. N. (1984). Misconceptions about the concept of natural selection by medical biology students. *Science Education, 68*, 493–503. https://doi.org/10.1002/sce.3730680412.
Campbell, N. A., & Reece, J. B. (2006). *Biologie* (6. Auflage). München: Pearson Education (p. 513). Deutschland.
Campos, R., & Sá-Pinto, A. (2014). Erratum: Early evolution of evolutionary thinking: teaching biological evolution in elementary schools. *Evolution: Education & Outreach, 7*. https://doi.org/10.1186/s12052-014-0005-4.

Catley, K. M., Lehrer, R., & Reiser, B. (2005). *Tracing a prospective learning progression for developing understanding of evolution*. Paper commissioned by National Academies Committee on Test Design for K-12 Science Achievement.

Catley, K. M., & Novick, L. R. (2009). Digging deep: Exploring college students' knowledge of macroevolutionary time. *Journal of Research in Science Teaching, 46*(3), 311–332. https://doi.org/10.1002/tea.20273.

Catley, K. M., Phillips, B. C., & Novick, L. R. (2013). Snakes and eels and dogs! Oh, my! Evaluating high school students' tree-thinking skills: An entry point to understanding evolution. *Research in Science Education, 43,* 2327–2348. https://doi.org/10.1007/s11165-013-9359-9.

Chi, M., Kristensen, A., & Roscoe, R. (2012). Misunderstanding emergent causal mechanism in natural selection. In S. Rosengren, S. Brem, & G. Sinatra (Eds.), *Evolution challengens: Integrating research and practice in teaching and learning about evolution* (pp. 145–173). New York: Oxford University Press.

Deadman, J., & Kelly, P. (1978). What do secondary school boys understand about evolution and heredity before they are taught the topics? *Journal of Biological Education, 12*(1), 7–15. https://doi.org/10.1080/00219266.1978.9654169.

Demastes, S. S., Settlage, J., & Good, R. (1995). Students' conceptions of natural selection and its role in evolution: Cases of replication and comparison. *Journal of Research in Science Teaching, 32*(5), 535–550. https://doi.org/10.1002/tea.3660320509.

Deniz, H., Donnelly, L. A., & Yilmaz, I. (2008). Exploring the factors related to acceptance of evolutionary theory among Turkish preservice biology teachers: Toward a more informative conceptual ecology for biological evolution. *Journal of Research in Science Teaching, 45*(4), 420–443. https://doi.org/10.1002/tea.20223.

Department for Education [DfE]. (2014). *National Curriculum.* Available at https://www.gov.uk/government/collections/national-curriculum.

Engel Clough, E., & Wood-Robinson, C. (1985). How Secondary Students Interpret Instances of Biological Adaptation. *Journal of Biology Education, 19*(2), 125–130. https://doi.org/10.1080/00219266.1985.9654708.

Evans, E. M. (2000). The emergence of beliefs about the origins of species in school-age children. *Merrill-Palmer Quarterly: A Journal of Developmental Psychology, 46*(2), 221–254.

Evans, E. M., Szymanowski, K., Smith, P. H., & Rosengren, K. S. (2005). *Overcoming an essentialist bias: From metamorphosis to evolution.* Atlanta, GA: In Biennial meeting of the Society for Research in Child Development.

Evans, E. M., Spiegel, A. N., Gram, W., Frazier, B. N., Tare, M., Thompson, S., et al. (2010). A conceptual Guide to Natural History Museum Visitors' Understanding of Evolution. *Journal of Research in Science Teaching, 47*(3), 326–353.

Ferrari, M., & Chi, M. T. H. (1998). The nature of naive explanations of natural selection. *International Journal of Science Education, 20*(10), 1231–1256.

Fiedler, D., Tröbst, S., & Harms, U. (2017). University students' conceptual knowledge of randomness and probability in the context of evolution and mathematics. *CBE-Life Sciences Education (LSE), 16*(2), 1–16, ar38. https://doi.org/10.1187/cbe.16-07-0230.

Flanagan, J. C., & Roseman, J. E. (2011). Assessing middle and high school students' understanding of evolution with standards-based items. In *Annual Conference of the National Association for Research in Science Teaching*. Orlando, FL.

Garvin-Doxas, K., & Klymkowsky, M. W. (2008). Understanding randomness and its impact on student learning: Lessons learned from building the Biology Concept Inventory (BCI). CBE – *Life Science Education, 7,* 227–233. https://doi.org/10.1187/cbe.07-08-0063.

Gelman, S. A. (2004). Psychological essentialism in children. *Trends in Cognitive Science, 8*(9), 404–409. https://doi.org/10.1016/j.tics.2004.07.001.

Gould, S. J. (1992). *Die Entdeckung der Tiefenzeit: Zeitpfeil und Zeitzyklus in der Geschichte unserer Erde* (p. 15). München: Carl Hanser.

Graf, D., & Hamdorf, E. (2011). Evolution: Verbreitete Fehlvorstellungen zu einem zentralen Thema In D. Dreesmann, D. Graf & K. Witte (Hrsg.), *Evolutionsbiologie: Moderne Themen für den Unterricht* (S. 32). Heidelberg: Springer AkademischerVerlag.

Graf, D., & Soran, H. (2011). Einstellung und Wissen von Lehramtsstudierenden zur Evolution – ein Vergleich zwischen Deutschland und der Türkei. In D. Graf (Ed.), *Evolutionstheorie – Akzeptanz und Vermittlung im europäischen Vergleich* (pp. 141–162). Berlin: Springer.

Grant, B. W. (2009). Practitioner research improved my students' understanding of evolution by natural selection in an introductory biology course. *Teaching issues and experiments in ecology, Vol. 6*: Research #4. Retrieved from http://tiee.ecoed.net/vol/v6/research/grant/.

Greene, E. D. (1990). The logic of university students' misunderstanding of natural selection. *Journal of Research in Science Teaching, 27*(9), 875–885. https://doi.org/10.1002/tea.3660270907.

Gregory, T. R. (2008). Understanding evolutionary trees. *Evolution: Education & Outreach, 1*(2), 121–137. https://doi.org/10.1007/s12052-008-0035-x.

Gregory, T. R. (2009). Understanding natural selection: Essential concepts and common misconceptions. *Evolution: Education & Outreach, 2*,156–175. https://doi.org/10.1007/s12052-009-0128-1.

Großschedl, J., Konnemann, C., & Basel, N. (2014). Pre-service biology teachers' acceptance of evolutionary theory and their preference for its teaching. *Evolution: Education and Outreach, 7*(18), 1–16. https://doi.org/10.1186/s12052-014-0018-z.

Ha, M., Haury, D. L., & Nehm, R. H. (2012). Feeling of certainty: Uncovering a missing link between knowledge and acceptance of evolution. *Journal of Research in Science Teaching, 49*(1), 95–121. https://doi.org/10.1002/tea.20449.

Halldén, O. (1988). The evolution of the species: pupil perspectives and school perspectives. *International Journal of Science Education, 10*(5), 541–552. https://doi.org/10.1080/0950069880100507.

Harms, U., Mayer, J., Hammann, M., Bayrhuber, H., & Kattmann, U. (2004). Kerncurriculum und Standards für den Biologieunterricht in der gymnasialen Oberstufe. In H.-E. Tenorth (Hrsg.), *Biologie, Chemie, Physik, Geschichte, Politik. Expertisen – im Auftrag der KMK* (S. 22–84). Weinheim: Beltz.

Illner, R. (2000). *Einfluss religiöser Schülervorstellungen auf die Akzeptanz der Evolutionstheorie* (Dissertation). Universität Oldenburg. Abgerufen von http://oops.uni-oldenburg.de/388/1/421.pdf.

Jensen, M. S., & Finley, F. N. (1995). Teaching evolution using historical arguments in a conceptual change strategy. *Science Education, 79*(2), 147–166. https://doi.org/10.1002/sce.3730790203.

Jensen, M. S., & Finley, F. N. (1996). Changes in students' understanding of evolution resulting from different curricular and instructional strategies. *Journal of Research in Science Teaching, 33*(8), 879–900. https://doi.org/10.1002/(SICI)1098-2736(199610)33:8%3c879:AID-TEA4%3e3.0.CO;2-T.

Jimenez-Aleixandre, M. P. (1992). Thinking about theories or thinking with theories?: A classroom study with natural selection. *International Journal of Science Education, 14*(1), 51–61. https://doi.org/10.1080/0950069920140106.

Jiménez-Aleixandre, M. P., & Fernández-Pérez, J. (1987). Selection or adjustment? Explanations of university biology students for natural selection problems. In J. D. Novak (Ed.), *Proceedings of the Second International Seminar on Misconceptions and Educational Strategies in Science and Mathematics* (vol. II, pp. 224–232). Ithaca, NY: Department of Education, Cornell University.

Johannsen, M., & Krüger, D. (2005). Schülervorstellungen zur Evolution - eine quantitative Studie. *Berichte des Instituts für Didaktik der Biologie, 14,* 23–48.

Jones, L., & Reiss, M. J. (Eds.). (2007). *Teaching about scientific origins: Taking account of creationism*. New York: Peter Lang.

Jungwirth, E. (1975). The problem of teleology in biology as a problem of biology-teacher education. *Journal of Biological Education, 9*(6), 243–246. https://doi.org/10.1080/00219266.1975.9654037.

Kalinowski, S. T., Leonard, M. J., & Andrews, T. M. (2010). Nothing in evolution makes sense except in the light of DNA. *CBE—Life Sciences Education, 9*, 87–97. https://doi.org/10.1187/cbe.09-12-0088.

Kampourakis, K. (2014). *Understanding evolution*. Cambridge: Cambridge University Press.

Kampourakis, K., Pavlidi, V., Papadopoulou, M., & Palaiokrassa, E. (2012). Children's teleological intuitions: What kind of explanations do 7–8 year olds give for the features of organisms, artefacts and natural objects? *Research of Science Education, 42*(4), 651–671. https://doi.org/10.1007/s11165-011-9219-4.

Kampourakis, K., & Zogza, V. (2008). Students' intuitive explanations of the causes of homologies and adaptations. *Science & Education, 17*(1), 27–47. https://doi.org/10.1007/s11191-007-9075-9.

Kampourakis, K., & Zogza, V. (2009). Preliminary evolutionary explanations: A basic framework for conceptual change and explanatory coherence. *Science & Education, 18*(10), 1313–1340. https://doi.org/10.1007/s11191-008-9171-5.

Kattmann, U. (1995). Konzeption eines naturgeschichtlichen Biologieunterrichts: Wie Evolution Sinn macht. *Zeitschrift für Didaktik der Naturwissenschaften, 1*, 29–42.

Kattmann, U. (2005). Lernen mit anthropomorphen Vorstellungen? - Ergebnisse von Untersuchungen zur Didaktischen Rekonstruktion in der Biologie. *Zeitschrift für Didaktik der Naturwissenschaften, 11*, 165–174.

Kattmann, U., Duit, R., Gropengießer, H., & Komorek, M. (1997). Das Modell der Didaktischen Rekonstruktion. *Zeitschrift für Didaktik der Naturwissenschaft, 3*(3), 3–18.

Kutschera, U. (2006). *Evolutionsbiologie* (2nd ed.). Stuttgart: Ulmer.

Lammert, N. (2012). *Akzeptanz, Vorstellungen und Wissen von Schülerinnen und Schülern der Sekundarstufe I zu Evolution und Wissenschaft* (Dissertation). Technische Universität Dortmund. Aufgerufen von https://eldorado.tu-dortmund.de/bitstream/2003/29476/1/Dissertation_Lammert.pdf.

Losh, S. C., & Nzekwe, B. (2011). The foundations: How education major influences basic science knowledge and pseudoscience beliefs. In *Atlanta Conference on Science and Innovation Policy* (pp. 1–16). IEEE. https://smartech.gatech.edu/bitstream/handle/1853//421-1500-2-PB.pdf Gesehen 21.07.2015.

MacFadden, B. J., Dunckel, B. A., Ellis, S., Dierking, L. D., Abraham-Silver, L., Kisiel, J., et al. (2007). Natural history museum visitors' understanding of evolution. *BioScience, 57*(10), 875–882.

Mayr, E. (1982). *The growth of biological thought: Diversity, Evolution, and Inheritance*. Cambridge: Harvard University Press.

McVaugh, N. K., Birchfield, J., Lucero, M. M., & Petrosino, A. J. (2011). Evolution education: Seeing the forest for the trees and focusing our efforts on the teaching of evolution. *Evolution: Education & Outreach, 4*, 286–292. https://doi.org/10.1007/s12052-010-0297-y.

Meikle, W. E., & Scott, E. C. (2010). Why are there still monkeys? *Evolution: Education & Outreach, 3*, 573–575. https://doi.org/10.1007/s12052-010-0293-2.

Meir, E., Perry, J., Herron, J. C., & Kingsolver, J. (2007). College students' misconceptions about evolutionary trees. *The American Biology Teacher, 69*(7), 71–76. https://doi.org/10.1662/0002-7685(2007)69%5b71:CSMAET%5d2.0.CO;2.

Meyer, J. H. F., & Land, R. (2005). Threshold concepts and troublesome knowledge (2): Epistemological considerations and a conceptual framework for teaching and learning. *Higher Education, 49*(3), 373–388. https://doi.org/10.1007/s10734-004-6779-5.

Meyer, J. H., & Land, R. (2006). Threshold concepts and troublesome knowledge: An introduction. In J. H. Meyer & R. Land (Eds.), *Overcoming barriers to student understanding: Threshold concepts and troublesome knowledge* (pp. 3–18). Abington, United Kingdom: Routledge.

Miller, J. D., Scott, E. C., & Okamoto, S. (2006). Public acceptance of evolution. *Science-New York then Washington-, 313*(5788), 765.

Monod, J. (1997). On the molecular theory of evolution. In M. Ridley (Ed.), *evolution* (p. 390). Oxforf: Oxford University Press.

Nadelson, L., Culp, R., Bunn, S., Burkhart, R., Shetlar, R., Nixon, K., & Waldron, J. (2009). Teaching evolution concepts to early elementary school students. *Evolution: Education and Outreach, 2*(3), 458–473. https://doi.org/10.1007/s12052-009-0148-x.

Nehm, R. H., Beggrow, E. P., Opfer, J. E., & Ha, M. (2012). Reasoning about natural selection: Diagnosing contextual competency using the ACORNS instrument. *The American Biology Teacher, 74*(2), 92–98. https://doi.org/10.1525/abt.2012.74.2.6.

Nehm, R. H., & Ha, M. (2011). Item feature effects in evolution assessment. *Journal of Research in Science Teaching, 48*(3), 237–256.

Nehm, R. H., Poole, T. M., Lyford, M. E., Hoskins, S. G., Carruth, L., Ewers, B. E., & Colberg, P. J. S. (2009). Does the segregation of evolution in biology textbooks and introductory courses reinforce students' faulty mental models of biology and evolution? *Evolution: Educations and Outreach, 2*, 527–532. https://doi.org/10.1007/s12052-008-0100-5.

Nehm, R. H., Rector, M. A., & Ha, M. (2010). "Force-Talk" in evolutionary explanation: Metaphors and misconceptions. *Evolution: Education and Outreach, 3,* 605–613. https://doi.org/10.1007/s12052-010-0282-5.

Nehm, R. H., & Reilly, R. (2007). Biology majors' knowledge and misconceptions of natural selection. *BioScience, 57*(3), 263–272.

Nehm, R. H., & Schonfeld, I. S. (2007). Does increasing biology teacher knowledge of evolution and the nature of science lead to greater preference for the teaching of evolution in schools? *Journal of Science and Teacher Education, 18,* 699–723. https://doi.org/10.1007/s10972-007-9062-7.

Nehm, R. H., & Schonfeld, I. S. (2008). Measuring knowledge of natural selection: A comparison of the CINS, an open-response instrument, and an oral interview. *Journal of Research in Science Teaching, 45*(8), 1131–1160. https://doi.org/10.1002/tea.20251.

Nehm, R. H., Schonfeld, I., & The, S. (2010b). Future of natural selection knowledge measurement: a reply to Anderson. *Journal of Research in Science Teaching, 47,* 358–362. https://doi.org/10.1002/tea.20330.

Neubrand, C., & Harms, U. (2017). Tackling the difficulties in learning evolution: Effects of adaptive self-explanation prompts. *Journal of Biological Education, 51*(4), 336–348. https://doi.org/10.1080/00219266.2016.1233129.

Newall, E. (2017). Evolution, insight and truth? *School Science Review, 99*(367), 61–66.

NGSS Lead States. (2013). *Next generation science standards: For states, by states*. Washington, DC: The National Academies Press.

Niebert, K., & Gropengießer, H. (2015). Understanding starts in the Mesocosm: Conceptual metaphor as a framework for external representations in science teaching. *International Journal of Science Education, 37*(5–6), 903–933. https://doi.org/10.1080/09500693.2015.1025310.

Novick, L. R., & Catley, K. M. (2006). Interpreting hierarchical structure: Evidence from cladograms in biology. In D. Barker-Plummer, R. Cox & N. Swoboda (Eds.), Diagrammatic Representation and Inference. *Diagrams 2006. Lecture Notes in Computer Science* (vol. 4045, pp. 176–180). Berlin Heidelberg: Springer.

Novick, L. R., Schreiber, E. G., & Catley, K. M. (2014). Deconstructing evolution education: The relationship between micro- and macro evolution. *Journal of Research in Science Teaching, 51*(6), 759–788. https://doi.org/10.1002/tea.21161.

Opfer, J. E., Nehm, R. H., & Ha, M. (2012). Cognitive foundations for science assessment design: Knowing what students know about evolution. *Journal of Research in Science Teaching, 49*(6), 744–777. https://doi.org/10.1002/tea.21028.

Pedersen, S., & Halldén, O. (1994). Intuitive ideas and scientific explanations as parts of students' developing understanding of biology: the case of evolution. *European Journal of Psychology of Education, 9,* 127–137. https://doi.org/10.1007/BF03173548.

Phillips, B. C., Novick, L. R., Catley, K. M., & Funk, D. J. (2012). Teaching tree thinking to college students: It's not as easy as you think. *Evolution: Education & Outreach, 5*, 595–602. https://doi.org/10.1007/s12052-012-0455-5.

Posner, G. J., Strike, K. A., Hewson, P. W., & Gertzog, W. A. (1982). Accommodation of a scientific conception: Toward a theory of conceptual change. *Science Education, 66*(2), 211–227.
Prinou, L., Halkia, L., & Skordoulis, C. (2008). What conceptions do Greek school students form about biological evolution? *Evolution: Education and Outreach, 1*(3), 312–317. https://doi.org/10.1007/s12052-008-0051-x.
Rector, M. A., Nehm, R. H., & Pearl, D. (2013). Learning the language of evolution: Lexical ambiguity and word meaning in student explanations. *Research in Science Education, 43*(3), 1107–1133. https://doi.org/10.1007/s11165-012-9296-z.
Reiss, M. J. (2011). How should creationism and intelligent design be dealt with in the classroom? *Journal of Philosophy of Education, 45,* 399–415.
Robbins, J. R., & Roy, P. (2007). The natural selection: Identifying and correcting non-science student preconceptions through an inquiry-based, critical approach to evolution. *The American Biology Teacher, 69*(8), 460–466.
Robson, R. L., & Burns, S. (2011). Gain in student understanding of the role of random variation in evolution following teaching intervention based on Luria-Delbruck Experiment. *Journal of Microbiology and Biology Education, 12*(1), 3–7. https://doi.org/10.1128/jmbe.v12i1.272.
Rosengren, K. L., Brem, S. K., Evans, E. M., & Sinatra, G. M. (Eds.). (2012). *Evolution challenges integrating research and practice in teaching and learning about evolution*. Oxford: Oxford University Press.
Ross, P. M., Taylor, C. E., Hudges, C., Kofod, N., Whitaker, N., Lutze-Mann, Kofod M., & Tzioumis, V. (2010). Threshold concepts in learning biology and evolution. *Biology International*, 47–52.
Rutledge, M. L., & Warden, M. A. (2000). Evolutionary theory, the Nature of Science & high school biology teachers: critical relationships. *The American Biology Teacher, 62*(1), 23–31. https://doi.org/10.1662/0002-7685(2000)062%5b0023:ETTNOS%5d2.0.CO;2.
Samarapungavan, A., & Wiers, R. W. (1997). Children's thoughts on the origin of species: A study of explanatory coherence. *Cognitive Science, 21*(2), 147–177.
Secretariat of the standing conference of the ministers of education and cultural affairs of the länder in the Federal Republic of Germany [KMK], 2005.
Settlage, J. (1994). Conceptions of natural selection: A snapshot of the sensemaking process. *Journal of Research in Science Teaching, 31*(5), 449–457. https://doi.org/10.1002/tea.3660310503.
Shtulman, A. (2006). Qualitative differences between naïve and scientific theories of evolution. *Cognitive Psychology, 52*(2), 170–194. https://doi.org/10.1016/j.cogpsych.2005.10.001.
Sinatra, G. M., Brem, S. K., & Evans, E. M. (2008). Changing minds? Implications of concetpual change for teaching and learning about biological evolution. *Evolution: Education and Outreach, 1*(2), 189–195. https://doi.org/10.1007/s12052-008-0037-8.
Southerland, S. A., Abrams, E., Cummins, C. L., & Anzelmo, J. (2001). Understanding students' explanations of biological phenomena: Conceptual frameworks or p-prims? *Science Education, 85*(4), 328–348. https://doi.org/10.1002/sce.1013.
Speth, E. B., Shaw, N., Momsen, J., Reinagel, A., Le, P., Taqieddin, R., & Long, T. (2014). Introductory Biology Students' Conceptual Models and Explanations of the Origin of Variation. *CBE – Life Sciences Education, 13,* 529–539.
Spindler, L., & Doherty, J. (2009). Assessment of the teaching of evolution by natural selection through a hands-on simulation. In *Teaching Issues and Experiments in Ecology, 6*. Retrieved from http://tiee.esa.org/vol/v6/research/spindler/pdf/spindler.pdf.
Strevens, M. (2000). The essentialist aspect of naive theories. *Cognition, 74*(2), 149–175. https://doi.org/10.1016/S0010-0277(99)00071-2.
Tamir, P., & Zohar, A. (1991). Anthropomorphism and teleology in reasoning about biological phenomena. *Science Education, 75*(1), 57–67. https://doi.org/10.1002/sce.3730750106.
Tibell, L. A. E., & Harms, U. (2017). Biological principles and threshold concepts for understanding natural selection - implications for developing visualizations as a pedagogic tool. *Science & Education, 26*(7), 953–973. https://doi.org/10.1007/s11191-017-9935-x.

Tracy, J. L., Hart, J., & Martens, J. P. (2011). Death and science: The existential underpinnings of belief in intelligent design and discomfort with evolution. *PLoS ONE, 6*(3), e17349. https://doi.org/10.1371/journal.pone.0017349.

Trend, R. D. (2001). Deep Time Framework: A preliminary study of U.K. primary teachers' conceptions of geologic time and perceptions of geoscience. *Journal of Research in Science Teaching, 38,* 191–221.

Van Dijk, E. M., & Kattmann, U. (2009). Teaching evolution with historical narratives. *Evolution: Education and Outreach,* 2(3), 479–489. https://doi.org/10.1007/s12052-009-0127-2.

Van Dijk, E. M., & Kattmann, U. (2010). Evolution im Unterricht: Eine Studie über fachdidaktisches Wissen von Lehrerinnen und Lehrern. *Zeitschrift für Didaktik der Naturwissenschaften, 16,* 7–21.

Weitzel, H., & Gropengießer, H. (2009). Vorstellungsentwicklung zur stammesgeschichtlichen Anpassung: Wie man Lernhindernisse verstehen und förderliche Lernangebote machen kann. *Zeitschrift für Didaktik der Naturwissenschaften, 15,* 287–305.

Yates, T. B., & Marek, E. A. (2014). Teachers teaching misconceptions: a study of factors contributing to high school biology students' acquisition of biological evolution-related misconceptions. *Evolution: Education and Outreach, 7*(1), 7.

**Ute Harms** is Director at the IPN - Leibniz Institute for science and mathematics education, Full Professor for Biology Education at the University of Kiel (Germany) since 2007, and Fellow of the Royal Society of Biology (Great Britain). She has a Ph.D. in cell biology and has worked as a high school teacher for several years. In 2000, she got her first Professorship for Biology Education at the Ludwig-Maximilians-University in Munich (Germany). From 2006 to 2007, she held a chair in Biology Education at the University of Bremen. Her main research interests are conceptual learning in biology and in science, focusing on evolution and energy, biology teacher education, biology-related competitions and transfer of contemporary topics in the life sciences to the public.

**Michael Reiss** is Professor of Science Education at UCL Institute of Education, University College London, Visiting Professor at the Universities of York and Kiel and the Royal Veterinary College, Honorary Fellow of the British Science Association, Docent at the University of Helsinki and a Fellow of the Academy of Social Sciences. After undertaking a Ph.D. and post-doctoral research in evolutionary biology and population genetics, he trained to be a science teacher and taught in schools for five years before returning to higher education. The former Director of Education at the Royal Society, his academic interests are in science education, bioethics and sex education and he has published widely on issues to do with creationism in schools.

# Evidence for the Success of a Quantitative Assessment Instrument for Teaching Evolution in Primary Schools in England

**Loredana L. Buchan, Momna V. Hejmadi and Laurence D. Hurst**

## 1 Introduction

There is growing recognition that young learners benefit from studying evolution when biology is first introduced in primary school (Fail, 2008; Weiss & Dreesmann, 2014) when they are most receptive to new ideas and are actively questioning how the world works (Nadelson et al., 2009). Primary education helps to provide the foundation for evolution understanding (Akerson & Donnelly, 2010) and to develop a deeper understanding of evolution as they progress through the school system (Wagler, 2012). In 2014 the Primary Science National Curriculum for England was altered to include the conceptual understanding of Evolution and Inheritance as a statutory requirement for all year 6 students.

However, it is also recognised that evolution is an extremely problematic and widely misunderstood topic (see Gregory 2009). As such, the introduction of this new content presents many challenges associated with teaching a complex science topic in primary schools, not least of which is the availability of age-appropriate resources for teaching evolution. Indeed, the bulk of the existing research focuses on the understanding of genetics and evolution in secondary school students, with very little known about the understanding of evolution in primary school children (Venville & Donovan, 2005; Venville, Gribble, & Donovan, 2005). More particularly, there is a notable dearth of experimental evidence for what 'works', in no small part due to a lack of assessment tools which is particularly acute for primary-aged children (Ha, Haury, & Nehm, 2012; Nadelson & Sinatra, 2008; Nehm & Schonfeld, 2008; Wiles & Alters, 2011). There is also a need to identify which teaching activities create the most effective learning experiences based on direct evidence (Beardsley, Bloom, & Wise, 2012; Glaze & Goldston 2015. Here we aim to address some of the above limitations and consider three issues.

L. L. Buchan (✉) · M. V. Hejmadi · L. D. Hurst
Department of Biology and Biochemistry, The Milner Centre for Evolution,
University of Bath, Bath BA2 7AY, UK
e-mail: l.buchan@bath.ac.uk



U. Harms and M. J. Reiss (eds.), *Evolution Education Re-considered*,
https://doi.org/10.1007/978-3-030-14698-6_2

## 1.1 Is It Possible to Develop a Fit-for-Purpose Quantitative Assessment Tool?

First, we ask whether it is possible to select and adapt assessment items to form a successful assessment instrument for use in primary schools? Most of the existing assessment tools for use in primary schools are phenomenographic in nature and for practical reasons have been implemented with relatively small sample sizes (Berti, Toneatti, & Rosati, 2010; Samarapungavan & Wiers, 1997; Solomon, 2002). Only one large-scale quantitative study of evolution education exists, based on a set of national standards-based multiple-choice items developed and field-tested with over 9000 students aged 11–18 as part of the AAAS Project 2061 (Flanagan & Roseman, 2011). As this wide age range affected the 'readability' and cognitive demand of some of the items, we sought to modify the instrument to match our younger mixed-age cohort.

## 1.2 Is It Possible to Teach Genetics and Evolution to Primary School Children?

We also address a second issue, whether it is possible to teach abstract concepts of genetics and evolution to primary children. There are numerous reasons as to why evolution is so hard to understand (Ferrari & Chi, 1998). First, understanding is linked to reasoning ability. The lower the reasoning ability of a child then the greater the number of alternative conceptions they hold and the harder it is for them to change their ideas after teaching (Oliva, 2003; Williams & Cavallo, 1995). Second, conceptual change is needed for some students to understand evolution and this is extremely difficult to achieve. Challenges to conceptual change can be divided into three categories: (a) basic or developmental constraints that are present from infancy and early childhood (Alters, 2005; Evans, 2008; Kelemen & Rosset, 2009); (b) prior knowledge or experiences that reinforce default ways of thinking, so-called naïve theories (Carey, 1985; Geary, 2006; Sinatra, Brem, & Evans, 2008); and (c) emotional/motivational reactions that make students reluctant to entertain the possibility of change (Brem, Ranney, & Schindel, 2003; Thagard & Findlay, 2009). Therefore, effective teaching of evolution is not just a matter of 'bolting on' new knowledge but it requires the 'un-teaching' and correction of alternative conceptions (Sinatra et al., 2008).

Third, whether 9–11-year-olds are able to understand evolution needs careful consideration. Children are rarely taught about genetics and DNA until they are around 15, as the concepts are thought to be too abstract and inappropriate for younger children (Engel Clough & Wood-Robinson, 1985). However, they are exposed to the same concepts through the media, comics, games and movies (Nelkin & Lindee, 1995). Carey (1985) proposed that young students are incapable of understanding natural selection because they have not yet developed the formal reasoning skills

**Table 1** Schematic outlining the work phase activities of the four SoW

| Scheme of work | Lesson 1 | Lesson 2 | Lesson 3 | Lesson 4 |
|---|---|---|---|---|
| | Variation and genetics | Natural selection in peppered moths | Geological time | Study of homology and common ancestry |
| 1 | Quantitative investigation of variation | Hunting moths[a] | Toilet roll timeline | Trilobites[c] |
| 2 | Quantitative investigation of variation | Hunting moths[a] | Toilet roll timeline | Pentadactyl limb[d] |
| 3 | Quantitative investigation of variation | PowerPoint[b] | Toilet roll timeline | Trilobites[c] |
| 4 | Quantitative investigation of variation | PowerPoint[b] | Toilet roll timeline | Pentadactyl limb[d] |

[a]Campos and Sá-Pinto (2013)
[b]Kelemen, Emmons, Seston Schillaci, and Ganea (2014)
[c]Wagler (2010)
[d]Nadelson et al., (2009)

needed. However, other studies show that children as young as five are able to grasp some ideas about genetics and natural selection given the correct type of instruction (Legare, Lane, & Evans, 2013; Venville & Donovan, 2007).

While there may be challenges, several authors have proposed possible primary level activities. These include classification by homologous structures (Chanet & Lusignan, 2008), modelling genes and DNA (Venville & Donovan, 2007), interpreting evolutionary trees (Ainsworth & Saffer, 2013) and the use of narrative texts to promote the understanding of evolution (Browning & Hohenstein, 2013). See Beardsley et al. (2012) for a partial summary of teaching intervention studies.

For our schemes of work (SoW), we adapted two different activities on homology/common ancestry and two on the predation of peppered moths (*Biston betularia*) by birds. We thus assembled different SoW (see Table 1) bringing together the most appropriate practical activities considering the age of our students, national curriculum requirements, resource availability and current pedagogical theory. While there are four SoW, we consider them en masse.

## *1.3 What Predicts Variation in Gain of Understanding?*

At the student level, we consider age, gender and ability as possible predictors of performance. Ability is an important variable, not least to examine whether our resources help the more able students without affecting the low ability group. As children get older, their ability to process abstract concepts improves (Piaget & Cook, 1952) and so they should be more able to understand evolution. Are ability and age important variables in predicting the change in understanding?

An a priori case for a possible role of gender is less straightforward. Some studies have shown that implicit stereotypes exist regarding science participation and performance (Nosek et al., 2009). There is also evidence that boys view science more positively than girls (Osborne, Simon, & Collins, 2003) and continue to outperform them in science in primary school (Gonzales et al., 2004). However, there is evidence suggesting that this gender gap is narrowing (Martin, Mullis, Foy, & Stanco, 2012). Our study is the first to consider whether a gender bias exists in evolution understanding in primary school children.

The final phase of analysis is at the class/school level, exploring possible candidate explanations. In particular, we consider whether teacher confidence, understanding and acceptance of evolution are predictors of performance. Teacher-level effects are important as the standard of evolution teaching is directly related to student understanding of the subject: unsatisfactory teaching is linked to the persistence of alternate conceptions (Alters & Nelson, 2002; Goldston & Kyzer, 2009; van Dijk & Kattmann, 2009). Teachers are expected to be 'experts' but, especially at primary level, may lack the confidence and skills to be able to teach evolution effectively. Poor understanding of the processes involved in evolution and the Nature of Science (NOS) equate to poor representation of the topic (Deniz & Donnelly, 2011; Nadelson & Nadelson, 2010; Papadopoulou, Stanissavljevic, Katakos, & Athanasiou, 2011). Several studies have reported that teachers do not feel adequately prepared to teach the subject (Aguillard, 1998; Griffith & Brem, 2004; Nadelson & Nadelson, 2010). Fewer primary school teachers are comfortable with teaching evolution (Fowler & Meisels, 2010) and have a lower acceptance of evolution compared to their secondary counterparts (Levesque & Guillaume, 2010; Losh & Nzekwe, 2011). The proportion of primary school teachers who hold a science degree is also significantly lower than their secondary colleagues, and consequently, it is quite common for primary school teachers to only be educated up to KS4 (GCSE or equivalent, i.e. age 16) standard in biology.

Teaching evolution is a controversial topic for some teachers (Allgaier, 2009; Berkman & Plutzer, 2010; Bowman, 2008). Secondary school teachers can allow religious and other beliefs to influence what they teach (Nehm & Schonfeld, 2007; Trani, 2004). Although England is a largely secular country, religiousness could still influence evolution teaching. Some teachers fear potential confrontation with students and parents (Beard, 1996) and may take the 'path of least resistance' by diluting the subject or even avoiding it altogether (Glaze & Goldston, 2015; Goldston & Kyzer, 2009). Alternatively, they devote minimal time to the topic in class (Berkman & Plutzer, 2011).

Importantly, teachers' alternative conceptions can be corrected by carefully planned interventions (Nehm & Schonfeld, 2007) but they need appropriate support to teach it in a conceptually sound way (Bandoli, 2008; Goldston & Kyzer, 2009). Providing primary school teachers with better training, resources and coping strategies will hopefully improve the standard of teaching (Glaze & Goldston, 2015; Griffith & Brem, 2004). Our analysis aims to help resolve where teacher-focus issues may lie and lead to the development of well-motivated teacher support.

## 2 Methodology

### *2.1 Development and Refinement of the Assessment Instrument*

In order to collect large-scale quantitative data, a paper and pencil assessment instrument was developed by selecting items from the AAAS science assessment website based on the research of Flanagan and Roseman (2011). Items were chosen for their relevance to the KS2 National Curriculum and appropriate cognitive demand. Each multiple-choice item had four alternative answers with common alternative conceptions acting as distractors. The selected fifteen items fell into five broad categories: homology/common ancestry; natural selection; variation; fossils; and geological time and extinction.

The assessment items were adapted to reduce reading difficulty and cognitive load while maintaining consistent item identity so that the assessment instrument could be used successfully in mixed-age upper primary classes. This was achieved by reducing the length and complexity of sentences and by using diagrams and tables of comparison. (See Fig. 1a, b for comparison.)

During the pilot phase, a full written version was allocated to each student to read and complete individually. This format was found to take too long, and concerns were raised about reading speed and concentration spans. At the pilot teachers' suggestion, the mode of delivery was altered to one in which the teacher read out the questions from a PowerPoint presentation while the students marked their responses on a grid as the test proceeded. To ensure consistency of delivery across schools, the teachers were trained to read the item out in full, allowing time for students to consider their answer before focusing on the key differences between the alternative answers emboldened in the text of the item (see Fig. 1b). From teacher feedback, this mode of delivery was much quicker to complete and helped to reduced reading problems.

Students were assessed at three different time points: pre-teaching to establish a priori knowledge, immediately after teaching to establish changes in understanding due to the teaching programme and 3–6 months later to evaluate retention. Demographic data collected from the students were confined to name, gender and date of birth. In order to avoid problems associated with disclosing formal science attainment scores, teachers were asked to identify the relative science ability of each student within their classes as being either high (top 1/3), middle or low (bottom 1/3). Other pertinent data were taken from recent Ofsted (the government Office for Standards in Education, Children's Services and Skills) reports and government databases.

**(a)**

Some organisms, such as a chimpanzee and a human, have many similarities. Others, such as a zebra and a worm, have fewer similarities. What is TRUE about the ancestors of these organisms?

(A) Chimpanzees and humans share a common ancestor with each other, but zebras and worms do not share a common ancestor with each other.

(B) Chimpanzees and humans share a common ancestor with each other, and zebras and worms share a common ancestor with each other, but chimpanzees and humans do not share a common ancestor with zebras and worms.

(C) Because chimpanzees, humans, zebras, and worms are separate species, none of them shares a common ancestor with any other.

(D) Chimpanzees, humans, zebras, and worms all share an ancient common ancestor.

**(b)**

Some organisms like a **human** and a **chimpanzee** (chimp) have **many similarities**. Others like a **lion** and a **worm** have **fewer**. The **circles** show organisms that **share a common ancestor**. Which answer is **TRUE** about **who shares a common ancestor?** Select **one** answer from the options below.

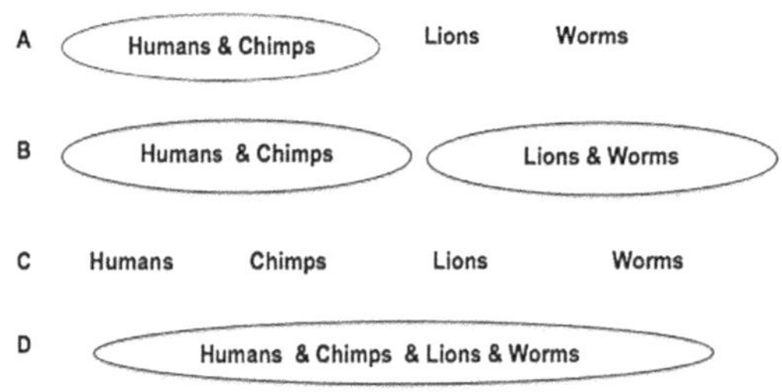


**Fig. 1** **a** Original version of question EN046004 from Flanagan and Roseman (2011). **b** Adapted version of the same question

## 2.2 *Teacher Questionnaire*

The teacher questionnaire was designed to assess acceptance and understanding of evolutionary theory and perceived confidence level related to teaching evolution. It also collected data on length of teaching experience, religiousness and highest qualification in biology. Assessment instruments used within this questionnaire were MATE to assess acceptance of evolution and CINS (Anderson, Fisher, & Norman, 2002) to assess understanding of natural selection, both were chosen for their appropriateness for teachers, contextual validity and ease of completion.

### 2.3 *Teaching Resource Development and Choice of Activities Based on Existing Educational Studies*

The needs of primary school teachers underpinned the development of the teaching materials. Detailed SoW were developed and adapted by liaison with pilot partner schools. Improvements in 'teachability' ensured the resources were endorsed and used by primary school teachers. All resources were fully differentiated and adjusted to be of the correct reading age (Kincaid, Fishburne, Rogers, & Chissom, 1975), so that they were suitable for all children in mainstream schools. Teacher information sheets and mark schemes were included and standardised comprehensive teacher training was given so that non-specialist primary teachers could teach the topic effectively, consistently and with confidence. The cost of the activities was minimised and used equipment that was easily available and suitable for use in the classroom.

Four different SoW were developed to enrich the conceptual understanding of natural selection and evolution. The lesson order (variation, natural selection/microevolution, geological time and macroevolution) was kept constant to facilitate a greater understanding of evolution within a constructivist framework. This best practice restriction accords with a recent randomised trial test that indicated that teaching genetics before evolution markedly improves evolution understanding, with no detriment to genetics understanding, compared to teaching evolution then genetics in secondary school (Mead, Hejmadi, & Hurst, 2017). The lesson on geological time scales was included to improve understanding of macroevolution and help the students visualise the vast periods of time involved (Dodick & Orion, 2003).

The lessons were structured according to the enquiry-based learning method (Van de Walle, 1990), the standard school lesson format used in schools. Each lesson consisted of three separate components: a starter activity to introduce the lesson and establish prior knowledge, a work phase activity and a plenary to consolidate. The starters and plenaries were carefully selected and modified from pre-existing teaching resources. They were identical in each of the four SoW to ensure the different work phase activities were embedded within the same conceptual framework. The work phase activities for lesson 1 and lesson 3 were also identical in each SoW, and there were two alternative work phase activities for lessons 2 and 4. This made a total of four different pathways through the teaching materials. This arrangement also allowed the impact of the work phase activities of lessons 2 and 4 to be evaluated separately (see Table 1).

The work phase activities for lessons 2 and 4 were based on relevant existing small-scale, mostly untested, research studies appropriate for use in primary schools and needing minimal resourcing. In lesson 2, the peppered moth was chosen as an appropriate exemplar species showing natural selection in action (Cook & Saccheri, 2012; Majerus, 2009). Two alternative activities were developed around moth predation to establish whether a student-centric 'hunting' activity was more effective than a seemingly more passive teacher-centric PowerPoint activity. The 'hunting' activity was based on part of Campos and Sá-Pinto (2013) study, students acting as birds using forceps to 'hunt' paper moths on white and newspaper environments.

Several rounds of timed predation were carried out to show differential survival and increased proportion of mimetic colours. The alternative activity was based on the picture story-book intervention of Kelemen et al., (2014). A PowerPoint presentation explaining the process of natural selection in peppered moths was developed from the description of the *pilosas* story-book. The two alternative activities for lesson 4 both involved the same learning experiences but were developed to establish whether homology and common ancestry were more understandable if based upon extinct or extant species. Both involved identification of homologous structures and actual model making. The extinct example chosen was various trilobite species (Wagler, 2010), while mammalian pentadactyl limbs formed the basis of the extant example (Nadelson et al., 2009).

### *2.4 School Recruitment Process*

All primary and middle schools within a 50-mile radius of Bath were invited to participate in the study. This distance allowed individual face-to-face contact and teacher training. In total, 17 schools (nine primary and eight middle) took part in this study, from a mixture of urban and rural settings. The sample was collected from 41 separate classes, taught by 37 different teachers. Data were collected from over 1000 students, 1151 completing the pre-test, 988 students completing both pre-test and post-test and 331 completing the retention test.

The project hinged upon the cooperation of primary school teachers and their adherence to the prescribed teaching package assigned at random. Each school was issued with a detailed scheme of learning together with the resources and received standardised comprehensive teacher training to ensure consistency of delivery.

## 3 Results and Analyses

### *3.1 The Assessment Instrument Is Fit for Purpose*

Before addressing the question of the impact of teaching on student understanding, it is necessary to appraise the utility of our mode of assessment. We consider four metrics to determine whether the instrument is fit for purpose.

First, we examine the internal consistency of the instrument using Cronbach's coefficient $\alpha$. This gave improving scores with successive student assessment session ($\alpha = 0.66$, pre-test; $\alpha = 0.71$, post-test; $\alpha = 0.87$, retention).

Second, if the students just guessed the answers randomly, we expect that their paired post-test and pre-test scores would be uncorrelated. However, we find a significant positive correlation between student pre- and post-teaching scores ($\rho = 0.443$, $P < 0.001$, Spearman's rank correlation).

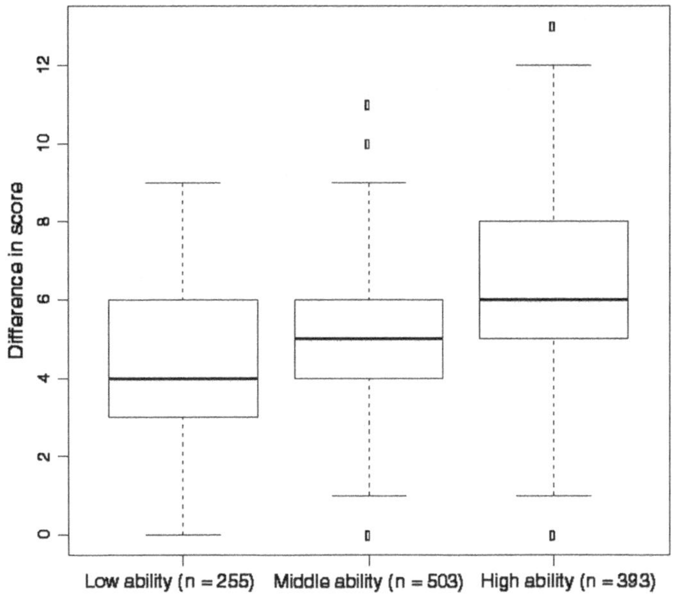


**Fig. 2** Pre-test score stratified by relative ability in science (maximum = 15)

Third, we would also expect higher ability students to possess greater a priori knowledge of the topic and consequently achieve higher marks before formal instruction while also providing evidence of criterion-related validity. We find a significant positive correlation between student ability and pre-test score ($\rho = 0.344$, $P < 0.001$, Spearman's rank correlation) and a significant difference between the pre-test scores in all three ability groups ($P < 0.001$, Kruskal–Wallis rank sum test with a Dunn post hoc), with the medians in the expected directions—higher ability students having higher pre-teaching scores (Fig. 2).

Fourth, the assessment items need to be accessible and discriminatory. The adapted assessment items were found to be easier to read and more appropriate for this age range compared to the original version, with a mean Flesch reading ease score = 70.04 and Flesch–Kincaid reading grade level = 5.88, compared to 63.21 and 7.17, respectively (Kincaid et al., 1975). The assessment instrument was of appropriate difficulty (Kaplan & Saccuzzo, 1997), as demonstrated by the mean percentage of correct responses (36.5% pre-teaching rising to 52.8% post-teaching). It also allowed clear discrimination between students shown by increasing mean item discrimination index values in successive tests (0.31 ± 0.12 pre-test, 0.34 ± 0.06 post-test, 0.36 ± 0.12 retention). We therefore conclude that the assessment instrument is fit for purpose for use with 9- to 11-year-old students.

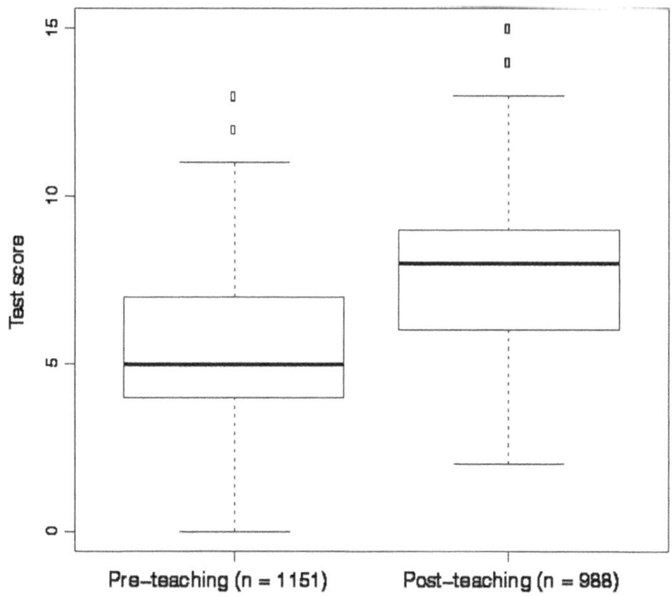


**Fig. 3** Test scores before and after teaching (unpaired)

## *3.2 Teaching Interventions Significantly Improve Student Performance*

Given that the mode of assessment is useful, we ask whether the teaching interventions improve student understanding. This we can address in two modes. First, we can consider all pre- and post-test scores in an unpaired manner and find that the mean student test score increased significantly by 16.3% between the pre-test (5.48 $\pm$ 2.13) and post-test (7.92 $\pm$ 2.38) ($P < 0.001$, Wilcoxon rank-sum test; Fig. 3).

As this analysis doesn't control for the performance of any given student, our second mode of analysis considers the distribution of the change in score values for all students who took both the pre-test and post-test assessment (Fig. 4). When pre-test and post-test scores were analysed in a paired manner, the mean student score increased significantly by 16.1% between pre-test (5.51 $\pm$ 2.15) and post-test (7.92 $\pm$ 2.38) (P < 0.001, Wilcoxon signed-rank test).

Additionally, all four SoW significantly improved student performance ($P = 2.82 \times 10^{-6}$, Kruskal–Wallis rank sum test), the relative utility of which will be assessed in further work.

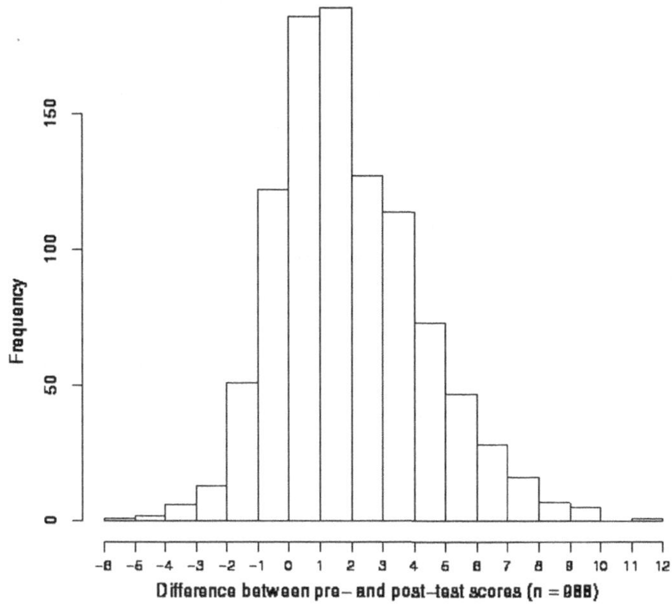


**Fig. 4** Histogram showing the distribution of the change in score values for students taking both pre-test and post-test ($n = 988$)

## *3.3 Evidence for Longer-Term Retention and for Waning*

A common feature of many teaching interventions is that they lead to short-term improvements in understanding, but such improvements are subsequently lost. To address the issue of longer-term retention, we consider a limited sample ($n = 320$) of pupils who took all three tests: pre-test, post-test and retention (mean $131 \pm 73$ days after the post-test). Can we detect evidence for retention and waning?

If there is some degree of retention, we expect that the retention scores should be significantly higher than the pre-test scores. A highly significant difference between pre-test and retention test scores was found ($P < 0.001$, Kruskal–Wallis rank sum test with Nemenyi's post hoc). Second, if there is a waning effect, whereby over time gains made are gradually lost, we also expect to see that post-teaching scores are higher than retention scores. We find this also to be the case ($P = 0.004$, Kruskal–Wallis rank sum test with Nemenyi's post hoc; Fig. 5). These results suggest that teaching interventions have some degree of long-term retention but also that understanding wanes over time.

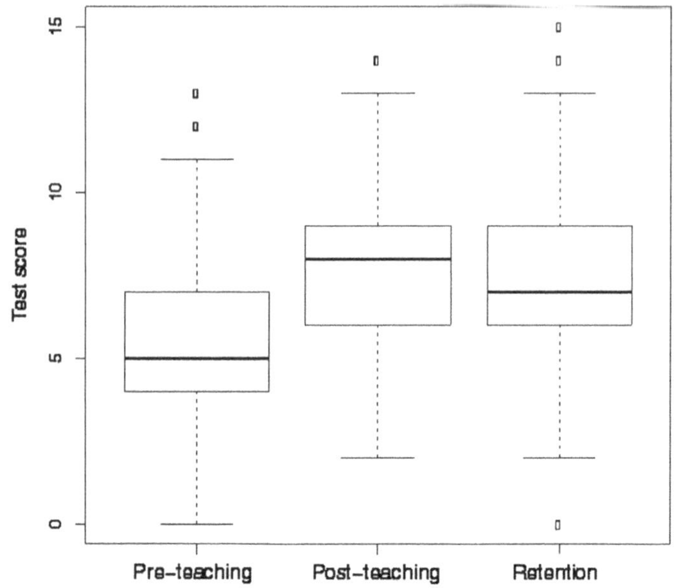


**Fig. 5** Pre-test, post-test and retention test scores for students completing all three tests ($n = 320$)

## *3.4 Science Ability and Gender Predict Response to Teaching*

If we consider the distribution of altered scores, while on average this change is positive there is a considerable variation to be explained. What might predict this variation? Some evidence suggests that primary school children see science as a masculine subject (Archer, 1992) and that existing gender stereotypes influence science aspirations and achievement in boys and girls (Ceci & Williams, 2007; Nosek et al., 2009). Do then we see evidence for differential improvement by gender? Likewise, we might expect that students showing higher ability in science relative to their peers might achieve a larger change in marks. Finally, it has been argued that older children may be better placed to grasp the abstractions of concepts such as natural selection and homology. Thus, we examine these three parameters.

At first sight, we find no evidence that any of them has any explanatory power. There was no significant difference in the difference between pre-teaching and post-teaching scores and gender ($P = 0.143$, Wilcoxon rank-sum test), relative ability in science ($P = 0.545$, Kruskal–Wallis rank sum test) with students of all ability making significant improvements in their scores, e.g. low ability students ($P = < 0.001$, Wilcoxon signed rank sum test), or student age ($\rho = -0.025$, $P = 0.442$, Spearman's rank correlation).

However, higher ability students achieve higher pre-test scores, and as there is a ceiling to the maximum score, we might expect less increase for students whose scores were already high. Indeed, change in score is negatively correlated with pretest score as expected by any ceiling effect ($\rho = -0.438$, $P < 0.001$, Spearman's rank correlation). Thus, we sought to adjust the change in score by correcting for the pre-test score. We consider the residuals of a LOESS regression model (a nonparametric technique that uses local weighted regression to fit a smooth curve through points in a scatter plot) of the change in score against pre-test score and then the distribution of these residuals as a function of gender, relative science ability and age.

Perhaps surprisingly, this analysis indicates no significant correlation between the residuals and student age within upper primary classes ($\rho = -0.047$, $P = 0.144$, Spearman's rank correlation; Fig. 6a). There is, by contrast, a significant difference between high, middle and low ability levels ($\rho = 0.182$, $P < 0.001$, Spearman's rank correlation; Fig. 6b), with higher ability students showing a greater corrected increase. The residuals stratified for gender also gave a significant result ($P = 0.011$, Wilcoxon test; Fig. 6c), with female students achieving higher corrected change scores; this remains significant after Bonferroni correction for three tests.

A multivariate regression model was performed to assess the combined effect of gender, age and ability. This was also found to be significant ($P < 0.001$, adjusted $R^2 = 0.033$), confirming that boys scored significantly lower than girls (coefficient $= -0.34$, $P = 0.01$), revealing significant increases in score with increasing science ability (coefficient $= 0.46$, $P < 0.001$) and that age was not significant (coefficient $= 0.001$, $P = 0.07$). However, only 3.3% of the variance was accounted with these factors alone.

## *3.5 Exploratory Class- and School-Level Analyses Suggest that Teacher Acceptance of Evolution Conditions Student Understanding*

All previous analyses performed at the level of the student come with the caveat that there may be pseudoreplication of data as pupils within any given class share the same teacher, thus introducing a component of non-independence between the students. For this reason, and to consider the importance of teacher-level effects, class-level analyses were carried out using mean class difference scores ($n = 40$) and data gathered from completed teacher questionnaires ($n = 33$), with some teachers teaching multiple classes.

Class-level analysis also enables us to study a series of possible confounding or predicting factors. In this exploratory analysis, only one significant factor was identified, this being a positive correlation between teacher acceptance of evolution and difference in mean class score ($\rho = 0.40$, $P = 0.02$, Spearman's rank correlation), mean $= 85.21 \pm 8.94$ (high acceptance), range 63–99 (moderate to very high acceptance (Rutledge & Sadler, 2007). As this is an exploratory, hypothesis-forming analysis, we did not perform multi-test control, but note that this one significant predictor fails to pass Bonferroni correction.

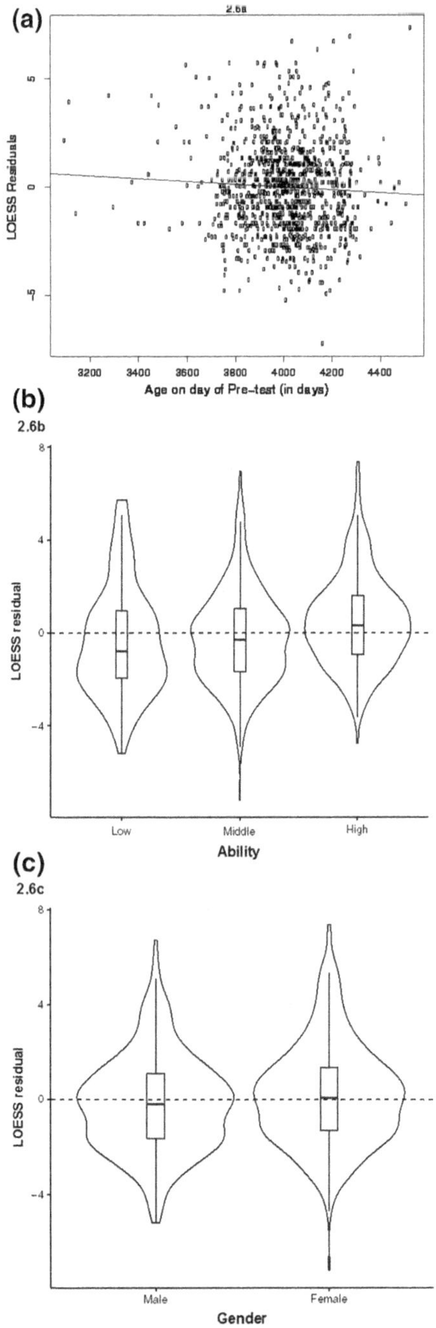


**Fig. 6** **a** Correlation between student age and LOESS residual. **b** Violin plot of LOESS residual scores stratified by relative student ability. **c** Violin plot of LOESS residual scores stratified by gender

School-level analyses were also carried out to compare performance between schools, using mean school difference scores ($n = 17$) and data gathered from recent Ofsted reports and government websites. Factors such as the school's index of multiple deprivation (IMD), type of school and % of students receiving free school meals were tested, and none were found to be significant.

## 4 Discussion

The assessment instrument and its mode of delivery are appropriate for use with 9- to 11-year-olds. It is of appropriate difficulty, allowing clear discrimination between students, and has acceptable internal consistency. Modification of 'readability' enables students to access and understand the assessment items, while its novel mode of delivery makes it quicker to complete and mitigates poor literacy skills.

The evidence presented by this project suggests that 9- to 11-year-olds have the cognitive ability to successfully understand the concepts of natural selection and evolution, when provided with appropriate resources and teaching instruction. There is evidence to suggest that the SoW are effective as there is a significant improvement in student performance in the post-teaching and retention tests compared to their pre-test scores; however, the magnitude of this improvement wanes with time. Importantly, all abilities reported significant increases in their scores after instruction ($P = <0.001$, Wilcoxon signed rank sum test).

We were unable to find any evidence that participating teachers were significantly hindered in their teaching of the topic by their own understanding of evolution, religiousness, highest biology qualification, formal evolution education, gender or years of experience. Only their acceptance of evolution seems to have a significant effect on class performance. All teacher characteristics will be investigated further and cross-checked with a second tranche of data currently being collected. A detailed analysis of the different SoW will also be carried out to help identify if certain teaching packages were more effective than others. In addition, qualitative feedback will be used to enhance and enrich our conclusions, our aim being to provide an evidential basis for establishing best practice.

The results come with many possible caveats and limitations. Although the project was offered to all primary and middle schools within a wide geographical area, participation was self-selecting. This bias is not a great concern because the study is not attempting to make generalisations about the entire teaching population. Instead, it is an exploratory study examining the effectiveness of different SoW delivered by highly motivated teachers, demonstrated by the time commitment they dedicated to the topic.

The fidelity of the results relied on the professionalism of the participating teachers in adhering to the SoW provided and standardised test delivery, after receiving identical training and follow-up. Additionally, the retention sample may have been biased towards completion by more motivated teachers and was also dependent on the scheduling of the topic in the individual schools, with only those teaching the topic early in the year being able to fit it in before the end of the academic year.

We also note that due to the cognitively progressive nature of the concept the order of the four lessons in each SoW could not be altered: variation, natural selection, geological time and finally evolution. Therefore, the relationship between lesson order and understanding was not studied.

**Acknowledgements** We would like to thank the dedicated teachers and their students who participated in this study, enabling the development of resources and collection of such high-quality data. Without their help, this research would not have been possible. We are also grateful to the Evolution Education Trust as the funder of this research and to Jonathan Milner for his support and continued interest in evolution education.

## References

Aguillard, D. W. (1998). *An analysis of factors influencing the teaching of biological evolution in Louisiana public secondary schools* (Vol. 60-03). ProQuest dissertations and theses; Thesis (Ph.D.), Louisiana State University and Agricultural & Mechanical College, 1998, Dissertation Abstracts International.

Ainsworth, S., & Saffer, J. (2013). Can children read evolutionary trees? *Merrill-Palmer Quarterly, 59*(2), 221–247.

Akerson, V., & Donnelly, L. A. (2010). Teaching nature of science to K-2 students: What understandings can they attain? *International Journal of Science Education, 32*(1), 97–124.

Allgaier, J. (2009). Scientific experts and the controversy about teaching creation/evolution in the UK Press. *Science & Education, 19*(6–8), 797–819. https://doi.org/10.1007/s11191-009-9195-5.

Alters, B. J. (2005). *Teaching biological evolution in higher education: Methodological, religious, and nonreligious issues*. Jones & Bartlett Learning.

Alters, B. J., & Nelson, C. E. (2002). Perspective: Teaching evolution in higher education. *Evolution, 56*(10), 1891–1901.

Anderson, D. L., Fisher, K. M., & Norman, G. J. (2002). Development and evaluation of the conceptual inventory of natural selection. *Journal of Research in Science Teaching, 39*(10), 952–978.

Archer, J. (1992). Gender stereotyping of school subjects. *The Psychologist, 5*(2), 66–69.

Bandoli, J. H. (2008). Do state science standards matter? Comparing student perceptions of the coverage of evolution in Indiana & Ohio public high schools. *The American Biology Teacher, 70*(4), 212–216.

Beard, M. N. C. E. N. J. (1996). Better biology teaching by emphasizing evolution & the nature of science. *The American Biology Teacher, 58*(6), 332–336.

Beardsley, P. M., Bloom, M. V., & Wise, S. B. (2012). Challenges and opportunities for teaching and designing effective K-12 evolution curricula. In *Evolution challenges: Integrating research and practice in teaching and learning about evolution* (p. 287).

Berkman, M., & Plutzer, E. (2010). *Evolution, creationism, and the battle to control America's classrooms*. Cambridge University Press.

Berkman, M., & Plutzer, E. (2011). Defeating creationism in the courtroom, but not in the classroom. *Science, 331*(6016), 404–405.

Berti, A. E., Toneatti, L., & Rosati, V. (2010). Children's conceptions about the origin of species: A study of Italian children's conceptions with and without instruction. *The Journal of the Learning Sciences, 19*(4), 506–538.

Bowman, K. L. (2008). The evolution battles in high-school science classes: Who is teaching what? *Frontiers in Ecology and the Environment, 6*(2), 69–74. https://doi.org/10.1890/070013.

Brem, S. K., Ranney, M., & Schindel, J. (2003). Perceived consequences of evolution: College students perceive negative personal and social impact in evolutionary theory. *Science Education, 87*(2), 181–206.

Browning, E., & Hohenstein, J. (2013). The use of narrative to promote primary school children's understanding of evolution. *Education 3-13* (ahead-of-print), 1–18.

Campos, R., & Sá-Pinto, A. (2013). Early evolution of evolutionary thinking: Teaching biological evolution in elementary schools. *Evolution: Education and Outreach, 6*(1), 25.

Carey, S. (1985). *Conceptual change in childhood.* Cambridge, MA: Bradford.

Ceci, S. J., & Williams, W. M. (2007). Why aren't more women in science. In *Top researchers debate the evidence*. Washington, DC: American Psychological Association.

Chanet, B., & Lusignan, F. (2008). Teaching evolution in primary schools: An example in French classrooms. *Evolution: Education and Outreach, 2*(1), 136–140. https://doi.org/10.1007/s12052-008-0095-y.

Cook, L., & Saccheri, I. (2012). The peppered moth and industrial melanism: Evolution of a natural selection case study. *Heredity, 110*(3), 207–212.

Deniz, H., & Donnelly, L. A. (2011). Preservice secondary science teachers' acceptance of evolutionary theory and factors related to acceptance. *Reports of the National Center for Science Education, 31*(4).

Dodick, J., & Orion, N. (2003). Cognitive factors affecting student understanding of geologic time. *Journal of Research in Science Teaching, 40*(4), 415–442.

Engel Clough, E., & Wood-Robinson, C. (1985). Children's understanding of inheritance. *Journal of Biological Education, 19*(4), 304–310.

Evans, E. M. (2008). Conceptual change and evolutionary biology: A developmental analysis. In *International handbook of research on conceptual change* (pp. 263–294).

Fail, J. (2008). A no-holds-barred evolution curriculum for elementary and junior high school students. *Evolution: Education and Outreach, 1*(1), 56–64.

Ferrari, M., & Chi, M. T. (1998). The nature of naive explanations of natural selection. *International Journal of Science Education, 20*(10), 1231–1256.

Flanagan, J. C., & Roseman, J. E. (2011). *Assessing middle and high school students' understanding of evolution with standards-based items.* Paper presented at the National Association of Research in Science Teaching Annual Meeting, Orlando, FL.

Fowler, S. R., & Meisels, G. G. (2010). Florida teachers' attitudes about teaching evolution. *The American Biology Teacher, 72*(2), 96–99.

Geary, D. C. (2006). Evolutionary developmental psychology: Current status and future directions. *Developmental Review, 26*(2), 113–119.

Glaze, A. L., & Goldston, M. J. (2015). US science teaching and learning of evolution: A critical review of the literature 2000–2014. *Science Education, 99*(3), 500–518.

Goldston, M. J., & Kyzer, P. (2009). Teaching evolution: Narratives with a view from three southern biology teachers in the USA. *Journal of Research in Science Teaching, 46*(7), 762–790.

Gonzales, P., Guzmán, J. C., Partelow, L., Pahlke, E., Jocelyn, L., Kastberg, D., & Williams, T. (2004). Highlights from the trends in international mathematics and science study (TIMSS), 2003. NCES 2005-005. US Department of Education.

Gregory, T. R. (2009). Understanding natural selection: Essential concepts and common misconceptions. *Evolution: Education and Outreach, 2*(2), 156–175.

Griffith, J. A., & Brem, S. K. (2004). Teaching evolutionary biology: Pressures, stress, and coping. *Journal of Research in Science Teaching, 41*(8), 791–809.

Ha, M., Haury, D. L., & Nehm, R. H. (2012). Feeling of certainty: Uncovering a missing link between knowledge and acceptance of evolution. *Journal of Research in Science Teaching, 49*(1), 95–121.

Kaplan, R., & Saccuzzo, D. (1997). Psychological testing: Principles, applications and issues (4e éd.). Pacific Grove: Brooks: Cole Publishing (1re éd. 1982).

Kelemen, D., Emmons, N. A., Seston Schillaci, R., & Ganea, P. A. (2014). Young children can be taught basic natural selection using a picture-storybook intervention. *Psychological Science, 25*(4), 893–902. https://doi.org/10.1177/0956797613516009.

Kelemen, D., & Rosset, E. (2009). The human function compunction: Teleological explanation in adults. *Cognition, 111*(1), 138–143.
Kincaid, J. P., Fishburne, R. P., Jr., Rogers, R. L., & Chissom, B. S. (1975). *Derivation of new readability formulas (automated readability index, fog count and flesch reading ease formula) for navy enlisted personnel*. Retrieved from.
Lawson, A. E., & Thompson, L. D. (1988). Formal reasoning ability and misconceptions concerning genetics and natural selection. *Journal of Research in Science Teaching, 25*(9), 733–746.
Legare, C. H., Lane, J. D., & Evans, E. M. (2013). Anthropomorphizing science: How does it affect the development of evolutionary concepts? *Merrill-Palmer Quarterly, 59*(2), 168–197.
Levesque, P. J., & Guillaume, A. M. (2010). Teachers, evolution, and religion: No resolution in sight. *Review of Religious Research*, 349–365.
Losh, S. C., & Nzekwe, B. (2011). Creatures in the classroom: Preservice teacher beliefs about fantastic beasts, magic, extraterrestrials, evolution and creationism. *Science & Education, 20*(5–6), 473–489.
Majerus, M. E. (2009). Industrial melanism in the peppered moth, *Biston betularia*: An excellent teaching example of Darwinian evolution in action. *Evolution: Education and Outreach, 2*(1), 63–74.
Martin, M. O., Mullis, I. V., Foy, P., & Stanco, G. M. (2012). *TIMSS 2011 International results in science*. ERIC.
Mead, R., Hejmadi, M., & Hurst, L. D. (2017). *Teaching genetics prior to teaching evolution improves evolution understanding but not acceptance*. Biology: Public Library of Science.
Nadelson, L., Culp, R., Bunn, S., Burkhart, R., Shetlar, R., Nixon, K., & Waldron, J. (2009). Teaching evolution concepts to early elementary school students. *Evolution: Education and Outreach, 2*(3), 458–473. https://doi.org/10.1007/s12052-009-0148-x.
Nadelson, L., & Nadelson, S. (2010). K-8 educators perceptions and preparedness for teaching evolution topics. *Journal of Science Teacher Education, 21*(7), 843–858.
Nadelson, L., & Sinatra, G. M. (2008). Educational professionals' knowledge and acceptance of evolution. *Evolutionary Psychology*.
Nehm, R. H., & Schonfeld, I. S. (2007). Does increasing biology teacher knowledge of evolution and the nature of science lead to greater preference for the teaching of evolution in schools? *Journal of Science Teacher Education, 18*(5), 699–723.
Nehm, R. H., & Schonfeld, I. S. (2008). Measuring knowledge of natural selection: A comparison of the CINS, an open-response instrument, and an oral interview. *Journal of research in science teaching, 45*(10), 1131–1160.
Nelkin, D., & Lindee, S. (1995). *The DNA mystique*. WH Freeman & Company.
Nosek, B. A., Smyth, F. L., Sriram, N., Lindner, N. M., Devos, T., Ayala, A. … Gonsalkorale, K. (2009). National differences in gender–science stereotypes predict national sex differences in science and math achievement. *Proceedings of the National Academy of Sciences, 106*(26), 10593–10597.
Oliva, J. M. (2003). The structural coherence of students' conceptions in mechanics and conceptual change. *International Journal of Science Education, 25*(5), 539–561.
Osborne, J., Simon, S., & Collins, S. (2003). Attitudes towards science: A review of the literature and its implications. *International Journal of Science Education, 25*(9), 1049–1079.
Papadopoulou, P., Stanissavljevic, J., Katakos, E., & Athanasiou, K. (2011). *Acceptance and understanding of evolution theory: A comparative study of Greek and Serbian teachers*. Paper presented at the Science learning and Citizenship, E-book Proceedings of the ESERA 2011 Conference, Lyon, France. http://lsg.ucy.ac.cy/esera/e_book/base/index.html.
Piaget, J., & Cook, M. (1952). *The origins of intelligence in children* (Vol. 8). New York: International Universities Press.
Rutledge, M. L., & Sadler, K. C. (2007). Reliability of the measure of acceptance of the theory of evolution (MATE) instrument with university students. *The American Biology Teacher, 69*(6), 332–335.

Samarapungavan, A., & Wiers, R. W. (1997). Children's thoughts on the origin of species: A study of explanatory coherence. *Cognitive Science, 21*(2), 147–177.
Sinatra, G. M., Brem, S. K., & Evans, E. M. (2008). Changing Minds? Implications of conceptual change for teaching and learning about biological evolution. *Evolution: Education and Outreach, 1*(2), 189–195. https://doi.org/10.1007/s12052-008-0037-8.
Solomon, G. E. (2002). Birth, kind and naïve biology. *Developmental Science, 5*(2), 213–218.
Thagard, P., & Findlay, S. (2009). Getting to Darwin: Obstacles to accepting evolution by natural selection. *Science & Education, 19*(6–8), 625–636. https://doi.org/10.1007/s11191-009-9204-8.
Trani, R. (2004). I won't teach evolution; it's against my religion. And now for the rest of the story…. *The American Biology Teacher, 66*(6), 419–427.
Van de Walle, J. A. (1990). *Elementary school mathematics, teaching developmentally*. ERIC.
van Dijk, E. M., & Kattmann, U. (2009). Teaching evolution with historical narratives. *Evolution: Education and Outreach, 2*(3), 479–489.
Venville, G., & Donovan, J. (2005). Searching for clarity to teach the complexity of the gene concept.
Venville, G., & Donovan, J. (2007). Developing year 2 students' theory of biology with concepts of the gene and DNA. *International Journal of Science Education, 29*(9), 1111–1131.
Venville, G., Gribble, S. J., & Donovan, J. (2005). An exploration of young children's understandings of genetics concepts from ontological and epistemological perspectives. *Science Education, 89*(4), 614–633.
Wagler, R. (2010). A missing link: K-4 biological evolution content standards. *Evolution: Education and Outreach, 3*(3), 443–450.
Wagler, R. (2012). Assessing "the framework" for kindergarten through fifth grade biological evolution. *Evolution: Education and Outreach, 5*(2), 274–278.
Weiss, M., & Dreesmann, D. C. (2014). Aspirations and expectations: Comparing scientist and teacher views as a source of ideas for teaching evolution. *Universal Journal of Educational Research, 2*(5), 421–431.
Wiles, J. R., & Alters, B. (2011). Effects of an educational experience incorporating an inventory of factors potentially influencing student acceptance of biological evolution. *International Journal of Science Education, 33*(18), 2559–2585.
Williams, K. A., & Cavallo, A. M. (1995). reasoning ability, meaningful learning, and students' understanding of physics concepts: Relating students' reasoning ability and learning styles to their physics misconceptions. *Journal of College Science Teaching*, 311–314.

**Loredana L. Buchan** has over 25 years of experience in education, teaching science in a large secondary school in Bath. During that time she held a variety of roles including head of KS3 science, PGCE mentor, head of year and head of biology. Currently, she is researching at the Milner Centre for Evolution at the University of Bath and is in the final year of her Ph.D. Her large-scale mixed-method study aims to develop effective Evolution and Inheritance teaching interventions for use by non-science specialist teachers in UK primary schools.

**Momna V. Hejmadi** Momna's research interests range from understanding cellular signalling in hypoxia to the pedagogy of learning. She is particularly interested in understanding how teaching genetics and evolution may link to the understanding and acceptance of evolution among school pupils. She also explores the role of teachers, family and peers in influencing acceptance of evolution across different demographic groups. She was awarded the UK Higher Education Academy National Teaching Fellowship in 2015, along with a number of awards for teaching excellence. Her projects have been funded by the UK Higher Education Academy, Centre for Bioscience, JISC OER phase 1 and 2 schemes, British Council, BBSRC and Milner Education Trust.

**Laurence D. Hurst** of the University of Bath is a Professor of Evolutionary Genetics, Director of the Milner Centre for Evolution, Director of the Genetics and Evolution Teaching Project and the President of the Genetics Society. He was educated at Cambridge, Harvard and Oxford. He has been elected a member of EMBO, a Fellow of the Academy of Medical Sciences and a Fellow of the Royal Society. He was awarded the Scientific Medal of the Zoological Society and the Genetics Society Medal. His biological research attempts to understand the role of errors in gene and genome evolution and to translate this into enhanced diagnostics and therapy. His education research uses large-scale trials to improve the teaching of genetics and evolution.

# Developing a Cross-Curricular Session about Evolution for Initial Teacher Education: Findings from a Small-Scale Study with Pre-service Primary School Teacher

**Berry Billingsley, Manzoorul Abedin, Keith Chappell and Chris Hatcher**

## 1 Introduction

The Science National Curriculum for primary schools in England has included since 2014 a number of objectives relating to evolution and inheritance for year 6 (10–11 years old). Rationales given for this change emphasised that the theory of evolution is a key concept which is fundamental to biology and scientific literacy and that it is important for children to begin learning about evolution at this younger age to help them gain a good level of understanding for when they study it in more depth later on (Borgerding, Klein, Ghosh, & Eibel, 2015). Evolution is widely perceived, however, as a science topic that presents multiple challenges for teachers (Sanders & Ngxola, 2009). Most of the research to date consists of studies with secondary school teachers and secondary school pre-service teachers and this indicates concerns about tensions around subject knowledge, personal conflicts with evolution, and expectations of resistance from students and/or their parents on the basis of religion. It seems reasonable to presume that primary school teachers and pre-services would also experience many of these tensions.

There are aspects of primary teaching and of the backgrounds, interests and expertise of primary school teachers that differ from teaching science in secondary school. Primary teachers and pre-service primary teachers are unlikely to have completed a

B. Billingsley (✉) · M. Abedin · K. Chappell
Faculty of Education, Canterbury Christ Church University, Canterbury, UK
e-mail: Berry.Billingsley@Canterbury.ac.uk

M. Abedin
e-mail: manzoorul.abedin@canterbury.ac.uk

K. Chappell
e-mail: k.chappell403@canterbury.ac.uk

C. Hatcher
Loughborough University, Loughborough, UK
e-mail: c.hatcher@lboro.ac.uk



U. Harms and M. J. Reiss (eds.), *Evolution Education Re-considered*,
https://doi.org/10.1007/978-3-030-14698-6_3

degree in science before their teacher training, in contrast with a typical secondary school science teacher. In addition, the primary school teacher typically teaches many curriculum subjects and in England this frequently includes teaching science and religious education (RE). Cross-curricular teaching is far more common in primary schools and there is a greater tendency to make time for teaching that adapts and responds to children's questions, concerns and interests.

The study described in this chapter was motivated by an interest in whether pre-service primary school teachers would appreciate an opportunity to explore ideas about evolution in a session that bridges their science and RE teacher education modules. Prior to the cross-curricular study, which is the main focus of this chapter, we gathered data from other cohorts of primary pre-service teachers during the two previous years to discover their attitudes to teaching evolution. We also conducted interviews with selected pre-service teachers. These interviews highlighted that pre-service primary school teachers are particularly concerned about the possibility that their school students with a religious faith will feel uncomfortable with, or conflicted by, the science. Pre-service teachers were also asked their attitudes towards different organisational formats, both within their own course and within their approaches to primary teaching—such as a cross-curricular session, a session of RE and a session of science. Responses by focus groups, by individuals in whole class sessions and in individual interviews indicated a mix of positions. Many of those interviewed said they would appreciate a cross-curricular session within their own programme but would be resistant to teaching a session using a cross-curricular format. This was typically said to be because of the risk of upsetting students with a religious faith. The design and statements in our survey were informed by this preliminary work.

## 2 Review of Literature

We begin this review of some of the existing literature by discussing perceptions and perspectives on single subject and cross-curricular teaching. The linking of subjects or disciplines for curriculum organisation is variously described as an integrated, interdisciplinary, multidisciplinary, transdisciplinary, blended, cross-curricular, cross-disciplinary, thematic or a topic-based process. The term 'cross-curricular learning' is used to describe the application of skills, knowledge and attitudes of different disciplines to a single experience, theme or idea; it also incorporates the interdisciplinary dimension of linking subjects to develop conceptual insight into particular phenomena, which, for the purposes of this study, is evolution as a teaching topic (Barnes, 2015). In the context of the National Curriculum in England, the Rose Report in 2009 emphasised the place and value of cross-curricular teaching across the curriculum. Foreseeably, however, using a cross-curricular approach to teach an area that is conceptually challenging and that is widely seen as contentious could create an overload of questions and possibilities for participants. These possibilities informed the current study and the design of our cross-curricular session for pre-service primary teachers.

Previous research highlights that school pupils frequently hold misconceptions relating to the mechanism of evolution, which is poorly understood (Kalinowski, Leonard, Andrews, & Litt, 2013). Existing research also reminds us that teachers and pre-service teachers may themselves have limited evolution content knowledge and hold misconceptions about evolution content (Dodick, Dayan, & Orion, 2010; Kim & Nehm, 2011; Nehm, Kim, & Sheppard, 2009). Previous studies also indicate that teachers often experience negative responses to their teaching from students, parents, community, colleagues and clergy (Bramschreiber, 2014; Chuang, 2003; Fowler & Meisels, 2010). Perhaps not surprisingly, some teachers report feeling very distressed about the prospect, or their experiences of, teaching about evolution (Griffith & Brem, 2004; Sanders & Ngxola, 2009). There are also teachers, however, who have not seen any students responding negatively while learning about evolution (Hanley, Bennett, & Ratcliffe, 2014).

Few published studies have investigated the stances taken by pre-service and in-service primary teachers about the teaching of evolution. Our research in this area has indicated that a majority of teachers are positive about the prospect of teaching evolution while at the same time expressing need for support with developing classroom activities, improving subject knowledge and coming up with strategies to ensure positive experiences for children with a religious faith (Billingsley & Abedin, 2016).

## 3 Purpose of the Research

To date, research has revealed that there are several challenges, which teachers and pre-service teachers experience or anticipate in relation to teaching about evolution. The present study builds on this existing research to explore pre-service primary teachers' perceptions of a cross-curricular teaching session in their teacher education programme. The aim of the session was to provide pre-service primary teachers with a space in which they could explore the relationships between science and religion prior to their regular science session on evolution in the programme where they would be developing pedagogies and subject knowledge relating to science.

Before discussing the cross-curricular teacher education session for primary pre-service teachers, this chapter first reports on a baseline survey with 158 pre-service teachers. We then describe and discuss data gathered before and after a cross-curricular session ($n = 45$), regarding participants' subject knowledge, perception and attitude to teaching about evolution.

# 4 Methodology

## *4.1 Design of the Survey*

Informed by the preliminary work, the aims of the baseline survey were to find out pre-service teachers' attitudes to teaching about evolution, planned approach to teaching evolution, subject knowledge of science and the relationship between religion and the nature of science.

The design of the survey instrument was informed by a series of studies in schools by the LASAR (Learning about Science and Religion) project team (see for example Billingsley, 2004; Billingsley, Brock, Taber, & Riga, 2016; Billingsley, Taber, Riga, & Newdick, 2011, 2012; Taber, Billingsley, Riga, & Newdick, 2011). The findings of these studies highlighted that school students have few if any opportunities to discuss a range of stances on the relationship between science and religion. We have also found that in science lessons, school students tend to hold back questions that they perceive to have a religious aspect and that misperceptions in some students' scientific understanding may not be apparent to their teachers. The themes addressed in the survey also drew on our review of issues that apply in secondary school when teaching evolution, as we surmised that many of these issues were also likely to apply in primary teaching with respect to evolution. These issues are weak teacher subject knowledge and resistance to teaching because of a perceived conflict by the teacher and/or the students between evolution and religious beliefs about human origins. The questionnaire was structured to determine pre-service teacher views on teaching evolution and subject knowledge before introducing the relationship between science and religion in order to not confound answers related to subject knowledge. Statements within the questionnaire included accurate subject knowledge as well as common misconceptions about the theory of evolution. These questions act as a tool to measure the impact of the cross-curricular session on subject knowledge. During the development of the questionnaire, we ran pilot studies with groups of teachers and pre-service teachers who did not participate in the final study. This included a pilot survey with postgraduate pre-service primary teachers and a pilot survey with primary school teachers attending a professional development workshop. Via these pilot studies, we honed the wording of the statements to reduce ambiguity. We also consulted with the project's Advisory Board which included senior academics in biology, ethics and theology, based in England and overseas.

The survey instrument consisted of 21 statements with a five-point Likert scale response section (strongly agree, agree, neither agree or disagree, disagree, strongly disagree). The survey was administered online and the instructions and design meant that participants could skip any question that they did not want to answer. Pre-service teachers also had the option of a space to explain their responses (labelled 'Comment if you'd like to'). These arrangements were to prevent participants from feeling pressured to give a response if they were reluctant or unsure about how to answer. The instructions also explained that participants' names would not be used in any reports. The surveys were provided to pre-service teachers using a computer lab during time slots organised within their taught sessions. Pre-service teachers were given the option to complete but not submit the survey if they wished. Questions to discover participants' religiosity and level of science qualification were placed at the end of the survey to avoid influencing how participants responded to the statements.

### *4.2 Sample*

The sample for the baseline survey consisted of 158 pre-service primary school teachers on a three-year undergraduate course. Those participating in the baseline survey were pre-service teachers on each of three iterations of the programme. The baseline survey was conducted in the first year of the three-year programme before participants had attended any teaching about evolution on their course.

## 5 Baseline Survey Findings

Analysis of the responses by this cohort of participants indicated that 43% were from comprehensive schools, 17% from academy comprehensives, 12% from private schools, 6% from academies and 4% each from colleges, sixth form colleges, grammar and state grammar schools. The remaining 6% came from technology colleges, grammar academies and British Military schools. Out of these schools, 77% were non-Church schools and the remaining 23% were Church schools.

We also noted that just under half, 45%, of 135 respondents of the total population identified themselves as Christians while the second largest group, about 27%, indicated that they did not have a religion ('none'); about 12% of the participants indicated they were atheist and 13% agnostic, 3% as Muslims and 1% as Hindus.

Just over a quarter (27%) of the pre-service teachers had a GCSE (qualification to age 16) or equivalent in general science, 55% had GCSE or equivalent in biology and another 18% had an A Level (qualification to age 18) in biology. The response rate to the baseline survey was good as 96% of pre-service teachers answered all questions except two. The principal findings are shown in Table 1. We have collapsed the categories for agree (agree/strongly agree and for disagree (disagree/strongly disagree).

In general, participants perceived that evolution is an important topic for primary children to learn. About 80% of participants agreed that 'Evolution is an important idea for children in primary school to learn about', and 75% agreed that they were 'glad that evolution will be taught in primary school'. Though positive on the inclusion of evolution in primary teaching, pre-service teachers revealed that they received very little teaching about evolution in their own school education. Some typical comments were:

> Very brief lessons

**Table 1** Findings of the baseline survey

| | Agree (%) | Neither agree nor disagree (%) | Disagree (%) |
|---|---|---|---|
| Evolution is an important idea for children in primary school to learn about in science | 81 | 16 | 3 |
| I am glad that evolution will be taught in primary school | 75 | 22 | 3 |
| Parents should be informed that a lesson on evolution will take place and can remove their child | 42 | 28 | 30 |
| I am looking forward to teaching evolution | 54 | 39 | 7 |
| I am concerned about teaching evolution | 41 | 33 | 26 |
| It will be important to take into account children's religious beliefs | 86 | 11 | 3 |
| Evolution is a theory and not a fact | 49 | 29 | 22 |
| I have an adequate understanding of evolution to teach at this level | 23 | 40 | 37 |
| Evolution says that humans evolved from monkeys | 47 | 34 | 19 |
| I think children are likely to ask questions about religion | 62 | 21 | 17 |
| The Church of England does not accept evolution | 38 | 48 | 14 |
| Christians believe in a six-day Creation | 67 | 18 | 15 |
| Christianity teaches that the universe was created in six days of 24 hours followed by a day of rest | 65 | 24 | 11 |
| Evolution says that life evolved over billions of years from simpler creatures | 84 | 14 | 2 |
| Darwin is the originator of the theory of evolution | 80 | 17 | 3 |
| The theory of evolution is in conflict with a belief in Creation | 70 | 22 | 8 |
| Darwin's theory was controversial because it contradicted religious teaching | 83 | 14 | 3 |
| Fossils are evidence for the theory of evolution | 79 | 16 | 5 |
| I will tell children they have a choice about whether to accept evolution | 76 | 20 | 4 |
| Children with religious beliefs are unlikely to accept evolution | 25 | 70 | 5 |
| Evolution is a very well supported explanation for how life came to be | 73 | 22 | 5 |

> Quite basic overview during GCSE, on the survival of the fittest and adaptations of living things
>
> I don't actually remember doing evolution in school until it was touched upon very briefly in GCSE biology

In response to the survey statement: 'I have an adequate understanding of evolution to teach at this level', only 23% agreed/strongly agreed. This said, 80% of the pre-service teachers agreed or agreed strongly that it was important for children in primary school to learn about evolution. In addition, their confidence in the validity of the theory was high. In response to the statement 'Evolution is a very well supported explanation for how life came to be', about three quarters (73%) agreed. Markedly fewer, but still a majority of 54%, agreed that they were looking forward to teaching evolution.

We found that 76% of the pre-service teachers agreed with the statement that 'I will tell children they have a choice about whether to accept evolution'. Some further explanation can be drawn from their other responses. Thus, while the large majority accepted the theory of evolution for themselves, there was less agreement about the level of acceptance they might find in students and still less agreement on whether it is appropriate to attempt to move students closer to acceptance—for example, 'Because evolution is subjective not everybody believes it'. A student further explained in a comment 'Evolution is a theory which they can choose to believe in or not' and 'I believe that evolution is how humans came to be on earth, but then again I am an atheist, so just because I believe it doesn't mean that the children I teach should'. Another expressed the view that 'Because there are different theories and religious beliefs on the concept of evolution and children cannot be forced into one idea'. Another pre-service teacher said 'Religious views are very different and oppose the view of evolution. Man developed from earlier creatures'. Another wrote 'The contradictions between Christianity and Darwin's theory. That as time changed animals and plants had to adapt to their surroundings in order to survive'.

A majority of 70% agreed or agreed strongly that the theory of evolution is in conflict with a belief in creation. In addition, 38% of the pre-service teachers agreed or agreed strongly that the Church of England does not accept evolution. These perceptions likely create a level of pressure for teachers teaching evolution, and 86% of participants agreed or strongly agreed that 'It will be important to take into account children's religious beliefs'. A quarter agreed or strongly agreed that children with a religious belief are unlikely to accept evolution.

## 6 The Cross-Curricular Teacher Education Session

The cross-curricular session on teaching science and religious education took place during the pre-service primary teachers' undergraduate programme. The cohort who attended the session was those on the third iteration of the course and as such was a subgroup of the full cohort of pre-service teachers involved in the baseline study. In

England, primary school teachers teach a range of subjects including science and RE. Each cohort, including this cohort, also attended a teacher education session on evolutionary biology as part of their science teaching module later in their programme. For those taking part in the cross-curricular teacher education session, the pre-session survey was administered in the week before the session and the post-session survey was administered at the end of the session. The data we report correspond to the pre-service teachers who participated in the pre-session baseline survey, session and the post-session survey.

The design of the session took into consideration the findings from the baseline survey responses from previous cohorts. We intended that the session would provide a forum in which pre-service teachers could voice and explore their own and other ways to conceptualise the relationships between science and religion. In addition, we wanted to encourage pre-service teachers to shift from the position that children should be offered a choice between science and religion to the position that science and religion are not necessarily incompatible. Thirdly, we aimed to address some common misconceptions relating to evolution and to enhance participants' science subject knowledge.

The first section of the presentation invited pre-service teachers to give their perceptions of how the media typically describe the relationship between science and religion. The discussion turned then to the notion that a school teacher can resist and critique perspectives that appear in the media, and participants then examined and shared examples of ways that the relationship is described in scholarship. The session then drew participants' attention to particular areas of confusion or gaps that were common in survey responses and sought to address these. One part of the presentation examined the misperception that the Church of England does not accept evolution. Pre-service teachers were also introduced to the idea that Darwin's work built not only on his own observations but also on other scholars' research and reflections.

## 7 Findings from the Pre- and Post-Studies of the Cross-Curricular Teacher Education Session

In the following sections, we discuss the survey data gathered before and after the cross-curricular teacher education session using comments to add detail to the quantitative findings.

The first question (which was open) on the post-questionnaire asked pre-service teachers what they perceived to be the key ideas in the session. The responses indicated that the session had successfully moved many students forward from a perception of necessary conflict. Comments included:

> We should provide children with role models that represent a variety of scientists' views so that they feel that they don't have to choose science or religion.

> That evolution is not something that should be taught as a conflicting idea to religion but that both ideas can exist alongside each other.
>
> Teaching children a 'balanced view' isn't as simple as I first read, and in doing so I could be influencing the children toward the idea that a decision has to be made as to whether they hold a scientific or religious view toward evolution.
>
> The idea that you can have both a religious (Christian) view and a science orientated view. I have both and I previously hadn't known of anyone who has both, other than my family so I didn't really know if my view was accepted.

### (a) **Attitudes to cross-curricular pedagogy**

In the post-session survey, pre-service teachers ($n = 45$) were asked if their teaching would be cross-curricular (science and RE) or single subject and why. The majority of students indicated that they were in favour of a cross-curricular session in their own approaches to teaching evolution (71%, $n = 32$), while a small number of pre-service teachers favoured single subject teaching or were unsure. In some of these cases, the cross-curricular session was in addition to a single-subject lesson. Six (13%) of the pre-service teachers were unsure and 7 (16%) felt that teaching about evolution should only take place in a science session. The comments selected below illustrate some of these positions:

> Single subject. The two should not be mixed. BELIEFS (RE) should not be intertwined with FACT (Science).
>
> I will teach them separately because I feel the combined teaching of them encourages children into making a choice.
>
> Cross-curricular, because different perspectives can help answer different questions.
>
> Probably both cross curricular and single subject. Some questions are best answered in isolation and others with other considerations.
>
> I think it would be a good idea to teach them separately and together so that they are represented equally and then have another lesson to discuss possible contradictions or how they complement each other.
>
> I would teach science first, and then follow up with the RE, giving all the theories and ideas. I would then put them both together showing how it is not a question of either or.

### (b) **Changes discerned in the data between the pre- and post-cross-curricular session surveys**

In this section, we compare the before and after data and discuss changes in the pre-service teachers' positions. Firstly, on the importance of teaching evolution, the proportion who agree increased from 73 to 89% following the study. About half of the participants both before and after (48 and 50%, respectively) indicated that they were 'looking forward to teaching evolution'. The proportion who agreed with the statement 'I have an adequate understanding of evolution to teach at this level' increased substantially from 24% in the pre-session survey to 53% in the post-session survey. We also found an increase in the proportion who agree with the statement that 'Children are likely to ask questions about religion'; post-cross-curricular session, this proportion was 84%, an increase of 17% from the pre-session level.

Pre-service teachers were divided in their opinions on whether 'Parents should be informed that a lesson on evolution will take place and can remove their child'—while more than 40% of them agreed in both the pre- and post-session surveys, half of them (50%) disagreed in the post-session survey, which was an increase from 38% at the pre-session stage.

We noted that there is a very slight increase in the proportion who agree that 'Not all scientists support evolution' with the before and after figures being 55 and 57%, respectively; for 'Evolution says that life evolved over billions of years from simpler creatures', agreement increased from 82 to 95%; for 'Fossils are evidence for the theory of evolution', the percentage who were in agreement increased from 66 to 86% and for 'Evolution is a very well supported explanation for how life came to be' the percentage in agreement increased from 67 to 75%. The number agreeing that 'The theory of evolution is in conflict with a belief in Creation' fell considerably from 73 to 41%.

Similarly, pre-service teachers' agreement with 'Christians believe in a six-day Creation' and 'Most Christians reject evolution' decreased by 13% (from 74 to 61%) and 14% (39 to 25%) following the session. About half of the pre-service teachers (48%) agreed that 'The Bishops of the Church of England do not accept evolution' in the pre-session survey. The figure reduced to 12% in the post-session survey.

## 8 Discussion

The concerns raised by the pre-service teachers in this study about teaching evolution are similar to those reported by other studies (see for example Sanders & Ngxola, 2009). Pre-service teachers said that they felt they lacked sufficient subject knowledge and they were also of the view that children would ask questions about religion. Findings from the data gathered here indicate that pre-service teachers are concerned to ensure that their students will have positive experiences of learning about evolution. At the same time, the perceptions held by a substantial majority (70%) were that the theory of evolution is in conflict with a belief in creation and a quarter of survey participants agreed or agreed strongly that children with a religious belief are unlikely to accept evolution. We note that three quarters of these pre-service teachers identified that they would tell children that they could choose what to believe.

Our intention in the design of the cross-curricular teacher education session was that pre-service teachers would have opportunities to consider other ways to conceptualise the relationship between science and religion and also to consider presenting any choice to students as a choice between 'conflict or not' rather than a choice between science and religion. The post-session data indicate that these aims were met and also that a significant proportion of the pre-service teachers following the cross-curricular session felt that they would also use this strategy with their students.

One unintended outcome of the session was the possibility of a slight increase (from 55 to 57%) in the numbers of the pre-service teachers who supposed that 'Not all scientists accept evolution'. There was an opportunity in follow-up interviews to find out more about what led to this. We found that some pre-service teachers had misunderstood the meaning of the phrase 'theistic evolution', which had been included at some points during the session.

With these points in mind, this leads us to offer a number of recommendations. The science of evolution is conceptually challenging and so too is the reasoning that underpins an appreciation that science and religion are not necessarily incompatible. Key concepts for evolution include variation, natural selection and adaptation, each of which can be understood/misunderstood in terms of conscious agency in addition to the manner in which evolutionary scientists use them. There is certainly potential to explore these notions with pre-service primary school teachers more fully to examine where potentially confusing notions are impacting on the understanding and acceptance of evolutionary theory by teachers and potential improvements in teaching. These may relate to religious notions and/or common usage of terms. There are many additional terms associated with these activities and we recommend a glossary and care by teacher educators as well as teachers when defining and using these terms.

With regard to perceptions of ways to relate science and religion, it is interesting to note the number of comments by pre-service teachers that seem to indicate a fairly passive acceptance of the notion of conflict. There may be the potential in sessions about religious education for pre-service teachers to introduce other science topics that are less commonly associated with conflict and explore how these relate to religious ideas prior to tackling the more specific concerns encountered in teaching evolution. In a similar way to the problems encountered in teaching evolution, the terminology used in science and religion discussions is often technical and involves the specific use of terms with different common usage. Useful work could be carried out in examining alternative ways to present concepts that would enable teachers and students to be more comfortable and to avoid misunderstanding and ambiguity.

Overall, we see an advantage with delivering the teaching for pre-service primary teachers in two sessions where the first is a cross-curricular teacher education session. The central aim of the first session is to develop pre-service teachers' own confidence and understanding. This includes ensuring that pre-service teachers appreciate that science and religion are not necessarily incompatible, countering misperceptions and establishing some key aspects of subject knowledge. We recommend that a second session is focused on developing classroom activities to develop and consolidate participants' understanding of evolution and ways to teach it.

## 9 Limitations and Suggestions for Further Research

For this study, the cross-curricular session was delivered to the full cohort on the programme in one iteration of the course. Further research could include creating a comparison group who only receive the science education teaching session that arises later in the programme to compare with this group who received a cross-curricular session in their first year. Another limitation is that this study was conducted only with pre-service teachers on an undergraduate teacher education programme and it would be interesting to discover whether those attending postgraduate courses respond in similar ways.

## Appendix

Pre-session survey

| Answer options | 5 (agree strongly) | 4 | 3 | 2 | 1 (disagree strongly) | Response count |
|---|---|---|---|---|---|---|
| Evolution is an important idea for children in primary school to learn about in science | 18 | 15 | 9 | 3 | 0 | 45 |
| I am glad that evolution will be taught in primary school | 15 | 18 | 12 | 0 | 0 | 45 |
| Parents should be informed that a lesson on evolution will take place and can remove their child | 12 | 6 | 10 | 8 | 9 | 45 |
| I am looking forward to teaching evolution | 9 | 12 | 18 | 4 | 1 | 44 |
| I am concerned about teaching evolution | 2 | 10 | 15 | 12 | 5 | 44 |
| It will be important to take into account children's religious beliefs | 29 | 10 | 3 | 1 | 1 | 44 |

(continued)

(continued)

| Answer options | 5 (agree strongly) | 4 | 3 | 2 | 1 (disagree strongly) | Response count |
|---|---|---|---|---|---|---|
| Evolution is a theory and not a fact | 20 | 10 | 7 | 2 | 6 | 45 |
| I have an adequate understanding of evolution to teach at this level | 4 | 7 | 17 | 10 | 7 | 45 |
| Not all scientists support evolution | 12 | 12 | 15 | 4 | 1 | 44 |
| Evolution says that humans evolved from monkeys | 11 | 13 | 11 | 3 | 7 | 45 |
| I think children are likely to ask questions about religion | 14 | 16 | 6 | 8 | 1 | 45 |
| The Bishops of the Church of England do not accept evolution | 8 | 13 | 17 | 5 | 1 | 44 |
| Christians believe in a six-day Creation | 19 | 14 | 4 | 7 | 1 | 45 |
| Evolution says that life evolved over billions of years from simpler creatures | 24 | 12 | 7 | 0 | 1 | 44 |
| Darwin is the originator of the theory of evolution | 17 | 20 | 4 | 1 | 2 | 44 |
| The theory of evolution is in conflict with a belief in Creation | 14 | 18 | 8 | 2 | 2 | 44 |
| Darwin's theory was controversial because it contradicted religious teaching | 24 | 14 | 4 | 1 | 1 | 44 |
| Fossils are evidence for the theory of evolution | 15 | 14 | 9 | 3 | 3 | 44 |
| I will tell children they have a choice about whether to accept evolution | 27 | 6 | 8 | 2 | 1 | 44 |
| Most Christians reject evolution | 5 | 12 | 18 | 6 | 3 | 44 |
| Evolution is a very well supported explanation for how life came to be | 12 | 17 | 10 | 2 | 3 | 44 |

Post-session survey

| Answer options | 5 (agree strongly) | 4 | 3 | 2 | 1 (disagree strongly) | Response count |
|---|---|---|---|---|---|---|
| Evolution is an important idea for children in primary school to learn about in science | 19 | 20 | 5 | 0 | 0 | 44 |
| I am glad that evolution will be taught in primary school | 18 | 16 | 9 | 1 | 0 | 44 |
| Parents should be informed that a lesson on evolution will take place and can remove their child | 8 | 11 | 3 | 8 | 14 | 44 |
| I am looking forward to teaching evolution | 7 | 15 | 18 | 3 | 1 | 44 |
| I am concerned about teaching evolution | 0 | 17 | 12 | 11 | 4 | 44 |
| It will be important to take into account children's religious beliefs | 22 | 12 | 9 | 1 | 0 | 44 |
| Evolution is a theory and not a fact | 21 | 11 | 8 | 3 | 1 | 44 |
| I have an adequate understanding of evolution to teach at this level | 5 | 18 | 19 | 2 | 0 | 44 |
| Not all scientists support evolution | 9 | 16 | 18 | 1 | 0 | 44 |
| Evolution says that humans evolved from monkeys | 1 | 0 | 4 | 9 | 30 | 44 |
| I think children are likely to ask questions about religion | 14 | 23 | 7 | 0 | 0 | 44 |
| The Bishops of the Church of England do not accept evolution | 1 | 4 | 25 | 9 | 4 | 43 |
| Christians believe in a six-day Creation | 11 | 16 | 10 | 5 | 2 | 44 |

(continued)

(continued)

| Answer options | 5 (agree strongly) | 4 | 3 | 2 | 1 (disagree strongly) | Response count |
|---|---|---|---|---|---|---|
| Evolution says that life evolved over billions of years from simpler creatures | 24 | 18 | 2 | 0 | 0 | 44 |
| Darwin is the originator of the theory of evolution | 11 | 19 | 11 | 2 | 1 | 44 |
| The theory of evolution is in conflict with a belief in Creation | 8 | 10 | 16 | 7 | 3 | 44 |
| Darwin's theory was controversial because it contradicted religious teaching | 12 | 20 | 9 | 3 | 0 | 44 |
| Fossils are evidence for the theory of evolution | 20 | 17 | 5 | 0 | 1 | 43 |
| I will tell children they have a choice about whether to accept evolution | 25 | 10 | 5 | 1 | 2 | 43 |
| Most Christians reject evolution | 1 | 10 | 14 | 16 | 3 | 44 |
| Evolution is a very well supported explanation for how life came to be | 13 | 20 | 9 | 1 | 1 | 44 |

## References


Barnes, J. (2015). *Cross-curricular learning 3–14*. Sage.

Billingsley, B. (2004). *Ways of approaching the apparent contradictions between science and religion*. (Ph.D., University of Tasmania).

Billingsley, B., & Abedin, M. (2016). *Primary children's perspectives on questions that bridge science and religion: findings from a survey study in England*. Presented at BERA Conference 2016. Leeds, United Kingdom.

Billingsley, B., Brock, R., Taber, K. S., & Riga, F. (2016). How students view the boundaries between their science and religious education concerning the origins of life and the universe. *Science Education*. https://doi.org/10.1002/sce.21213.

Billingsley, B., Taber, K. S., Riga, F., & Newdick, H. (2011). *Teaching and learning about science and religion*. Paper presented at the ASE (Association for Science Education) Annual Conference, Reading.

Billingsley, B., Taber, K., Riga, F., & Newdick, H. (2012). Secondary school students' epistemic insight into the relationships between science and religion: A preliminary enquiry. *Research in Science Education*, 1–18. https://doi.org/10.1007/s11165-012-9317-y.

Borgerding, L. A., Klein, V. A., Ghosh, R., & Eibel, A. (2015). Student teachers' approaches to teaching biological evolution. *Journal of Science Teacher Education, 26*(4), 371–392.
Bramschreiber, T. L. (2014). Teaching evolution: Strategies for conservative school communities. *Race Equality Teaching, 32*(1), 10–14.
Chuang, H. C. (2003). Teaching evolution: Attitudes & strategies of educators in Utah. *The American Biology Teacher, 65*(9), 669–674.
Dodick, J., Dayan, A., & Orion, N. (2010). Philosophical approaches of religious Jewish science teachers toward the teaching of 'controversial' topics in science. *International Journal of Science Education, 32*(11), 1521–1548.
Fowler, S. R., & Meisels, G. G. (2010). Florida teachers' attitudes about teaching evolution. *The American Biology Teacher, 72*(2), 96–99.
Griffith, J. A., & Brem, S. K. (2004). Teaching evolutionary biology: Pressures, stress, and coping. *Journal of Research in Science Teaching, 41*(8), 791–809.
Hanley, P., Bennett, J., & Ratcliffe, M. (2014). The inter-relationship of science and religion: A typology of engagement. *International Journal of Science Education, 36*(7), 1210–1229.
Kalinowski, S. T., Leonard, M. J., Andrews, T. M., & Litt, A. R. (2013). Six classroom exercises to teach natural selection to undergraduate biology students. *CBE-Life Sciences Education, 12*(3), 483–493.
Kim, S. Y., & Nehm, R. H. (2011). A cross-cultural comparison of Korean and American science teachers' views of evolution and the nature of science. *International Journal of Science Education, 33*(2), 197–227.
Nehm, R. H., Kim, S. Y., & Sheppard, K. (2009). Academic preparation in biology and advocacy for teaching evolution: Biology versus non-biology teachers. *Science Education, 93*(6), 1122–1146.
Sanders, M., & Ngxola, N. (2009). Addressing teachers' concerns about teaching evolution. *Journal of biological education, 43*(3), 121–128.
Taber, K. S., Billingsley, B., Riga, F., & Newdick, H. (2011). To what extent do pupils perceive science to be inconsistent with religious faith? An exploratory survey of 13–14 year-old English pupils. *Science Education International, 22*(2), 99–118.

**Dr. Berry Billingsley** is associate professor of Science Education at the University of Reading where she leads the primary evolution research project and the LASAR (Learning about Science and Religion) research project. She teaches on a range of teacher education courses including undergraduate and postgraduate courses for future primary and secondary teachers.

**Dr. Manzoorul Abedin** is research fellow with LASAR at the Faculty of Education, Canterbury Christ Church University. He specialises in designing and presenting teacher education sessions and in addition, children's workshops on questions bridging science and religion.

**Keith Chappell** is associate research fellow at Canterbury Christ Church University, UK, within the LASAR (Learning about Science and Religion) project. He holds a Ph.D. in biology from the University of Hull, UK and another in theology from the University of Oxford, UK. Until recently he taught ecology and evolutionary biology at the University of Reading, UK. His interests relate to the relationships between science and religion and the power and limits of science.

**Chris Hatcher** is biologist at Loughborough University. He is currently conducting research on the evolutionary ecology of carnivorous plants. His research has inspired the development of unique teaching resources to support learning about evolution. He has designed and led student education and teaching strategy sessions on learning about evolutionary adaptation.

# Developmental Progression in Learning About Evolution in the 5–14 Age Range in England

**Terry Russell and Linda McGuigan**

## 1 Introduction

This chapter reports classroom-based research motivated by teachers' need for guidance in meeting the demands of the newly introduced subject matter of 'evolution and inheritance' in the national curriculum for ages 9–11. The first phase was conducted as a scoping study, characterised as using a design-based research (DBR) approach seeking practicable instructional design solutions. Eleven teachers working across the 5–11 years age range developed strategies supportive of developmental progression in understanding evolution. Five interrelated sub-domains were defined: 'deep time', 'fossils', 'variation', 'inheritance' and 'macroevolution'. This partitioning was a response to the complexity of the subject matter, the psycho-logic of pupils' developing understanding, specific curricular requirements and to ensure the classroom manageability of the research activities.

The first phase of the research was wide-ranging: a tight experimental design was neither practicable nor ethical, given the pressing requirement faced by teachers to deliver the curriculum. The second phase considered conceptual continuity across the primary to secondary transition. Twelve different teachers and their classes across the 9–14 years age range participated in this more focused enquiry into pupils' understanding of macroevolution. The concept of 'macroevolution' for the target age group carried similar assumptions to those of the Berkeley 'Understanding Evolution' website (https://evolution.berkeley.edu). In this 'big picture', we regarded pupils' insights into *speciation* and *common origin*, perhaps also *most recent common ancestor*, as major gains in understanding the fundamentals of evolutionary theory, while acknowledging the further formal development of the concept that would be

T. Russell (✉) · L. McGuigan
University of Liverpool, 126 Mount Pleasant, Liverpool L69 3GR, England, UK
e-mail: t.j.russell@liv.ac.uk

L. McGuigan
e-mail: l.mcguigan@liverpool.ac.uk

 
U. Harms and M. J. Reiss (eds.), *Evolution Education Re-considered*,
https://doi.org/10.1007/978-3-030-14698-6_4

possible later. At the very least, appreciating the macroevolutionary process would establish a basic scientific literacy with respect to evolutionary theory.

The second-phase research also included consideration of the *process* of pupils' science learning, with teachers encouraged to include science argumentation as an intervention strategy. Neither teachers nor pupils were assumed to be closely familiar with argumentation as it is currently described in the science education research literature (Duschl & Grandy, 2013; Kuhn, 2010; Mercier, 2011). All teachers engaged pupils in classroom discourse, adjusting the suggested procedure to their habitual practice. Following these sessions, students' own ideas and feelings about the value of class discussion in supporting their science learning were collected. In this manner, the research was structured to provide insights into both *intra*-psychological and *inter*-psychological cognitive strategies for understanding macroevolution. Metacognitive reflections on representational preferences required individuals to make personal choices that suited them, while social discourse—listening carefully, weighing evidence and articulating ideas overtly—exposed personal ideas to public scrutiny.

Particular learning design principles were incorporated into both phases of the research with the intention of facilitating deep and resilient learning. A *metacognitive* approach was encouraged: pupils were explicitly invited to think about their own and others' thinking. *Multimodality* was utilised, manifest in the use of alternative formats to encapsulate ideas. Multimodality embraces redundancy between representations as supportive of deeper understanding, rather than assuming that parsimony is more efficient and effective. The act of explicitation of internal representations and translation between different formats through *representational redescription* (Karmiloff-Smith, 1992) was promoted. Argumentation foregrounds the social aspect of conceptual change, but while pupils' ideas or claims are most usually conceived as formulated in speech, our research has defined a far wider range of modalities in which claims could be constructed, presented and subjected to argument. This perspective extends the nuances of meaning that can be exchanged and critiqued as claims, especially for younger learners. All these approaches were adopted within the overarching intention of constructing developmental learning progressions (DLPs). We assume contributing factors to DLPs to be (i) a cognitive-developmental maturational dimension as a limiting factor, (ii) relevant perceptual experience that provides grist to the mind/brain's mill and (iii) tailored instructional experiences. 'Instruction' includes culturally transmitted ways of organising thinking—language, mathematics, scientific models and other templates that structure understanding. DLPs are educational constructs, and while no single route to the goal of understanding is assumed, research can provide guidance that details both impediments and constructive scaffolding devices that take into account age-appropriate progression in understanding.

The authors took responsibility for setting the research agenda, while teachers (deemed to be the expert judges of the needs and capabilities of the children they taught) formatively assessed pupils' current understanding to inform intervention. The touchstone was the maximisation of applied, practical outcomes for teachers' practices, informed by robust field-testing in ecologically valid classroom contexts. The design outcomes comprised classroom-useable entities: intervention strategies in the form of specific modes of constructing knowledge for enhanced understanding. A summary overview of age-related pedagogical strategies, a curricular blueprint for teaching evolution and inheritance in the 5–11 age range, was one important outcome (Russell & McGuigan, 2019). The project produced data in the form of pupils' responses to classroom elicitation tasks, teachers' insights exchanged over a SharePoint facility, teachers' digital research diaries, researchers' classroom visit records, teacher group meetings and individual interviews with pupils.

## 2 Progression in Understanding in the Conceptual Sub-domains

Some outcomes, intervention strategies and developmental progressions in four of the first-phase research sub-domains are presented: variation, fossils, deep time and selective breeding. The more detailed consideration of macroevolution follows, where the combined outcomes of both phases of research are reviewed.

### *2.1 Variation*

Within-species variation is fundamental to evolution by natural selection: it is what makes changes over generations possible. Yet children tend to believe all living things within a species are the same (Evans, 2008). This 'essentialist' outlook sees all frogs, rabbits, oak trees, dandelions, etc., as identical. Superficially, this belief appears to be verified by observation. Mathematical tools enhance the developmental appreciation of how living things within the same species may vary (Lehrer & Schauble, 2012; McGuigan & Russell, 2015).

Teachers encouraged children's enquiries into differences in apparently identical living things, for example, tadpoles emerging from a mass of frogspawn. Children observed, drew, measured and recorded in tally charts numbers emerging from their eggs each day. By using a simple quantitative strategy, these 4–6-year-olds began to notice tadpole variability in the timing of their development.

Another unequivocally essentialist expectation was that if the seeds they planted at one time were given the same amount of water and sunlight, they would all grow to the same height and produce the same number of leaves, etc. Teachers asked children to observe, count and measure differences between collections of plants of the same kind. Younger children (around 5 years) growing sunflower seedlings employed their mathematics vocabulary to describe the heights of sunflower plants and used ordinal relations to sequence seedlings from shortest to tallest. With adult help, they were able to transform this arrangement into something resembling a pictogram, putting live plants of similar height into columns. They traced around the plants and in the air with their fingers to describe the curve, revealing a pattern that resembled a normal

**Fig. 1** A 'hill-shaped' distribution of plants

distribution. Drawing around pots of bean seedlings arranged on two axes to compare differences in the number of leaves (Fig. 1) revealed a pattern children decided to call 'hill shaped'. These multimodal interactions introduced early encounters with mathematical modelling and what will later be understood as normal distributions.

The 'hill-shaped' metaphor proved useful to children's appreciation of a recurring pattern across different data sets, such as hand span measurements. Towards the end of Year 6 (10–11-year-olds), children made careful measurements of one another's heights and, given help with the intervals on the x-axis, plotted their results to show variation as normally distributed. Further discussions considered likely patterns if height data were to be collected from older or younger children.

## 2.2 *Fossils*

Unlike the other conceptual domains discussed here, fossils are concrete objects rather than abstract ideas; the topic is also the subject of more specific reference in the national curriculum (Department for Education, 2013). Year 3 pupils are required to '*describe in simple terms how fossils are formed when things that have lived are trapped within rock*'. Fossils' importance is as sources of evidence for evolution, but the significance of this evidence can only truly be appreciated when interlinked with a notion of deep time. The topic is revisited in Year 6 when the requirement is to '*recognise that living things have changed over time and that fossils provide information about living things that inhabited the Earth millions of years ago*'.

Even very young children tend to be familiar with fossils. They revealed a keen interest in handling and observing them and imagining links with earlier life forms. Project teachers developed some practical multimodal strategies and productive science discourse approaches for moving children's thinking about fossil formation forward as a critical source of evidence for evolution. Practical experience of handling, observing closely and recording in drawings real fossils proved useful. Children's detailed drawings of fossils were coupled with their knowledge of current life forms that provided clues to species' identities. Imaginative thinking was encouraged as children considered what the former living animal or plant might have looked like and where it might have been found. Some ideas were recorded in sequenced drawings or as 3-D models, hypothesising the process of their formation.

Findings from children's personal library and Internet research formed the basis for claims to be exchanged and justified in science discourse activities. Children (9–11 years) revealed an appreciation of fossils forming over many thousands of years and some intuitive understanding of the replacement of the soft body parts with minerals from rock. The process of clay particle sedimentation in water was set up for observation. Knowledge voids in need of attention included a tendency to overlook plants as sources of fossils, the timescale of extinction, the process of fossilisation and the role of fossils as evidence of extinct species contributing to evolutionary theory.

## 2.3 Deep Time

Children's and adults' difficulty with understanding deep time is well recognised: Meir, Perry, Herron and Kingsolver (2007) found misunderstandings about how time is mapped onto phylogenetic trees. Time and change are the irreducible dimensions against which evolution can be described. Pupils' appreciation of the *range* of organisms that change and the *timescale* of modification was found to be limited. For instance, evolution of plants was barely mentioned and the time suggested for evolution to occur could be as short as a human lifespan. Terms such as 'mya' (millions of years ago) and the timescale of billions of years must become familiar if the essence of evolutionary change is to be grasped. Significantly, time is not a subject addressed directly in the science curriculum, the expectation being that the concept is acquired incidentally to other subjects.

Our developmental perspective suggested three aspects in need of attention: (i) familiarity with definitions of very large numbers; (ii) decisions about appropriate scales to apply; and (iii) the representational formats to use. A lack of familiarity with the definitions of 'million' and 'billion' was attended to by encouraging multimodal translations between written and spoken versions of both numbers and words. Using a simplified value for the age of the Earth of 4.5 billion years, teachers encouraged children in the 9–11 years age range to express this value in as many different ways as they were able: 'four point five billion', 'four and a half thousand million', '4,500,000,000' and so forth.

The approximation of 4.5 billion years as the age of the Earth simplified the range of distance scales suited to different contexts and age groups. A 45-mm line could fit easily on the page of a notebook, while a 450-mm line could scaffold discussion through an interactive whiteboard. A 4.5-m string could be stretched across a classroom with evolutionary events suspended as notes or drawings at measured points. A 45-m trail laid out in the schoolyard with signposted evolutionary events was manageable for even small children, while a 450-m trail required more stamina and planning for 10–11-year-olds. Teachers' selection of distance was governed by available space, the amount of evolutionary detail to be added and children's capability to handle each scale.

An alternative to the 2-D linear format used 450 sides of paper (225 sheets) in a ring binder, each side standing for about ten million years. This representational format for the passage of time proved accessible to children throughout the age range. Pupils used personal research and negotiated the selection of events for inclusion in their binders. Other formats to represent time include spirals and books in concertina form (Russell & McGuigan, 2014). Representational redescription was used by pupils to move between formats so as to consolidate understanding.

## 2.4 Inheritance and Selective Breeding

Selective breeding, the deliberate management of heritable features for transmission to offspring, was judged to be more immediately accessible as a step towards understanding evolutionary processes than would be launching directly into natural selection. Darwin was no doubt of similar mind in making the first chapter of 'Origin', 'Variation Under Domestication' (Darwin, 1859). The outcomes of selective breeding are likely to be familiar to pupils through their experience of pets. They are observable over much shorter timescales and are also controlled rather than natural selection's trial and error. The research found that primary children tend to think of offspring as identical to their parents or in receipt of equal characteristics from each parent (Russell & McGuigan, 2015a). Challenging these ideas led to the production of a number of imaginative activities, including making 3-D fantasy pets, both parents and progeny. The models provided the means to elicit beliefs about heritable features. Many younger children represented the offspring as identical to the adults, only smaller. Occasionally, there was an appreciation of differences and recognition that descendants might 'get features from mum or dad'. The characteristics of 3-D constructions could be treated as 'claims' and argued for and against, with evidence.

Selective breeding was approached through various contexts, the most engaging of which included breeding of assistance dogs and designer pets. To inform their designs, children collected data about different breeds: health, size, lifespan, hair shedding, behavioural traits and so on. Their thinking about how characteristics were passed from adult to offspring was revealed as they justified their choice of parent dogs in group and class discussion.

Selective breeding for desirable traits was more acceptable to children than the idea of preventing the propagation of undesirable traits. Faced with a disadvantageous trait, children's inclination was to accommodate the underdog by nurturing and training, rather than to prevent breeding. Culling was never countenanced. Pupils emphasised environmental effects, upbringing and experience on offspring. The disposition towards supporting disadvantaged individuals is strongly encouraged in schools' social cultures. This is in complete contrast to the reality of ruthless natural selection and the Malthusian inevitability of insufficient resources driving competition to survive. The evidence pointed to the merits of establishing an understanding of the process of selective breeding as both more accessible and more amenable to pupils' outlook, both logically and emotionally, prior to the introduction of natural selection. The potentially entrenched moral and affective conflict between pupils' compassion and the realities of struggles for survival is one that is probably best addressed explicitly through discussion.

# 3 Macroevolution

The first-phase research revealed little evidence of understanding across the 5–11 age range of gradual speciation over immense periods of time. Rather, assumptions about linear transformation, often confused with lifespan maturation or metamorphosis, were commonly offered as explanations for how change occurred. Evolution was understood as change, but it tended to be thought of as one species transforming into another—usually one more advanced or better equipped. Little awareness of the process of branching of species was encountered. The widely disseminated graphic in which a series from primitive to modern hominids is portrayed across the image (Fig. 5) seemed to be familiar to children. This image possibly contributes to pupils' assumptions about evolutionary transformations being linear, giving rise to puzzlement amongst 5–11-year-olds, as expressed in the following questions:

> Y4 (8–9 years) pupil: *Why are there still apes if apes have changed into people? Why haven't all the apes been used up?*

The second phase of the research explored the concept of macroevolution more closely. The fact that the sample extended across the primary to secondary transition allowed possibilities for ensuring continuity and progression across the two phases of education to be explored.

## *3.1 Introduction of Macroevolution Through Alternative 'Tree of Life' Representations*

The literature review conducted in Phase 1 (Russell & McGuigan, 2015b) noted that Catley, Phillips and Novick (2013) advocated the introduction of cladograms for US students as an essential component of understanding evolution. Moreover, UK research by Ainsworth and Saffer (2013) reports that children from age 8 to 9 years demonstrated success in accessing the *internal logical* aspects of cladograms. The latter research stopped short of making any claims for an appreciation of the role of speciation in macroevolutionary change in those young subjects. Zoos and museums are reported to use cladograms to communicate information about exhibits (Chua et al., 2012), so children visiting such exhibitions may well have encountered this informational device. Cladograms do not feature in Key Stages 2 or 3 mainstream science experiences in England.

The second phase of the research undertook a more detailed and structured enquiry to support the design of interventions facilitating a branching view of evolution. The interest was in the possibilities for establishing a sound notion of macroevolution and the use of cladograms to represent this concept in specific detail. The research intended to clarify what forms of prior or parallel scaffolding experience would help to make the target of introducing cladograms accessible to pupils and at what age. To this end, we acknowledge that our explorations occasionally employed modified approximations to the formal cladogram structures used by evolutionary biologists, in the interests of extending the possibilities of phylogenetic or 'tree thinking'. In addition to working with whole classes, more detailed responses were collected from a sub-sample of six pupils stratified by gender and overall science achievement (high, medium or low within their own class, $n = 72$) interviewed from each of the twelve participating classes.

Six representations of macroevolution, described in the following section, were the focus of the research. The researchers met with each teacher individually to discuss the materials to be introduced to children. Copies of the book, *One Smart Fish*, were provided, together with a section of a real tree branch, PowerPoint slides of Darwin's sketch, the hominid evolution image and a cladogram. The 3-D modelling activity was described as being valued but optional, to avoid imposing onerous management demands on teachers. Background notes were provided on each representation, but teachers were urged to employ their normal, age-appropriate, classroom techniques. As the intention was for pupils to make their own sense of the representations, teachers were asked not to undertake expository teaching. The fact that all the representations were of evolution was to be made clear, with pupils asked to reflect on what worked for them in supporting their understanding and to identify anything they found difficult to comprehend. Discussion between pupils was to be encouraged during this reflective process. Several secondary teachers used a 'circus' activity in which the materials were placed in stations around the laboratory and students moved between them in small groups, considering each representation in turn. Others tended to read the story to pupils, pass around the real branch and invite discussion of each of the other representations in turn. Six of the twelve classes found time to introduce the 3-D modelling activity after the other representations had been explored.

## 3.2 *Narrative Fiction: One Smart Fish*

Christopher Wormell's (2011) *One Smart Fish* is a picture storybook of about 500 words over just 30 pages. The story traces an anthropomorphic fish's move from the water to land, resonating with tetrapod evolution, ending with a colourful double-page graphic of the fanning diversification of animal species. Turned on end, the illustration reveals the Tree of Life form as a compact and accessible pictorial representation.

**Fig. 2** 'One Smart Fish' graphic

Many science educators are opposed to any hint of anthropomorphism; others welcome the communicative power of narrative fiction (Hopkins & Weisberg, 2017). Reassurance for sceptics should reside in the English curriculum's requirement for pupils to be able to distinguish between fact and fiction. While the author confirms there is no didactic intent in the book, the narrative and illustrations include two relevant evolutionary facts. Firstly, there is clear fossil evidence of the transition from the sea to the land, as, for example, the famous fossil *Tiktaalik* (Shubin, 2009). Secondly, the fanning double-page illustration of species evolution is a wonderful graphic representation of multiple species from common descent, incidentally, with the absence of any language demand on its comprehension (Fig. 2).

Teachers read the story to their classes of younger children. Older pupils were invited to consider the suitability of the story as an introduction to evolution for younger children.

## *3.3 A Section of a Tree Branch*

The etymology of 'clade' from the Greek 'clados' for 'branch' confirms the close proximity in imagery between the real section of tree branch and the abstract symbolic, tree-like form of the cladogram. The heuristic utility of the tree branch is that it offers a 3-D metaphor that can be held in the hands and rotated. Once the branching metaphor is explained, various evolutionary journeys can be projected mentally onto the form. The stem represents common descent and each branching node a site at which most recent common ancestors (MRCAs) split into different species; the tips of twigs can be thought of as being in present time, while any that are cut or

**Fig. 3** Real tree branch

broken can stand for extinction. There is power in its simplicity in challenging linear assumptions about how evolution progresses. Teachers recognised branches as an invaluable zero-cost resource that could make an invaluable contribution to pupils' (and their own) conceptualisation of macroevolution (Fig. 3).

### *3.4 Darwin's Tree of Life Sketch*

The tree of life sketch Darwin drew in his notebook describes his thinking about the interrelationships between diverse organisms and the descent of all living things from a common ancestor (Darwin, 1837). Darwin's early struggle to articulate his growing awareness of the origin of species might have had a resonating appeal to pupils facing their own sense-making challenge (Fig. 4).

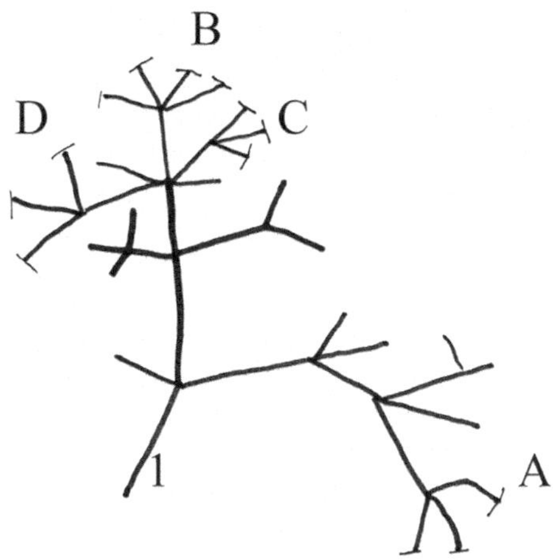


**Fig. 4** Darwin's Tree of Life Sketch

### *3.5 Hominid Evolution in Pictorial Form*

The ubiquitous hominid evolution image in Fig. 5 has been reproduced in many variants in books, magazines and on clothing. Pupils tended to be familiar with this image in some form or other. Following the customary reading of text and images from left to right (in Western cultures), the temptation is to interpret the image as a linear transformation model in which an individual monkey or ape might transform into a human.

**Fig. 5** Hominid evolution aka 'Ascent of Man'

## 3.6 Simple Cladogram of Hominid Evolution

The cladogram format presents information in a compact schematic form about the pathways and timescale of species change. Pupils' interest in human evolution influenced the choice of subject matter. The liberty was taken of leaving some lines unlabelled to convey the sense of the incompleteness of the fossil record (Fig. 6).

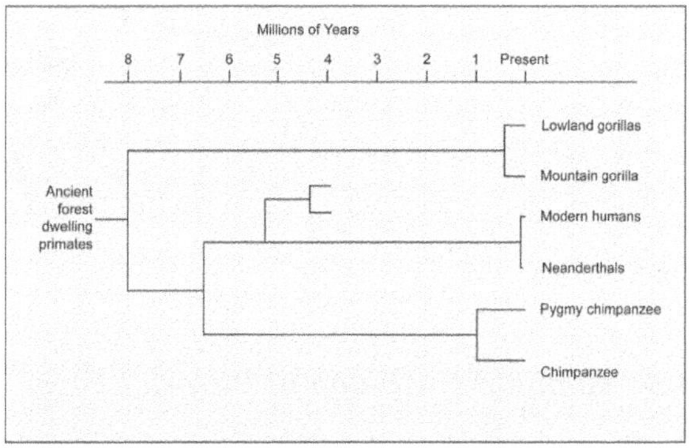


**Fig. 6** Cladogram of hominid evolution

**Fig. 7** Constructing a 3-D cladogram

### 3.7 *Pupils' 3-D Modelling of a Cladogram*

The invitation to model their ideas about evolution in three dimensions, informed by the various formats encountered, was a deliberate stimulus for representational redescription. In this case, the 2-D cladogram schematic was a major influence in the translation into a 3-D model, made to scale and with species labels attached. Construction materials were provided, with modelling decisions needing to be agreed within small collaborative groups of about four pupils. This discussion and negotiation proved to be a useful rehearsal for classroom argumentation. Only half the classes found time to engage in this task, though others were enthused to do so at a later date (Fig. 7).

## 4 The Role of the Cladogram in Understanding Evolution

The six sub-sample pupils from each class were withdrawn, one at a time, from their normal science sessions, for a discussion that followed a structured protocol. Interviews lasted 25–35 min and were audio-recorded for later transcription and analysis. The focus reported here was in response to being asked, '*Looking back over the various ways of showing evolution, which aspects of each representation did you find helpful?*', followed by, '*And which aspects caused you difficulties in understanding?*'. Here, we consider pupils' reactions to the cladogram, bearing in mind that they had discussed with peers and reflected upon the set of materials but had not been in receipt of direct instruction. Responses can thus be thought of as likely to be conservative with respect to the finer points of the cladogram, but valid in revealing pupils' intuitive sense-making and dispositions towards the format.

Almost all (89%) expressed a positive view that the cladogram had supported their understanding (Table 1). Many comments were generally affirmative but non-specific: '*It's helpful*'. Where probing stimulated an elaborated response, the aspects referred to positively were the *timescale*, the depiction of *branching or speciation* and clarification of *what was living and what was extinct.* Only four pupils in the entire sample (three being the youngest in the sample) did *not* find the cladogram to be helpful to their understanding of evolution. A further four pupils, again from the youngest group, were equivocal about its usefulness.

### 4.1 *Timescale*

The cladogram was the sole format that included a timescale, its usefulness being commented upon by one-third of the 9–10-year-olds, but more than half of each of the older age groups. Additionally, many mentioned the *absence* of any scale of time in relation to the tree branch and Darwin's sketch. The sense is thus that, having been

**Table 1** Helpfulness of the cladogram to pupils' understanding of evolution

| | Overall response to cladogram | | | Aspects of cladogram found to be helpful | | | Problems with cladogram | |
|---|---|---|---|---|---|---|---|---|
| | Useful | Equivocal | Not useful | Timescale | Branching, speciation | Clarifies alive and extinct | Lines confusing complicated | Lacks labels |
| Y5 (age 9–10) $n = 12$ | 42[a] (5) | 33 (4) | 25 (3) | 33 (4) | 8 (1) | 17 (2) | 50 (6) | 17 (2) |
| Y6 (age 10–11) $n = 18$ | 100 (18) | – | – | 61 (11) | 72 (13) | 11 (2) | 17 (3) | 28 (5) |
| Y7 (age 11–12) $n = 18$ | 94 (17) | – | 6 (1) | 50 (9) | 50 (9) | 33 (6) | 6 (1) | 17 (3) |
| Y 8–9 (age 13–14) $n = 24$ | 100 (24) | – | – | 71 (17) | 50 (12) | 13 (3) | 42 (10) | 21 (5) |
| Total $n = 72$ | 89 (64) | 6 (4) | 6 (4) | 57 (41) | 49 (35) | 18 (13) | 28 (20) | 21 (15) |

[a]Percentages, raw numbers in brackets

exposed to the complete set of representations, expectations of a need for a timescale had been raised and might even have been overlooked as self-evident in commenting on the cladogram.

## 4.2 *Speciation*

Comments suggesting the helpfulness in the fact of the cladogram showing branching occurred with a relatively low frequency amongst the youngest age group, but were made by half or more of older pupils. This was encouraging; an aspiration of the intervention was to increase cognisance of speciation as a key aspect of macroevolution. It is acknowledged that, even in response to direct questions referring to the cladogram, the affordances of other representational formats in raising awareness of speciation would be expected to be influential. Certainly, the *One Smart* Fish narrative had an impact on pupils' grasp of speciation, and so too did Darwin's sketch and the 3-D modelling activities. But it was also the case that the cladogram served to clarify interpretations associated with other formats:

> Y6 Boy: *In this one* [the cladogram] *you can see they all evolve from primates. Some people say that humans evolve from monkeys but in this you can see that we all evolved from the same thing but we didn't evolve from monkeys. Monkeys evolve from the same thing as humans.*

### *4.3 Complicated or Confusing Aspects of the Cladogram*

The presentation of some lines of descent lacking any labels—intended to convey the incomplete nature of the fossil evidence—attracted comments, including suggestions that the cladogram was overly complicated or simply confusing. Yet teachers noted that the cladogram, including its lack of labels, promoted much discussion, and suggested that omitting some labels might have been useful.

Six per cent of the sample, all from the older age ranges, suggested the need for illustrations of the species described—the feature that had appealed in the hominid evolution image.

> Y8 Girl: *Some animals have become extinct and it doesn't say what animals and how long ago they became extinct. There is no pictures.*

The following insight suggests that the cladogram might not offer an instant revelation: a level of personal effort needed to be invested:

> Y6 Boy: *Yes, it was very clear. Well, not really clear, but once I had a proper look I really understood it, because it shows points in time. You can see when a species went extinct and when, how many million years ago, that that species was introduced or went extinct. It's not very clear to someone who hasn't looked in-depth. Like, it's hard to understand for some people but for others it's easy to understand when they really put their mind to it.*

## 5 Understanding Cladograms in the Context of Other Representations

The various formats were intended to enable an entrée to 'tree thinking' for younger pupils, but also to scaffold an accessible approach to cladograms for all students. The developmental aspiration was to ease pupils into a more accurate way of thinking about macroevolution, to nurture an appropriate orientation towards evolutionary change as early as realistically possible. The extent to which the important evolutionary connections in each representation were common to all, or capable of being translated across any pair of them, was probed, with interviewees being asked, 'Do you see any similarity between the cladogram and …', followed by each of the other representations: Darwin's Tree of Life, the hominid evolution graphic and the tree branch. About half the pupils in age bands 11–14 years and three quarters at age 9–10 confirmed a similarity between the Darwin sketch, hominid evolution and branch representations. This was felt to be a positive endorsement of using multiple representations. Interestingly, the percentage commenting on a *lack* of similarity between the cladogram and the hominid evolution graphic increased with age. This outcome was interpreted as a positive awareness of the contrast between the linear and the branching conception of evolution. The graphic immediacy of the hominid evolution image was being challenged or rejected by the older pupils in the sample. While there is insufficient direct evidence for the project intervention procedures to claim credit for students' rejection of the linear model, the outcomes are consistent with the intended shift in thinking. Our retrospective view favours an interpretation that emphasises the various formats working in concert to reinforce tree thinking. None should be dismissed as too simple a prop for older students.

> Y6 Teacher: *I think comparing all of the different representations helped to draw out all of the key ideas about evolution, as no one model perfectly shows them all.*

The apparently simplistic tree branch was seen to have value even with older students:

> Y8 Teacher: *Have used it with Y10* [14–15 years] *biology group who "all" had an "Ahhhh!" moment.*
>
> Y9 Teacher: *I think using the tree branch as a model was an excellent tool, the tactile nature of the branch encouraged students to use it as they were talking about it. Students, who struggled with the cladogram, had fewer troubles with the tree branch.*
>
> Y8 Teacher: *Comparing representations is a very good approach, especially when combined with allowing pupils to peer question each others' views. It allowed pupils to access a model that was most pertinent to their way of understanding. It is a technique that would work for many topics and allows independent working.*

## 6 Argumentation and Classroom Discourse to Support Understanding

Opportunities to exchange views on the perceived merits or shortcomings of each representation complemented pupils' personal reflections on what was valued in each. The project promoted pupils' expression of ideas as claims supported with evidence and encouraged them to listen critically and respond thoughtfully. This inter-psychological strategy emphasised the essential role of discourse practices in science (Kuhn, Hemberger, & Khait, 2016). The project's attention to attentive listening and use of reasoning was commented upon as resonating with participating schools' wider curricular priorities. Though the project was not resourced to offer professional development support for argumentation, there was a shared commitment to using class discussion to promote science understanding that drew on familiar practices.

The project provided teacher guidance materials to help prepare pupils to strengthen their arguments by using research which gathered supporting evidence. A brief article on managing classroom science argumentation (Russell & McGuigan, 2016a) and a protocol for researching and engaging in argument, building on work by Lehrer and Schauble (2012), were developed. The latter advised on breaking down the functions of listening carefully to others' utterances and assembling evidence in support of beliefs, clearly expressed as knowledge claims. Teachers used various techniques to induct pupils into argumentation, including some in their existing repertoire. Behaviours relevant to a science discussion were negotiated, defined, reviewed and reflected upon; pupils' attention was drawn to these agreed criteria during discourse and the same principles used by teachers to evaluate the discussion. Several teachers modelled ways of expressing and responding to each others' ideas with sensitivity and respect: '*I agree with your idea, but…*', or '*Following on from your idea, I think …*', and so on. These phrases gave licence to the pupils to respond to each other directly, without teacher mediation. The importance of pupils bringing considered positions to the argumentation sessions was recognised. Techniques included sharing emerging research findings to stimulate further library and Internet searches that might strengthen their arguments, ranking evidence according to its perceived importance and pairing reasons with evidence in support of a case.

## *6.1 Pupils' Views About the Value of Peer Discourse to Their Understanding of Science*

A seven-criterion rating scale on science discourse was developed to probe students' affective and cognitive perceptions of their science classroom discussion experiences, a relatively neglected area of enquiry (Mercer & Dawes, 2014). This scale was used initially during the sub-sample interviews and later was administered in questionnaire form to the entire sample. Judgements were scaled 1–5, where 5 was high. Criteria and mean ratings by age group are summarised in Table 2.

Responses of the youngest age group were less consistent than those of older students, suggesting that many of the 9–10-year-olds were probably being stretched to the limits of their capabilities in discussing evolution. Overall, the four older age groups rated the criterion, '*Discussion helps my personal science learning*' highest. It seems reasonable to accept this outcome as a strong endorsement of including discourse sessions as an integral component of science learning. Equally consistent in this upper age range was the absence of surprise at the ideas expressed by their peers. Ideas that might have been extraordinary to a visiting observer were taken in their stride by participating pupils. Furthermore, students tended not to rate highly the science accuracy of the ideas that were shared. It seems reasonable to infer that the act of sharing ideas was what pupils appreciated as supportive of their learning; they were not surprised overall by others' ideas nor were they particularly impressed with others' scientific accuracy. The benefit may reside, at least in part, in discussion acting as a sounding board to clarify one's own understanding in the face of alternative points of view.

The interview sub-sample conversations generated qualitative comments that illuminated the rating scale evidence. The open questions were posed: '*Do you think a class discussion where you debate or argue with evidence is a good way to learn sci-*

**Table 2** Pupils' views of class discussion (mean values on scale 1–5 where 5 is high)

| Year groups | | | | | | |
|---|---|---|---|---|---|---|
| | Y5 ($n$ = 40) | Y6 ($n$ = 63) | Y7 ($n$ = 80) | Y8 ($n$ = 83) | Y9 ($n$ = 45) | All ($n$ = 311) |
| Discussion helps my personal science learning | 4.1 | 4.4 | 4.3 | 4.3 | 3.9 | 4.3 |
| Others' claims were found to be of interest | 4.3 | 4.2 | 4.0 | 4.1 | 3.5 | 4.0 |
| Reasons were given by others for their claims | 4.2 | 4.3 | 4.1 | 3.9 | 3.6 | 4.0 |
| Personal agreement with others' expressed ideas | 3.6 | 4.0 | 3.8 | 4.1 | 3.7 | 3.9 |
| Scientific accuracy of others' ideas | 3.5 | 3.8 | 3.7 | 3.6 | 3.3 | 3.6 |
| Evidence was provided by others for their claims | 3.9 | 4.0 | 3.8 | 3.7 | 3.2 | 3.8 |
| Others' ideas were found to be surprising | 3.9 | 3.7 | 3.2 | 3.1 | 3.1 | 3.3 |

*ence?*'. About 60% responded positively, 20% were ambivalent, while a further 20% expressed a negative view, the latter expressions tending to be linked with shyness or a fear of being 'wrong'.

> Y5B: *A little bit like, worried, because I could have got it wrong but I don't really know if I want to share with the class, my ideas. I don't really.*
>
> Y6B: *To be honest I told you a lot of ideas but I wouldn't really do that in front of a lot of people. To be honest I'm a bit shy to, I don't like sharing my ideas that much.*

Y6G: *Yes, I think it* really *helped me in multiple ways and I think it's a good way to help science and evolution really. It is a good way so I would recommend it. Because I was thinking of one idea, I didn't really think of any other ideas – I just got one idea stuck in my head. But when other people shared their ideas it started to make me think more and then help me understand evolution more because I didn't just think one solid answer in my head anymore and I was thinking more about evolution. So that helped a lot.*

Y8G: *I think that if you exchange ideas, then if you have an idea, someone could perhaps build on that and help your understanding. With my ideas I think people sort of helped me to sort of understand my own ideas a bit more.*

Y8G: *Me personally, I don't like doing that because I … I don't know … Because I'm not very confident with science anyway, I don't like to step forward and say my ideas because most of the time they could be wrong. But people that understand and people that are more confident to give their ideas and that helps me because I know what they're thinking and I know what they understand so I could reflect on that.*

Our previous research into class discourse with a younger age range has convinced us of the need to take a long, developmental view of students' engagement in dialogic practices and to prioritise the expression of ideas from the point of entry to school (Russell & McGuigan, 2016b). In this perspective, argumentation incorporates formative assessment. As one Y6 girl put it, '*my answers can help the teacher to see if I've been learning on the right track …*'. There is no doubt that teachers felt positively about this aspect of the project.

Y6 Teacher: *This has made me think carefully about how discussion/debating is something that should be modelled and practised throughout school, as it is such a powerful tool for learning. I can imagine that some practitioners would find the 'undefined/unknown journey/destination' of this type of exercise unsettling – it requires some subject knowledge on their part, as well as the ability to adapt and think on your feet! However, I think it is good for children to know that you don't know all the answers all of the time either! It made me listen more carefully to what the children were saying and gave me valuable insights into their understanding.*

## 7 Conclusions

Using multimodal representations in a reflective, metacognitive manner to elicit ideas and as the basis for evidence-based classroom discussion was, in some respects, a novel approach to teaching science for our collaborating teachers. Their feedback on the value of the strategy was overwhelmingly positive. All referred to the high quality of pupils' engagement and the positive contribution to learning.

The value of considering the affordances and limitations of different models and translating between them was widely appreciated in relation to understanding macroevolution. Other areas of the science curriculum would appear to be open to a similar approach. As an example, waveforms in physics could be approached in a similarly multi-representational and multimodal manner: 2-D and 3-D; in words; and using mathematics.

Our research points to the importance of both intra- and inter-psychological constructivist approaches. Pupils readily engage and empathise with their peers' struggles to articulate understanding. Pupils' perception of the science argumentation to which they were exposed was positive; they were undeterred if their peers' contributions revealed insufficient scientific evidence or accuracy. The critical factor seems to be the opportunity for personal metacognitive reflection vis-à-vis others' formulations. Not least, the formative assessment value of publicly expressed articulations of understanding was valued by teachers and recognised by some pupils.

The challenge to teachers of the introduction of evolution and inheritance to the statutory science curriculum in England was the motivation for the research reported here. Evidence from the two project phases confirms that Darwinian evolution can (because it can be made accessible) and should (because it provides a sound explanatory basis for future learning) permeate the biology curriculum progressively from the earliest years.

Currently, cladograms are not introduced until late in secondary biology education, if at all, yet research suggests that tree thinking in general and cladograms in particular are invaluable in developing the macroevolutionary perspective and addressing critical areas of current neglect. Firstly, they offer explicit means to support younger and older pupils' appreciation of deep time. Secondly, a pervasive view of linear transformation akin to the Lamarckian transmission of acquired characteristics is explicitly challenged. Thirdly, if cladograms were to be introduced from around the age of 10 years onwards, as our research deems viable, the familiar hiatus around the transfer between primary and secondary educations would be ameliorated.

The introduction of cladograms would appear to be made more accessible to pupils when accompanied by the various scaffolding alternative representations described, even for older students. The linear implications of the pervasive 'hominid evolution' or 'ascent of man' image appear to have been successfully counterbalanced by exposure to other representations.

Although current scientific thinking suggests evolutionary branches may tangle and bond via horizontal cross-breeding and transgenerational changes to the genome via epigenetic mechanisms (Heard & Martienssen, 2014), 'tree thinking' is nevertheless an important scaffolding metaphor for scientists and non-scientists. A basic understanding of the Tree of Life metaphor targeted at 9–14-year-olds and building on the strategies described here would be revolutionary if understood by the general population and have the potential to move the curriculum and understanding from the nineteenth to the twenty-first century.

**Acknowledgements** This project was funded by the Nuffield Foundation, but the views expressed are those of the authors and not necessarily those of the foundation. Further information can be found at http://www.nuffieldfoundation.org/evolution-and-inheritance.

## References

Ainsworth, S., & Saffer, J. (2013). Can children read evolutionary trees? *Merrill-Palmer Quarterly, 15*(2), 221–247.

Catley, K. M., Phillips, B. C., & Novick, L. R. (2013). Snakes and eels and dogs! Oh, my! Evaluating high school students' tree-thinking skills: an entry point to understanding evolution. *Research in Science Education, 43,* 2327–2348. https://doi.org/10.1007/s11165-013-9359-9.

Chua, K. C., Qin, Y. Q., Block, F., Phillips, B., Diamond, J., Evans, E. M., Horn, M. & Shen, C. (2012). FloTree: A multi-touch interactive simulation of evolutionary processes. In *Proceedings of the ACM International Conference on Interactive Tabletops and Surfaces,* pp. 290–302. ACM Press.

https://dl.acm.org/citation.cfm?doid=2396636.2396684. Accessed January 29, 2018.

Darwin, C. (1837). *Notebook B*. Cambridge University Library.

Darwin, C. (1859). *On the origin of species by means of natural selection*. London: John Murray.

Department for Education. (2013). Science programmes of study: key stages 1 and 2. *National Curriculum in England* © Crown copyright 2013 Reference: DFE-00182-2013.

Duschl, R. A., & Grandy, R. (2013). Two views about explicitly teaching the nature of science. *Science & Education, 22,* 2109–2139.

Evans, E. M. (2008). Conceptual change and evolutionary biology: A developmental analysis. In S. Vosniadou (Ed.), *International handbook of research on conceptual change* (pp. 263–294). New York: Routledge.

Heard, E., & Martienssen, R. A. (2014). Transgenerational epigenetic inheritance: myths and mechanisms. *Cell, 157*(1), 95–109. https://doi.org/10.1016/j.cell.2014.02.045.

Hopkins, E. J., & Weisberg, D. S. (2017). The youngest readers' dilemma: A review of children's learning from fictional sources. *Developmental Review, 43,* 48–70. https://doi.org/10.1016/j.dr.2016.11.001.

Karmiloff-Smith, A. (1992). *Beyond modularity: Developmental perspective on cognitive science*. Cambridge, MA: MIT Press.

Kuhn, D. (2010). Teaching and learning science as argument. *Science Education, 94,* 810–824. https://doi.org/10.1002/sce.20395.

Kuhn, D., Hemberger, L., & Khait, V. (2016). Dialogic argumentation as a bridge to argumentative thinking and writing. *Journal for the Study of Education and Development, 39*(1), 25–48. https://doi.org/10.1080/02103702.2015.1111608.

Lehrer, R., & Schauble, L. (2012). Supporting enquiry about the foundations of evolutionary thinking in the elementary grades. In S. M. Carver & J. Shrager (Eds.), *The journey from child to scientist* (pp. 171–207). Washington: APA.

McGuigan, L., & Russell, T. (2015). Using multimodal strategies to challenge early years essentialist beliefs. *Journal of Emergent Science, 9,* 34–41.

Meir, E., Perry, J., Herron, J. C., & Kingsolver, J. (2007). College students' misconceptions about evolutionary trees. *The American Biology Teacher Online, 69*(7), 71–76.

Mercer, N., & Dawes, L. (2014). The study of talk between teachers and students, from the 1970s until the 2010s. *Oxford Review of Education, 40*(4), 430–445.

Mercier, H. (2011). Reasoning serves argumentation in children. *Cognitive Development, 26*(3), 177–191. https://doi.org/10.1080/13546783.2013.819036.

Russell, T., & McGuigan, L. (2014). How long is a piece of string? 4.5 billion years perhaps! *Primary Science, 134* (The Association for Science Education).

Russell, T., & McGuigan, L. (2015a). Animals don't just grow feathers when they want to Research into inheritance and evolution. *Primary Science, 138* (The Association for Science Education).

Russell, T., & McGuigan, L. (2015b). Review of literature and resources: Evolution and inheritance, Nuffield Foundation http://www.nuffieldfoundation.org/pupils-understanding-evolution-inheritance-and-genetics.

Russell, T., & McGuigan, L. (2016a). Developing argumentation with the 4–11 age range: Research into inheritance and evolution. *Primary Science, 144*. (The Association for Science Education).

Russell, T., & McGuigan, L. (2016b). *Exploring science with young children: a developmental perspective*. London: Sage.

Russell, T., & McGuigan, L. (2019). Teaching and learning about evolution: A developmental overview. Primary Science. No.156. Jan/Feb 2019 pp. 33–35 Association for Science Education. Hatfield, U.K.
Shubin, N. (2009). *Your inner fish: A Journey into the 3.5-billion-year history of the human body* (1st Penguin edition). ISBN-10: 0141027584.
Wormell, C. (2011). *One Smart Fish*. (Red Fox edition). 978 1 862 30652 3.

**Terry Russell** is Emeritus Professor at the University of Liverpool, and Psychologist with a special interest in cognitive development as applied to the learning of science at all ages. He has worked in Southeast Asia, Africa and at King's College London and directed the Centre for Research in Primary Science and Technology at the University of Liverpool for over 20 years. He has directed national and international projects and published extensively. This chapter reflects his commitment to evidence-based activity that improves teachers' and pupils' educational experiences, where theory informs practice through practical and accessible applications.

**Dr. Linda McGuigan** has an established record of research into the development of science understanding. She is Honorary Senior Research Fellow at the University of Liverpool. She has conducted research into children's conceptual development in science, assessment and curriculum development. Attracting national and international interest, her work has been funded by national assessment and curriculum agencies. She has co-authored a number of books, reports, articles and digital and hard copy materials to support practice. Focusing on children's conceptual progression, she brings a developmental perspective to science learning, teaching and assessment.

# Teaching Evolution Along a Learning Progression: An Austrian Attempt with a Focus on Selection

**Martin Scheuch, Jaqueline Scheibstock, Heidemarie Amon and Helene Bauer**

## 1 Introduction and Defining the Problem

Since the nineteenth century, evolution theory with its key concepts of variation and selection has been a central backbone of the discipline of biology. First formulated by Charles Darwin and Alfred Russel Wallace, evolution theory has been further developed. In the 1930s, the theory expanded to include early genetics and population thinking and mathematically modelling of populations (Dobzhansky, 1973; Laland et al., 2014, 2015; Mayr, 1982). Since the millennium, an extended synthesis has tried to include life science research conducted since the first modern synthesis, for example epigenetics and evolutionary developmental biology (Laland et al., 2015; Pigliucci & Müller, 2010).

Evolution education should help students gain a deeper understanding of the scientific background of biological phenomena. The aim of this type of biology education is not to get a grasp solely on phenomena in biology, which leads to fact-based and rote learning, but to learn biology in the context of the central theory of evolution. Explanations of biological phenomena with evolution in the background are the key to making sense of the mechanisms of life. Scientific reasoning in the light of evolution is therefore one possible way to increase the acceptance of evolutionary theory in Austria (Eder, Turic, Milasowszky, Adzin, & Hergovich, 2011). In a current

M. Scheuch (✉) · J. Scheibstock · H. Amon · H. Bauer
Universität Wien ZLB – Didaktik der Naturwissenschaften Österreichisches Kompetenzzentrum für Didaktik der Biologie – AECC-Bio, Porzellangasse 4/2/2, Vienna 1090, Austria
e-mail: martin.scheuch@agrarumweltpaedagogik.ac.at

J. Scheibstock
e-mail: jaqueline.scheibstock@univie.ac.at

H. Amon
e-mail: heidemarie.amon@univie.ac.at

H. Bauer
e-mail: helene_bauer@gmx.at



U. Harms and M. J. Reiss (eds.), *Evolution Education Re-considered*,
https://doi.org/10.1007/978-3-030-14698-6_5

position paper of the German Academy of Sciences about evolution education in schools and at universities (Nationale Akademie der Wissenschaften Leopoldina, 2017), the prime concern is to introduce evolution as a basis for scientific literacy to learn about life on earth.

To support these overall goals in biology education, two lines of research in evolution education help to further develop more effective teaching and learning for students. First, research about students' conceptions contributes to the knowledge about preconceptions and the difficulties in learning evolution concepts. Two compilations collected a lot of this research (Hammann & Asshoff, 2015; Kattmann, 2015). Second, a more recent endeavour has arisen in science education with respect to learning progressions (LPs). These focus on scientific content and try to establish coherent, revising and deepening teaching and learning over several years of schooling with important stepping stones in comprehension (Duncan & Rivet, 2013).

In Austria, evolution is only rarely mentioned in the state curriculum. This absence is compounded by inconsistencies in school lessons and textbooks (Eder, Seidl, Lange, & Graf, 2018; Scheuch, Amon, Hoffmeister, Scheibstock, & Bauer, 2017; Scheuch & Wäger, 2018). To tackle this deficiency, Jelemenská, Amon and Wenzl (2010) have developed a teaching and learning sequence at three different grades at lower and upper secondary levels. The process of development was situated in coaching sessions with the teachers and a biology education researcher (Jelemenská, 2012; Jelemenská et al., 2010) and used topics of the state curriculum as starting points in the respective grades. This book chapter presents a long-term study on students' developing conceptions aiming to further develop LPs with a focus on different forms of selection in evolution.

## 2 Evolution as a Central Theme in School Along a Learning Progression

In the last two decades, a new branch of science education research has developed; the educational theory on LPs attempts to link curricular demands and research on students' conceptual learning (Duschl, Maeng, & Sezen, 2011). One tradition preceding the LP research of the German-speaking science education community is the model for *educational reconstruction* (Duschl et al., 2011; Kattmann, Duit, Gropengießer, & Komorek, 1997). In this triangular model, the knowledge about students' conceptions is linked with the analysis of the content as a baseline, and as the third corner a learning environment is established relating the knowledge stemming from the two first cornerstones (ibid.). In planning LPs there are also similar cornerstones linked to each other (Duschl et al., 2011): (1) the *formulation of big ideas* of the discipline which are not only important for learning in the respective discipline, but also for research in this scientific field; (2) *scientific practices* which help to engender an epistemological stance which helps to make new knowledge more accessible to learners and (3) scaffolding *the learning process of the students.*

Learners need educational material based on their everyday conceptions in order to be able to construct more scientifically accurate conceptions (Hammann & Asshoff, 2015; Kattmann, 2015; Kattmann et al., 1997; Krüger, 2007; Riemeier, 2007). To frame this in a long-term sequential arrangement, it needs lower and upper anchors (Duschl et al., 2011, p. 151): where to start with the big ideas and basic phenomena and everyday experience related to them, and where to go with many more abstract concepts related to the big ideas as more complex learning goals (p. 152). Research on the students' learning gives us additional anchors along the individual LPs and informs researchers about growing ideas, even if they are, in some way, still misrelated to the scientific upper anchors (Duncan & Rivet, 2013).

To improve evolution education, teaching and learning should be organized along a well-researched LP based on students' conceptions. This means that evolution should already be introduced at an early grade level and presented via several examples of everyday biological phenomena (a recent example is given by Wyner & Doherty, 2017). In subsequent grades, the learner comes back to the initial idea, repeating it, applying it to new phenomena, building upon it and further differentiating it with newly integrated concepts. Thus, the previous knowledge is activated and can be used to tackle new questions. This instructional design and long-term planning of biology education helps learners to extract more abstract principles for conceptual learning and enables them to transfer their knowledge into new problem-solving situations. With this idea of a curriculum in mind, evolution should be a recurrent theme during the whole course of schooling so that the basic ideas in biology can fully be developed and applied by learners (Gropengießer, 2010).

LPs were also an important background for the development of the 'Next Generation Science Standards' (NGSS) in the USA (Duncan & Rivet, 2013). These standards can stand as an example for our attempts to improve the evolution LPs related with the problems arising from the Austrian State curriculum (see below).

## *2.1 Next Generation Science Standards*

A promising development has been achieved in the USA with the implementation of the Next Generation Science Standards[1] with a consistent set of learning objectives. Against the background of LPs, these learning objectives of the science subjects were planned from kindergarten to grade 12. At first, the so-called disciplinary core ideas[2] were identified and then the LPs were formulated stepwise, for the subsequent grades. One example is 'natural selection' in the field of evolution education. This is the example of a lower anchor (grade 3–5 in the US): '*Sometimes the differences in characteristics between individuals of the same species provide advantages in surviving, finding mates, and reproducing*'.

[1]http://www.nextgenscience.org/ (29.1.2018) Next Generation Science Standards of USA.

[2]http://nstahosted.org/pdfs/ngss/20130509/MatrixOfDisciplinaryCoreIdeasInNGSS-May2013.pdf (29.1.2018) Disciplinary Core Ideas in biology.

In planning such curricula, the content is the starting point, but the abilities and the perspectives of the learners are taken into consideration to develop the sequence. It is very important for learners to contextualize new knowledge in the light of previously learned ideas.

## 2.2 Austrian State Curriculum and Teaching Evolution

State curricula are defined for all the school types separately: primary (6–10 years), lower secondary (10–14 years) and many different upper secondary school types (14–18/19 years). There are no consistently planned LPs, e.g. from primary school to a standard secondary school or to an upper secondary vocational school. In primary school, evolution and its key processes are not mentioned. The history of development of the earth is not even a topic, although the interest of this age group in dinosaurs and fossils is great (Dawson, 2000). In lower secondary school curricula,[3] evolution is firstly mentioned in the 7th grade (12–13 year-olds) in connection with the '*history of the development of life*'. Secondly, evolution is mentioned in the state curriculum of upper secondary schools in grade 12 (17–18 year-olds) as '*evolutionary mechanisms, chemical and biological evolution, theories of evolution*' as well as 'the *history of human development*' and '*evolution as a basis for diversity of organisms and for the change of ecosystems, organs, and cellular structures*'.[4] Additional big ideas, which can be equated with the disciplinary core ideas of the NGSS, were included in 2016 in the biology curriculum of upper secondary schools (15–18 year-olds) to support competence-based biology teaching. Out of seven big ideas, two have links to evolution. The first one is '*reproduction*', in which the role of reproduction for evolutionary processes is included. The second one ("*variability, kinship, history, and evolution*") is dedicated to mechanisms and phenomena related to evolution on the whole. These innovations are intended to strengthen teaching and learning about evolution in biology lessons in Austrian schools; however, these big ideas are not assigned to specific places in the curriculum; therefore, the responsibility for including them lies in the hands of the teachers. Numerous studies have shown that students have problems with understanding and applying mechanisms of evolution (Kampourakis, 2014; Wandersee, Good, & Demastes, 1995; Weitzel & Gropengiesser, 2009). Thus, it is important to assist teachers in determining progressions that enhance student learning in evolution.

---

[3]https://www.ris.bka.gv.at/GeltendeFassung.wxe?Abfrage=Bundesnormen&Gesetzesnummer=10008568 (29.1.2018) „Gesamte Rechtsvorschrift für Lehrpläne – allgemeinbildende höhere Schulen, Fassung vom 28.1.2018".

[4]https://www.ris.bka.gv.at/Dokument.wxe?Abfrage=BgblAuth&Dokumentnummer=BGBLA_2016_II_219 (29.1.2018) 219. „Verordnung der Bundesministerin für Bildung, mit der die Verordnung über die Lehrpläne der allgemein bildenden höheren Schulen geändert wird".

## 3 Learning Progression Based on Students' Conceptions: Evolution and Selection

In 2009, a biology teacher and a biology educator started to develop an LP for evolution focused on artificial, natural and sexual selection (Jelemenská, 2012; Jelemenská et al., 2010) for grades 8 (13–14 year-olds) to 10 (15–16 year-olds). The starting point was the exhaustive literature on students' conceptions about evolution. The LP designed for three grades (13–16 year-olds) includes several learning activities focusing on selection: firstly, breeding as an example of artificial selection; secondly, changing beak size within a population of Darwin's finches by varying environmental conditions (dry and moist years) as an example of natural selection; and finally, examples of the different forms of sexual selection, like female choice in the peacock. The aim of the grade-specific, interconnected teaching and learning sequences is to help students to develop a coherent knowledge about central concepts of evolution, especially natural selection. The rationale behind this LP is to develop the concept of selection from an everyday understanding of breeding to the most complex sexual selection in the context of behavioural biology. For this LP, the developers looked for topics in the respective grades in the state curriculum, where examples of evolution education could be realistically included.

The 8th grade sequence (for ~14 year-olds) includes three lessons aimed at introducing the concepts 'variation', 'artificial selection' and 'natural selection' to the students[5]: in the first lesson, software for modelling the breeding process of dogs confronts the students with artificial selection. The students have to select some dogs for mating. It is possible to repeat the selection for several generations and to create new dog breeds. The aim of this section is to enable the students to realize that they need different dogs at hand for a selective process, so they can use this diversity or variation during the breeding process in an artificial population. This is followed by a short lecture about Darwin's life, given by the teacher, which includes his voyage on board *The Beagle* and his later experiences and thoughts concerning pigeon breeding, in order to underline the fact that variation is always given, being the prerequisite of natural (and artificial) selection. In addition, students are led towards Darwin's stay on the Galapagos Islands and his observations there, to make them familiar with his thinking. This acts as a basis for later activities. In the second lesson, an activity about his observations of birds is conducted. Students are asked to develop their own criteria to sort European native birds on the basis of similarities and estimated relationships before getting a set of Darwin finches from the Galapagos Islands and sorting them on the same basis. The students are following Darwin's path as noted by him: '*Seeing this gradation and diversity of structure in one small, intimately related group of birds, one might really fancy that from an original paucity of birds in this archipelago, one species had been taken and modified for different ends*' (Darwin, 1845; Keynes, 2001). During the sorting process, the students also discuss similarities and differences. Via a PowerPoint presentation, the teacher presents information

[5]The authors are willing to provide the whole sequence of 8th grade upon request via e-mail.

to the students about how Darwin's finches reached the islands as well as about the selective processes associated with their foraging behaviour as it relates to differences in size and strength of their bills. In the third lesson, the students are given a learning task on the foraging behaviour of two species of ground finches and the selective process in dry and moist years. They then discuss their work and thus come to understand the change in traits within the population. The goal of this task is to make the students apply or transfer their knowledge about variation and natural selection to a specific situation which was not discussed in detail before. The concepts of both 'variation' and 'natural selection' are revised and deepened. To further improve their understanding, the differences between artificial and natural selection in terms of time and directedness are also discussed. This intervention is carried out for the purpose of enabling the students to formulate their ideas and to find differences as well as commonalities between these two processes.

The 9th grade (~15 years-old) sequence consists of four lessons. During the first lesson, students revise the tasks used in grade 8 on artificial selection and natural selection. However, a new factor is discussed in order to activate past knowledge built by students, namely for them to discuss the potential effect of changing environmental factors on the mean occurrence of traits in a population over generations. In the following sequence, investigations on the role of variation in a population as a prerequisite for selection are repeated, to deepen the students' understanding. An additional task about Darwin's finches continues the learning about the environment. The students must interpret data on the population density of two species of ground finches co-occurring on one island due to possible dependency on dry weather. They are asked to extract data from a graph to fill in a table of the average length of bills. This exercise aims at training their ability to interpret diagrams, which is relevant for understanding variation in a population. After this activity, an example of blackcaps (*Sylvia atricapilla*) is given to the students to transfer and to deepen their conceptual knowledge about natural selection. The blackcap is an example where recent evolutionary processes can be observed. It is widely known that this bird species from Central Europe arrives in spring for breeding and leaves in autumn when food becomes scarce. Global warming already shows effects on the migration behaviour of birds. Since the 1960s, it has been observed that the blackcaps do not only arrive from the south-west but also from the north-west since an increasing number of individuals stay in Great Britain (GB) for the winter (Pennisi, 2005). The context being responsible for this new behaviour is that the increased winter-feeding in GB is triggering the beginning of speciation due to a different timing of migration and therefore the beginning of reproductive isolation and initial morphological variation. In the final discussion, commonalities and differences in terms of selection between the examples of dog breeding (artificial selection: working with favoured traits), Darwin's finches (natural selection: traits selected by environment) and the blackcaps (natural selection: changes in migration behaviour) are elaborated.

During the 10th grade (~16 years-old), four lessons are incorporated. The curricular framework of this grade is sexuality and animal behaviour. The development of conspicuous traits like long and colourful tail feathers through sexual selection is introduced. Darwin already realized that some traits like eye-catching plumage or huge antlers cannot be explained by natural selection based on environmental factors only. He modified his theory to include the concept of sexual selection. Sexual selection can be used to challenge the students' idea that selection always leads to optimal adaptation. The 10th grade lessons start with a transfer of their knowledge about natural selection to a simulation which is focused on the peppered moth within its habitat in order to revise the concept of natural selection and link it to the two aforementioned grades. Regarding this simulation (presented in class via video projector), the students work together with the teacher and make suggestions about which colours of moths are more likely to be eaten by predators. This serves as a classical example, where visual predator selection is the foremost driver of melanism; in the meantime, other factors have been identified as well (e.g. Cook & Saccheri, 2013). Again, a new example should enable the transfer of previously learned mechanisms to explain phenomena and data.

The first example of sexual selection (introduced by the teacher in a PowerPoint presentation) is about the antlers of deer. In class, students are encouraged to think about the reasons for the development of such a cumbersome structure. The subsequent activity proceeds with photographs about widowbirds, peacocks and native barn swallows as similar examples. The two following tasks deal with experiments on the manipulation of the tail length of male widowbirds and barn swallows (Andersson, 1982; Møller, 1990). Students get brief descriptions and graphs about tail length and mating success and are asked to advance a hypothesis on the basis of the data about how differences in tail length have occurred and about how tail length affects the mating success. Then, the different forms of female choice, as these occur in the bird examples, and male competition, in the deer example, are worked through in the form of a teacher–learner interaction, presenting and discussing several pictures of sexual dimorphisms in deer and bird populations. Finally, students transfer their knowledge to human sexuality and discuss which factors (e.g. body odour) might play a role in finding a partner and they can reflect on evolutionary principles. This final step is made to meet the students' interests in human-centred topics and show them the role evolution can play in their lives.

These three sections, in three different grades and with a focus on selection, are our main research focus. We want to trace the students' learning within and between the sections.

## 4 Research and Development Questions

Our overarching developmental aim is based on the results of the interview study to further develop and expand this LP. The questions of this study are dedicated to the longitudinal learning of the students in the LP: How do they learn within this LP? Which conceptions about selection and variation have been developed?

## 5 Research Design, Methodology and Methods

This is a longitudinal interview study with students before and after the teaching sequences, focusing on the students' conceptions and concept development. Throughout the whole study, we follow a qualitative research paradigm (Silverman, 2016) with an open mind regarding the chosen methods following a pragmatic position. This means that the way data are collected and analysed may depend on (already be set up following) a pre–post-design, but which questions are asked in the interviews depends on the interviews conducted in lower grades and on the exhaustive, up-to-date literature research.

This study started with a master thesis by Scheibstock (2014), where pre–post-interviews in the 8th grade (14 year-olds) were conducted. Based on the *case study* approach by Yin (2009), the development of students' conceptions regarding *variation* and *natural selection*, two evolutionary key concepts, were investigated. To choose four pupils, a *group discussion* about the topic of evolution was made with the whole class (Lamnek, 2005). The criteria were the different personalities, knowledge, opinion and discussing behaviour. Following the same research design in every grade, the students were questioned about evolution before (pre) and after (post) the teaching sequence. The interviews were *problem-centred*, *open* and *guideline-based* (Baalmann, Frerichs, Weitzel, Gropengießer, & Kattmann, 2004; Duschl et al., 2011; Fenner, 2013). Generally, in the post-interview in the 8th grade and in the interviews in the upper secondary grades, relevant statements of the previous interviews were included for clarification (Helldén, 2005). This is an example of the guidelines of both interviews (pre- and post-) in the 8th grade:

> Introductory question
>
> It is known that the present-day long-necked giraffes had short-necked giraffes as ancestors which lived in woods. How would you explain the evolution of the long-necked giraffes? (cf. Johannsen & Krüger, 2005)
>
> natural selection
>
> What happens to a species if the environment is changing? Who or what exactly changes in the course of time?

Additionally, in all grades, *participatory observation* in every learning sequence was conducted by the researcher, presuming that the interviewees would make relevant statements concerning their conceptual development of variation or natural selection (Lamnek, 2010). Thus, a process-based approach was added as suggested by Riemeier (2007).

As a method of analysis, the *qualitative content analysis* by Gropengießer (2005) was used to *transcribe*, *redact* and *explain* the pre- and post-interviews by Scheibstock. Afterwards, the pre- and post-interviews were compared student per student in order to track their individual LPs. In 2016, when the same students were interviewed again in the 10th grade, they were asked questions about a new concept (sexual selection) which is the theoretical frame of the teaching sequence in the 10th grade (for organisational reasons, these students had not been able to do the evolution section in grade 9; they therefore did not work further with the data on the Darwin finches and did not learn about the blackcap example).

# 6 Results of the Development of the Students' Conceptions

In general, the interviews in the 8th and 10th grades show that the students are able to develop a more scientific conception of *natural selection* as they progress through the LPs' curriculum (Table 1). Most of the students' conceptions that were found in our interviews are similar to what previous studies have found. Gregory (2009) made an extensive literature review on conceptions for natural selection. Zabel and Gropengiesser (2011) looked for learning progress and created a conceptual landscape to depict the development of students' understanding of variation and natural selection. In the following results, we present the LPs of our students. We highlight those results that are new to the research from our perspective.

The following student conceptions were found in the pre-interview in the 8th grade:

- goal-oriented adaptive behaviour
- adaptive physical change
- the outsider theory (one begins evolving, others imitate)
- intentional genetic transmutation
- problem to define and work with the terms 'species' and 'population'.

**Table 1** Overall learning progressions between the four interviews in 8th and 10th grades

8th grade teaching sequence: 3 lessons *variation* & *selection* by example of dog breeding & Darwin's finches — 2 years → 10th grade teaching sequence: 5 lessons *natural* & *sexual selection* by example of birds' tail feathers & non-adaptive traits

| | Pre-interview in 8th grade (→ Teaching sequence 1) | | Post-interview in 8th grade | | Late Post- & Pre-interview in 10th grade (→ Teaching sequence 2) | | Post-interview in 10th grade |
|---|---|---|---|---|---|---|---|
| → Variation → | individuals within a population are different | = | individuals within a population are different | ≠ | individuals within a species are alike, but not identical | ! | problem to interpret the scientific terms "species" & "population"; individuals are different |
| | | | | * | variation as a prerequisite for changes in traits / selection | = | variation as a prerequisite for selection |
| | animals show variation because humans also do | ≠ | animals show variation because humans and other animals also do | ≠ | animals show variation because other animals and humans also do | | |
| | | | | * | nature gives rise to variation | | |
| | | | | * | variation means genetic diversity | | |
| → Natural selection → | goal-oriented adaptive behaviour | ≠ | selection by the environment | = | selection by the environment | ≠ | selection by the environment and/or the (mating) partner |
| | adaptive behaviour (does not always lead to adaptation) | ≠ | better adapted species due to "survival of the fittest" (scientific notion) | ! | "survival of the fittest" means goal-oriented adaptive behaviour | ≠ | better adapted species due to "survival of the fittest" (scientific notion); individuals cannot change |
| | adaptive physical change (adaptation by usage) | ≠ | sexual reproduction as the key to evolution; incremental adaption over generations | ! | sexual reproduction as goal-oriented adaptive behaviour | ≠ | sexual reproduction as the key to evolution; incremental adaption over generations |
| | need to adapt due to environmental changes in order to survive | = | need to adapt due to environmental changes in order to survive | = | need to adapt due to environmental changes in order to survive | = | need to adapt due to environmental changes in order to survive |

[≠] means **conception changes**; [=] means **no development**—in other words: conception stays the same; [*] means **new conception**—this conception was not yet found; [!] means **break/problem**—often because (new) more sophisticated conceptions are mixed with (old) less sophisticated conceptions

All of the students, prior to the 8th grade evolution sequence, imagined that organisms sense their need to adapt and act intentionally in order to survive. The students believed that the will to survive induces gradual physical changes in organisms due to a repetitive use of beneficial traits. Subsequently, these changes are undergone by all individuals of a population, which results in further speciation. One student also mentioned the fact that one individual might be different compared to other members of the same species, which enables it to cope with the changing environment. All others then try to follow its example and induce changes in their body aiming at developing similar favourable traits:

> Maybe, one – so to speak – newcomer, maybe one was different, [...] or recognizes [...] but everyone recognizes somehow, but this one thinks about it and drags the others along somehow, or like that. But, anyhow, the whole species has changed, because [...] there are no giraffes with short necks any longer.

Zabel (2009) also found the outsider model, but the social aspect of the pioneer and the followers seems to be new.

Furthermore, two of the four students questioned suggested that genetic information (material) is transformed. In detail, they think that more adaptive traits become dominant, while less adaptive traits become recessive. When organisms note instinctively their need to adapt, their body changes the genetic material.

An important finding was the fact that—for one student—the meanings of the two terms 'species' and 'population' were not clear enough. As this fact represents another barrier for understanding evolutionary concepts, we wanted to know if the other interviewees had the same problems. Hence, for the pre-interview in the 10th grade, we prepared the guideline accordingly.

The following student conceptions were found in the post-interview in the 8th grade:

- speciation due to 'survival of the fittest'
- selection by the environment
- sexual reproduction as the key to evolution
- increasing adaptation over generations
- the need to adapt due to environmental changes
- sexual reproduction as goal-oriented adaptive behaviour.

The post-interviews in the 8th grade, conducted after the teaching sequence, proved that the students' conceptions had changed to a large extent and that they now possessed a more scientific conception of *natural selection*. In particular, they were convinced that speciation is based on the 'survival of the fittest', which we used as a category because it was used by the students themselves during the interviews. It is not the concept as Spencer, and later Darwin, used the phrase (Gregory, 2009) and which has been as being a misleading concept (ibid.). We guess that this term came in through another additional textbook used, since the concept was not integrated in the LP materials. It means, in the students' view, that if the environment changes, only those individuals which can adapt to these changes better will survive. These are the only ones to be able to mate and reproduce sexually. Consequently, their genetic information is passed onto following generations. Less adapted organisms will not survive and will therefore not contribute to the gene pool of the next generation, i.e. pass on their old genes for short necks. At this point one aspect has become clear: although this conception about natural selection is not at all scientifically correct (not death or survival, but the choice and frequency of the genes which are passed on determine the gene pool of the next generation), it seems to be an important *stepping stone* for the students in progressing (Duncan & Rivet, 2013; Duschl et al., 2011). The intervention supports them to reconstruct their old conceptions about evolution.

Another more sophisticated conception developed by the students is 'selection by the environment'. They understand that the environmental conditions determine which traits of organisms enable specific individuals to cope better. Also, students have realized that changes in organisms cannot be achieved by the will of an individual, but only through sexual reproduction:

> Everything came into being (was created) with the ability to change if necessary – through mating, gene alteration and mutation. Mutation results from mating. Males and females have different genes. When they mate, descendants with new genes are created.

This student thinks that changes are based on gene alteration and mutations, which can only arise from sexual reproduction. Therefore, succeeding generations have new genes. Besides, he/she is convinced that every organism is able to change if necessary.

All of the interviewees understood sexual reproduction as the key to evolution which is triggered by environmental changes. This conviction makes them reconstruct their everyday concept, which is why the inclusion of sexual reproduction functions as another *stepping stone* (Duncan & Rivet, 2013; Duschl et al., 2011). Although not perfectly correct, this is effective for scientific reasoning: all students were finally convinced that only one poorly adapted individual cannot achieve optimal adaptation throughout its life. Even though much progress regarding the development of more sophisticated conceptions can be noticed, some unscientific conceptions persisted. For example, see the first sentence in the quote above; here are some other examples:

> A: If the environment changes, the giraffes reproduce.
>
> B: The environment changes. In order to survive, the giraffes change too.
>
> C: If the environment changes, the species also changes. This means, if one thing evolves, others must also evolve.

Ergo, all of the students held on to the conviction that organisms sense the need to adaptation which is triggered by environmental changes. Two of them stated that these changes initiate sexual reproduction aiming at the creation of descendants with favourable traits. So, goal-directed thinking persists and new knowledge is linked to this underlying conviction.

The following student conceptions were found in the pre-interview in the 10th grade:

- speciation due to 'survival of the fittest'
- variation as a prerequisite for selection
- selection by the environment

- different environmental conditions cause different speciations
- increasing adaptation over generations
- sexual reproduction as the key to evolution and as goal-oriented adaptive behaviour
- the need and intention to adapt due to environmental changes
- difficulties in understanding the scientific terms 'species' and 'population'.

The findings of the pre-interview in the 10th grade show that scientific conceptions of the evolutionary key concept *natural selection*, which the students were able to acquire in the 8th grade, are still available to them to a great extent. They still trace the change of a species back to the concept of the 'survival of the fittest'. They even express a 'new' scientific conception, which had not yet been noticed before: 'variation as a prerequisite for selection'. Furthermore, all of the students still explain that selection is decisively influenced by environmental conditions. They point out that different conditions require different adaptations in the respective organisms. Additionally, after two years the students are still convinced that adaptation happens over generations due to sexual reproduction. Although this is not completely correct, it can be seen as another stepping stone. Subsequent teaching and learning sequences can build on this aspect and provide opportunities for the learners to further reconstruct their knowledge about evolution.

The following conception is less scientific and is persistent: organisms sense a certain need to adapt due to changes in the environment in order to survive as a species. Although students develop more sophisticated conceptions in general, one of the three students interviewed, who compared to the others tends to think more scientifically, seems to have problems explaining what the 'survival of the fittest' really means. Even though this student had already explained this concept in a scientific way in the 8th grade, he now confuses the following two contradictory concepts: 'intention to adapt' and 'selection by the environment'.

All of the students have problems in defining, using correctly and distinguishing between the scientific terms 'species' and 'population', which are important for an understanding of evolution. This fact was concluded from an intervention conducted during the interview, where the students were asked to group several pictures of three species of rabbits. All students show the same results: individuals looking similar are defined as a species and the term 'population' is assigned to all the species together as an overarching group. Therefore, the use of this term during the instruction does not make any sense to the students and improvements have to be planned to cope with this point.

The following student conceptions were found in the post-interview in the 10th grade:

- speciation due to the 'survival of the fittest'
- variation as a prerequisite for selection
- at the beginning everyone is different in order that selection can take place
- selection by the environment
- different environmental conditions cause different speciations
- sexual reproduction as the key to evolution
- increasing adaptation over generations

- the environment affects the formation of new traits
- the need to adapt due to environmental changes
- difficulties in understanding the scientific terms 'species' and 'population'.

After the teaching sequence in the 10th grade, the findings of the post-interview showed that the two conceptions 'speciation due to the 'survival of the fittest'' and 'variation as a prerequisite for selection' were still available to the students. One of the students noted that individuals are different at the beginning in order that selection can take place. On the other hand, all of the students were convinced that the environment itself has a selective function, with different environmental conditions causing a different speciation. As in the pre-interview in the 10th grade and the post-interview in the 8th grade, they still understood that adaptation happens over generations due to sexual reproduction being the key to evolution (as a stepping stone idea). A new conception was 'The environment affects the formation of new traits'. The need (by organisms) for adaptation induced by changes in the environment could neither be exchanged for, nor further developed into, another more scientific concept. Moreover, as this teaching sequence did not specifically aim at defining the scientific terms 'species' and 'population', we assume that no development has taken place in this regard.

## 7 Discussion and Outlook

### *7.1 Study on Students' Learning Progressions*

The first interview took place before the students attended the teaching sequence (8th grade) about *natural selection* for the first time. Findings about their conceptions correspond highly to findings of other studies on the same topic (Baalmann et al., 2004; Gregory, 2009; Hammann & Asshoff, 2015; Johannsen & Krüger, 2005; Kattmann, 2015; Wandersee et al., 1995; Zabel, 2009; Zabel & Gropengiesser, 2011). As evolutionary processes are not experienced in the students' everyday lives, pre-Darwinian conceptions seem to be a common attempt to explain these natural phenomena. Students without or with very little scientific understanding have a tendency to think of evolution as an intended act, or a necessity. Some imagine that organs further develop through repetitive use, similar, but not identical, to Lamarck's ideas of evolution (Kampourakis & Zogza, 2007) or other historical phases of scientific thought (Ha & Nehm, 2014). These newly grown or strengthened organs are passed on to the next generation, which is why descendants are better adapted (Scheibstock, 2014). Furthermore, everyday conceptions of evolutionary processes are based on teleological convictions, like 'goal-oriented adaptive behaviour' and 'adaptive physical change', according to which changes during one's lifetime are possible (Scheibstock, 2014).

Regarding the (newly acquired) students' conceptions of 'selection by the environment' and/or 'survival of the fittest', which occurred after the first teaching sequence, it has to be pointed out that even if they are not perfectly correct scientific conceptions, they seem to be important '*stepping stones*' (Duncan & Rivet, 2013; Duschl et al., 2011). Critical to the 'survival of the fittest' seems to be the notion about thinking of an individual in the context of this phrase; this point even expands the critique of Gregory (2009) about this concept. With these stepping stones, it is possible for students to accept the fact that it is not a goal-oriented adaptive behaviour which is pivotal to evolutionary change, but *natural selection*, whereby environmental conditions and the survival or death of specific individuals determine whose genes will be passed on to following generations. Another example of a stepping stone in understanding could be the idea of 'goal-oriented adaptive behaviour'. It seems that this idea is developed further into the conviction that sexual reproduction is used as a response to environmental changes with the aim of creating descendants with favourable traits due to speciation. After the first teaching sequence (and two years later), students no longer think that goal-oriented actions lead to evolutionary changes during one's lifetime, but see 'sexual reproduction' as an mechanism intended to cope with environmental changes, for the next generation. This is an example of *conceptual reconstruction* in progress, but it is not yet completed (Krüger, 2007) because the role of sexual reproduction is learnt, but embedded in the previous teleological thinking.

After two years, in the pre-interviews for the 10th grade, a new conception can be found: 'variation as a prerequisite for selection'. This suggests that the first teaching sequence in the 8th grade (whose thematic frame is based on the concepts of 'variation' and 'natural selection') serves to introduce these specific evolutionary concepts to learners in an effective way as they accept them and even combine them. Besides, it is interesting to note that one of the students thinks that individuals are different at the beginning in order that selection can take place. Obviously, this student considers 'variation' as a temporary state, which enables the process of selection and which disperses as soon as all poorly adapted individuals no longer exist.

The fact that one of the 10th grade students still mentions conceptions that were already believed to have been overcome only proves that conceptual reconstruction can go backwards as well as forwards. Furthermore, this corresponds to the findings of Wandersee et al. (1995), who state that it takes years to understand the whole concept of evolution as many complex and emotionally afflicted concepts need to be accepted by the students. This underlines the demand for theory-based and well-evaluated LPs, which consider the students' teleological conceptions and enable scientific reasoning. And those should start in the lower grades.

## 7.2 Further Development of the Learning Progression

Although carefully planned, the LP curriculum still has holes and pitfalls. Research helps to identify some of them, to fill them and to enable a growing LP to reach even more students.

One problem of our LP is rooted in the fact that the material was aligned to the state curriculum in Austria and not solely on previous research about evolution learning. This curriculum was not designed like the NGSS (see above) but in case of evolution presents itself as jagged, and therefore, students cannot develop their ideas consistently. As a consequence, the LP developers looked for topics in the respective years to which evolution education topics could be added. Finally, the breeding was included in grade 5 (already an extension of the LP—see below in the next paragraph) and grade 8 (domesticated animals), Darwin finches (grade 8) and blackcap evolution (ecology) in grade 9; sexual selection was introduced in grade 10, because in this year ethology is a big topic. In grade 12, the state curriculum reserves nearly a third of the year for teaching evolution. This strategy was chosen to make it easier to include evolution into regular biology lessons over the years and fulfil the state curriculum at the same time (cf. McComas, 2016). This makes it easier for teachers to include the developed materials into their lessons, but does disrupt the idea of an ideal LP.

The results of the students' interviews in the 8th and 10th grades have revealed that students have problems distinguishing between 'species' and 'population'. Knowing these terms is crucial for the understanding of the mechanisms of evolution. To improve the whole LP, we planned an additional sequence of two lessons in grade five (10–11-year-olds). In the state curriculum, mammals and pets have to be covered in this grade. Therefore, the breeding of dogs is already introduced at this level with a card sorting task. The students receive a number of cards, each with a picture of an individual dog, and are asked to sort them into breeds, with a special emphasis on the definitions of population (the respective dog breed), the species (all dogs worldwide) and individuals (whether it is chosen for breeding as an individual or not). This is also an extension of the LP due to the fact that Austrian textbooks fail to build up evolution concepts consistently, especially the population concept (Scheuch & Rachbauer, in prep.; Scheuch & Wäger, 2018).

Another issue is the missing learning experience of the students with the sequence in grade 9. One concept is absent in all other learning sequences, and this is reproductive isolation. This concept would have been included in the blackcap example, but unfortunately it was not possible to teach in that respective grade.

We hope that this enlargement of the LP will help students to improve their understanding of the different levels where evolution takes place. However, due to the results of this study, another question turns up: is breeding with an intentional aim by humans, as it is a directed and therefore goal-oriented activity, helpful as a starting point for learning about evolution? We have seen that the students stick to teleological thinking and even attach new scientific ideas to this basic conception; therefore, it could be counterintuitive to put an emphasis on breeding at the beginning of the LP. On the other hand, breeding is very close to everyday life, and other researchers like Russell and McGuigan (Chapter "Developmental Progression in Learning About Evolution in the 5–14 Age Range in England" in this volume) also propagate this idea as a basis for an LP. Anyway, the teacher has to be aware of this contrast between intended breeding, which easily builds on the students' previous conceptions, and natural selection, which is not goal-directed.

**Acknowledgements** We would like to thank Patricia Hoffmeister, who started this project at the University of Vienna. Many thanks also to the students who agreed to participate in the study. This project was supported by the Hochschuljubiläumsstiftung der Stadt Wien (Grant Nr. H-316715/2017).

## References

Andersson, A. (1982). Female choice selects for extreme tail length in a widowbird. *Nature, 299,* 818–820.

Baalmann, W., Frerichs, V., Weitzel, H., Gropengießer, H., & Kattmann, U. (2004). Students' conceptions about processes for adaptation—Results of an interview study. *Zeitschrift für Didaktik der Naturwissenschaften, 10,* 7–28.

Bauer, H. (2017). *Leitmotif evolution—Analysis of a learning progression with focus on coherence.* Vienna: University of Vienna.

Cook, L. M., & Saccheri, I. J. (2013). The peppered moth and industrial melanism: Evolution of a natural selection case study. *Heredity, 110,* 207.

Darwin, C. (1845). *Journal of researches into the natural history and geology of the countries visited during the voyage of H.M.S. Beagle round the world, under the Command of Capt. Fitz Roy* (2d ed.).

Dawson, C. (2000). Upper primary boys' and girls' interests in science: Have they changed since 1980? *International Journal of Science Education, 22*(6), 557–570.

Dobzhansky, T. (1973). Nothing in biology makes sense except in the light of evolution. *The American Biology Teacher, 35*(3), 125–129.

Duncan, R. G., & Rivet, A. E. (2013). Science learning progressions. *Science, 339*(6118), 396–397.

Duschl, R., Maeng, S., & Sezen, A. (2011). Learning progressions and teaching sequences: A review and analysis. *Studies in Science Education, 47*(2), 123–182.

Eder, E., Seidl, V., Lange, J., & Graf, D. (2018). Evolution-education in the German speaking countries. In H. Deniz & L. A. Borgerding (Eds.), *Evolution education around the globe* (pp. 235–260). Berlin: Springer.

Eder, E., Turic, K., Milasowszky, N., Adzin, K. V., & Hergovich, A. (2011). The relationships between paranormal belief, creationism, intelligent design and evolution at secondary schools in Vienna (Austria). *Science & Education, 20*(5), 517–534.

Fenner, A. (2013). *Students' conceptions concerning evolution theory and evaluation of a learning sequence on adaptation through selection.* Gießen.

Gregory, T. R. (2009). Understanding natural selection: Essential concepts and common misconceptions. *Evolution: Education and Outreach, 2*(2), 156–175.

Gropengießer, H. (2005). Qualitative content analysis in subject didactics research. In P. Mayring & M. Gläser-Zikuda (Eds.), *Practice of qualitative content analysis* (pp. 172–189). Weinheim, Basel: Beltz.

Gropengießer, H. (2010). Teaching biology. In H. Gropengießer, M. Beier, & J. Wolter (Eds.), *Biology* (Vol. 1). Stuttgart, Leipzig: Klett.

Ha, M., & Nehm, R. H. (2014). Darwin's difficulties and students' struggles with trait loss: Cognitive-historical parallelisms in evolutionary explanation. *Science & Education, 23*(5), 1051–1074.

Hammann, M., & Asshoff, R. (2015). *students' conceptions in biology education—Causes for learning difficulties.* Seelze: Klett-Kallmeyer.

Helldén, G. (2005). Exploring understandings and responses to science: A program of longitudinal studies. *Research in Science Education, 35*(1), 99–122.

Jelemenská, P. (2012). Teachers' conceptions about teaching and learning of evolution—A case study of a biology education coaching. *Zeitschrift für Didaktik der Naturwissenschaften, 18,* 229–259.
Jelemenská, P., Amon, H., & Wenzl, I. (2010). *What do you understand about evolution?—Planing evolution lessons based on students' conceptions.* Retrieved from https://www.imst.ac.at/imst-wiki/images/d/d5/1847-Langfassung-Jelemenska.pdf.
Johannsen, M., & Krüger, D. (2005). Students'concptions in evolution—A quantitative study. *Berichte des Institutes für Didaktik der Biologie, 14,* 23–48.
Kampourakis, K. (2014). *Understanding evolution.* Cambridge: Cambridge University Press.
Kampourakis, K., & Zogza, V. (2007). Students' preconceptions about evolution: How accurate is the characterization as "Lamarckian" when considering the history of evolutionary thought? *Science & Education, 16*(3), 393–422.
Kattmann, U. (2015). *Understanding students—Everyday conceptions in biology education.* Hallbergmoos: Aulis Verlag.
Kattmann, U., Duit, R., Gropengießer, H., & Komorek, M. (1997). The model of educational reconstruction—A framework for science education research and development. *Zeitschrift für Didaktik der Naturwissenschaften, 3*(3), 3–18.
Keynes, R. (2001). *Charles Darwin's beagle diary.* Cambridge: Cambridge University Press.
Krüger, D. (2007). Conceptual change-theory. In D. Krüger & H. Vogt (Eds.), *Theories in science education research: A handbook for teacher students and doctoral students* (pp. 81–92). Berlin, Heidelberg: Springer.
Laland, K., Uller, T., Feldman, M., Sterelny, K., Müller, G. B., Moczek, A., … Odling-Smee, J. (2015). The extended evolutionary synthesis: Its structure, assumptions and predictions. *Proceedings of the Royal Society B: Biological Sciences, 282*(1813).
Laland, K., Uller, T., Feldman, M., Sterelny, K., Müller, G. B., Moczek, A., … Strassmann, J. E. (2014). Does evolutionary theory need a rethink? *Nature, 514*(7521), 162–164.
Lamnek, S. (2005). *Group discussion—Theory and practice* (2nd ed.). Weinheim, Basel: Beltz, UTB.
Lamnek, S. (2010). *Qualitative research in social studies.* Weinheim, Basel: Beltz.
Mayr, E. (1982). *The growth of biological thought: Diversity, evolution, and inheritance.* Cambridge, Massachusetts, London: Harvard University Press.
McComas, W. F. (2016). Biology education is evolution education. *The American Biology Teacher, 78*(2), 91.
Møller, A. P. (1990). Effects of parasitism by a haematophagous mite on reproduction in the barn swallow. *Ecology, 71,* 2345–2357.
Nationale Akademie der Wissenschaften Leopoldina. (2017). *Teaching evolutionary biology at schools and universities.* http://www.leopoldina.org/uploads/tx_leopublication/2017_Stellungnahme_Evolutionsbiologie.pdf.
Pennisi, E. (2005). New migration route could lead to new species of bird. *Science, 310.*
Pigliucci, M., & Müller, G. B. (Eds.). (2010). *Evolution—The extended synthesis.* Cambridge, Massachusetts: MIT Press.
Riemeier, T. (2007). Moderate constructivism. In D. Krüger & H. Vogt (Eds.), *Theories in science education research: A handbook for teacher students and doctoral students* (pp. 69–79). Berlin, Heidelberg, New York: Springer.
Scheibstock, J. (2014). *Learning effects in biology lessons with the topic of evolution—Study on the development of students' conceptions with selection and variation.* Wien: University of Vienna.
Scheuch, M., Amon, H., Hoffmeister, P., Scheibstock, J., & Bauer, H. (2017). Evolution—A leitmotif for school. *Plus Lucis, 1,* 14–18.
Scheuch, M., & Rachbauer, S. (in prep.). Teaching evolution with Austrian biology textbooks. In libreriauniversitaria.it (Ed.), *New persectives in science education* (Vol. 8, 4p). Firenze/It: Editore Filodiritto.

Scheuch, M., & Wäger, M. (2018). Occurrence and representation of evolution in Austrian biology textbooks. In libreriauniversitaria.it (Ed.), *New perspectives in science education* (Vol. 7, 4p). Firenze/It: Editore Filodiritto.
Silverman, D. (Ed.). (2016). *Qualitative research*. London, Thousand Oaks, New Dehli, Singapore: Sage Publications.
Wandersee, J. H., Good, R. G., & Demastes, S. S. (1995). Research on biology education about evolution: An inventory. *Zeitschrift für Didaktik der Naturwissenschaften, 1*, 43–54.
Weitzel, H., & Gropengiesser, H. (2009). Development of conceptions on phylogenetic adaptation: Understanding learning obstacles and creating learning opportunities. *Zeitschrift für Didaktik der Naturwissenschaften, 15*, 287–305.
Wyner, Y., & Doherty, J. H. (2017). Developing a learning progression for three-dimensional learning of the patterns of evolution. *Science Education, 101*(5), 787–817.
Yin, R. K. (2009). *Case study research: Design and methods* (4th ed.). Thousand Oaks, California: SAGE.
Zabel, J. (2009). *The role of narration for understanding evolution* (Vol. 24) (Doctoral thesis). Carl von Ossietzky Universität Oldenburg.
Zabel, J., & Gropengiesser, H. (2011). Learning progress in evolution theory: Climbing a ladder or roaming a landscape? *Journal of Biological Education, 45*(3), 143–149.

**Dr. Martin Scheuch** is a trained vegetation ecologist with working experience in biodiversity and nature conservation. He was Postdoc at the University of Vienna and is currently Professor for Biology Education at the University College for Agricultural and Environmental Education in Austria. He teaches biology teachers and environmental educators. His research interest is pedagogical content knowledge (PCK) in biology. He is the leader of a project about a learning progression for evolution education and the development of the students' conceptions. He is interested in out-of-school learning in biology (outdoors, botanical gardens and museums), as well as in the context of citizen science and environmental education. In vocational education, he works with competency-based learning tasks for agricultural schools in Austria.

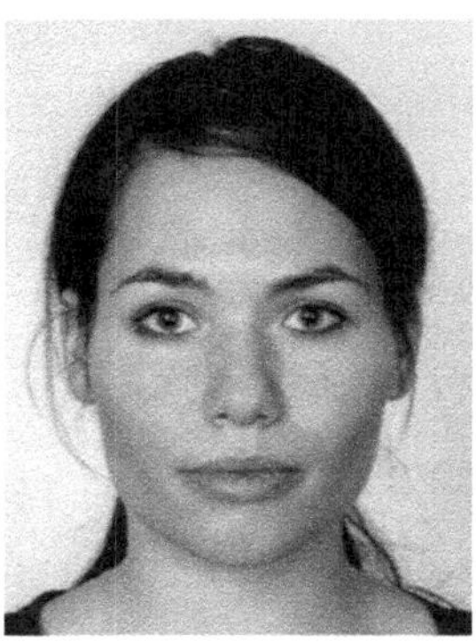

**Mag. Jaqueline Scheibstock** studied biology and French to become a teacher for lower and upper secondary schools at University of Vienna. The topic of her master thesis was students' conceptions about evolution—this is still in the centre of high interest to her. After teaching in secondary schools for a few years, she started to work in the biology education research at the Austrian Educational Competence Centre for Biology at the University of Vienna. She continues her work from the master thesis and currently makes a longitudinal study on the development of evolution concepts of students, which learn evolution in a carefully planned learning progression.

**Mag. Heidemarie Amon** has obtained the master's in biology education from the University of Vienna and has been working since 1987 as a teacher of biology in Vienna. She works part time at the Austrian Educational Competence Centre for Biology at the University of Vienna since 2008 in biology teacher education (pre-service and in-service). Her interests include the development of competency-based learning tasks for students. Together with Dr. Patricia Hoffmeister (Jelemenská), she started 2010 to develop teaching and learning sequences in different levels of lower and upper secondary schools to work with students' conceptions about evolution.

**Mag. Helene Bauer** is High School Teacher for biology and chemistry in Austria. In her master thesis, she dealt with the concept of evolution as a leitmotif in biology teaching, and the underlying pedagogical content knowledge (PCK) for teaching evolution. She identified six different big ideas in a previously developed learning progression and connected those with the students' beliefs and conceptions.

# Inequitable Foundations? Educational Equality in Evolution

**Jaimie L. Miller-Friedmann, Susan E. Sunbury and Philip M. Sadler**

## 1 Background

Evolution is unique among all life science topics. It is all-encompassing, incorporating portions or strands of almost every other field of life science into its framework. Evolution has also had a history unlike every other major scientific topic—socially, religiously, and politically. For many decades, there has been a debate in the USA about the teaching of the theory of evolution in the nation's schools. This debate is framed by the near-universal acceptance by the scientific community of evolution and natural selection as the organizing principles in modern biology and why it is important for all students to learn these principles. Opposed are people who question evolution itself, or the teaching of evolution and natural selection; others want students to also learn of alternative (mainly religious) viewpoints. Students are often exposed to ideas about evolution by family members, religious leaders, and even the media before entering a life science classroom. A 2014 survey found that "religious belief was the strongest determinant of people's views on evolution—much more so than education, socioeconomic status, age, political views, or region of the country" (Hill, 2014).

In a 2005 Gallup Poll survey (Table 1), a majority of Americans indicated that they would *not* be upset if creationism was taught in their schools, while fewer reported that they would not be upset if evolution was taught in schools. Additionally, 30%

J. L. Miller-Friedmann (✉)
Department of Education, University of Oxford, 15 Norham Gardens, Oxford OX2 6PY, UK
e-mail: friedmann@education.ox.ac.uk

S. E. Sunbury
Smithsonian Astrophysical Observatory, Cambridge, MA, USA
e-mail: ssunbury@cfa.harvard.edu

P. M. Sadler
Harvard University, Cambridge, MA, USA
e-mail: psadler@cfa.harvard.edu



U. Harms and M. J. Reiss (eds.), *Evolution Education Re-considered*,
https://doi.org/10.1007/978-3-030-14698-6_6

**Table 1** Results from a 2005 Gallup Poll eliciting public preference to teaching evolution versus creationism (Carlson, 2005)

| | Yes, I would | No, I wouldn't |
|---|---|---|
| If the public schools in your community taught the theory of creationism—that is, the idea that human beings were created by God in their present form and did not evolve from other species of animals—would you be upset, or not? | 22% | 76% |
| If the public schools in your community taught the theory of evolution—that is, the idea that human beings evolved from other species of animals—would you be upset, or not? | 34% | 63% |
| Would you be upset if evolution but NOT creationism were taught in the public schools in your community? | 30% | 70% |
| Would you be upset if creationism but NOT evolution were taught in the public schools in your community? | 18% | 82% |

of Americans indicated they would be upset if only evolution was taught, while 18% would be upset if only creationism was taught (Carlson, 2005).

These results indicate that, at the time, there was a national inclination toward conflating religious explanations for natural phenomena with state-run educational systems. Although the USA as a nation has endeavored to separate church and state, it would seem that the local governments not infrequently have had other preferences with regard to evolution and creationism. In the intervening years, a series of national standards have been formulated, all of which include the theory of evolution but not creationism. Clearly, there is some disagreement between what the government wants and what people want their children to be taught.

A majority of the states have adopted and adapted the biological evolution strands from the National Research Council's National Science Education Standards and now the Next Generation Science Standards, but many have included new requirements for critical thinking when teaching these standard-based materials as a way to permit the introduction of alternative viewpoints to evolution. Moreover, assessment of student understanding of these evolutionary concepts is now part of nearly all the high-stakes tests that states require for high school graduation. While this debate is often regarded as an issue unique to the USA, there is evidence that creationism is being introduced into classrooms across the globe (Blancke, Boudry, Braeckman, De Smedt, & Cruz, 2011; Borczyk, 2010; Kutschera, 2008; Moore & Kraemer, 2005). Even in countries where Ministers of Education support teaching the theory of evolution, creationist views are still held by members of the public and even teachers, including biology teachers (Clement, 2015; Curry, 2009; Miller, Scott, & Okamoto, 2006).

Why is it essential for students to be able to understand evolution and natural selection as called for in the NSES and NGSS (NRC, 1996; NGSS, 2013)? Over the years, scientists and educators have a powerful, twofold answer to that question: (i) Evolution and natural selection are the unifying and organizing principles for all modern biology, and (ii) studying this theory and mechanism requires that students become engaged with the nature of science, particularly what scientists mean when they talk about a theory. Evolution both underlies and unifies the varied fields within life science (Futuyama, 2000; Rutledge & Mitchell, 2002; Sager, 2008; Wiles, 2010; Wiles & Alters, 2011) and includes genetics, origins of life, the history of science, biochemistry, paleontology, and more. Theodosius Dobzhansky, the evolutionary biologist and geneticist, wrote that "Seen in the light of evolution, biology is, perhaps, intellectually the most satisfying and inspiring science. Without that light, it becomes a pile of sundry facts some of them interesting or curious, but making no meaningful picture as a whole" (Dobzhansky, 1973). This light, or underlying foundation, gives coherence to otherwise disparate topics lumped together under an umbrella heading of "life science." The interconnectedness of topics provides the pathways for the teaching and learning of all life science. Furthermore, evolution addresses and answers fundamental questions about the Earth—how did the different kinds of environments emerge, different species arise, life originate, and extinctions occur? It provides the best explanation for commonalities between the rich diversity of organisms on the planet. In essence, by tying all organisms together, evolution displaces anthropomorphism in the same way that a heliocentric model of the solar system replaced geocentrism (AAS, 1990).

Evolution and natural selection are objective and universally accessible; like all other scientific explanations, they are derived from repeated observation and experimentation. Evolutionists agreed on a modern, contemporary theory only after decades of argument. The debate involved common-sense explanations for phenomena on one side, and on the other experimental testing of the behavior of nature. While there is no longer any substantive debate within the scientific community on the guiding principles of evolution, like any science theory, it is continually open to modification. It exemplifies the gradual development of a fluid theory, incorporating new findings and the contributions of many different kinds of scientists all interested in furthering the understanding of evolution (NSES, 1998). The development of evolutionary theory is an ongoing example of how science functions, and the ways in which science is dependent upon many voices and contributions, remaining mutable and adaptable. This history contains essential lessons for every student, lessons that scientists and educators believe every person should understand.

Both federally funded agencies and national education associations have unequivocally held that the teaching and learning of evolution is essential to the nation's youth. The American Association for the Advancement of Science (AAAS, 2012) places the discussion of evolution first in its framework on life science, emphasizing the ways in which evolution unifies all other life science subjects. The National Research Council (NRC, 1996), National Science Teachers' Association (NSTA, 1997, 2013), National Academy of Sciences (NAS, 2004), and National Association of Biology Teachers (NABT, 2011) have all published literature insisting that evolution be accorded a place in the curriculum that mirrors its importance in the field of life science. The National Science Education Standards (NRC, 1996) clearly states that: "No standards should be eliminated from a category. For instance, 'biological evolution' cannot be eliminated from the life science standards" (p. 112).

But being able to teach and/or learn evolution from a scientific point of view has not always been easy in the USA. The Butler Law was established in Tennessee in 1925 and prohibited evolution from being taught in all schools in the state, including universities and normal schools (teaching colleges). John Scopes was arrested and charged with disobeying this law. He was found guilty in trial, confirming the prohibition on teaching evolution. The Butler Law remained until 1967, repealed then on a teacher's First Amendment rights to free speech. Other states enacted similar prohibitive laws in 1925. Perhaps this was in reaction to or in spite of the influence of the Scopes Trial, which sparked further interest in whether or not evolution was being taught in the classroom. Anti-evolution bills are still being introduced in state legislations across the country, even in states not known for religious fundamentalism. While most are struck down, several states including Texas, Louisiana, and Tennessee have passed bills which allow public school teachers to teach alternative conceptions to evolution. In other states, public funds are utilized to support private schools which teach creationism.

At the local level, there are concerns that many teachers and even entire school districts omit evolution from their curricula in favor of creationism (Long, 2012). This stems from fears that teaching evolution will conflict with students' religious beliefs (Yasri & Mancy, 2014) and even cause them discomfort when the topic of evolution is discussed at school. However, Lac, Hemivich, and Himelfarb (2010) found that only 10% of parents reported that their children felt uncomfortable when the topic of evolution was presented in the classroom.

Historically, teacher surveys have been conducted to elicit the degree to which evolution was taught, if at all. In the winter of 1939, there was a nationwide survey to which there were 3186 respondents; it found that evolution was being taught in less than half of the high schools in the USA (Riddle et al., 1942). A 1982 survey revealed that almost 75% of the nation's life science teachers placed moderate to strong emphasis on evolution (Ellis, 1983). A survey distributed to Minnesota teachers in 2003 revealed that 88% included evolution in their curriculum, while 12% excluded it; 20% included creationism in their curriculum (Moore & Kraemer, 2005). More recently, it was reported that only 28% of high school biology teachers regularly teach the concept of evolution as prescribed by the National Science Standards (Berkman & Plutzer, 2011).

While these surveys show that biology teachers do include evolution in their curriculum, there is a rise in the percentage of teachers who introduce non-scientific alternative points of view; in the Berkman and Plutzer survey, 13% of high school biology teachers advocate for creationism in their classrooms. More concerning is the 60% of biology teachers who do not take a stance on evolution and use a variety of educational techniques to avoid addressing the controversy (Berkman & Plutzer, 2011). In a follow-up study, it was revealed that Catholic school teachers were most comfortable discussing the differences between evolution and creationism (Mervis, 2015).

This sets evolution apart from all other life science standards. For almost every other unit in life science, there is no prohibition or qualification as to whether or not it will be taught, nor are alternative viewpoints presented to the students. The factors contributing to a de-emphasis of evolution in the classroom have varied from poor textbook coverage, community resistance to evolution, educational policy banning evolution, allowing teachers to teach evolution but with caveats, or requiring teachers to instruct their students in all religious, social and scientific viewpoints at the same time (Eglin, 1983; Rutledge & Mitchell, 2002; Shankar & Skoog 1993; Troost, 1967; Zimmerman, 1987). There are two contentious issues—natural selection as the mechanism for evolution, as opposed to creationism or intelligent design, and human evolution as a species of Animalia, as opposed to human exception from common ancestry with other species. These two topics are critical to learning evolution and to understanding the big picture—the interconnectedness of all life and the great diversity of species that have lived or now live on the Earth (NCSE, 2008).

There is general consensus that a problem exists in respect of equal access. Most states require students to pass a high-stakes exit exam that includes standards-based evolutionary concepts, yet students nationwide are taught the subject of evolution at unequal depths and in many instances as one of several viewpoints. Teaching and learning evolution has then become a matter of educational equality. The charge that all nation's students have equal opportunity and access to quality education is central to the most recent White House science commission report (PCAST, 2012). The standards in science have been adopted to ensure a minimum level of scientific literacy for every student, regardless of socioeconomic status (SES) or any other outside influence. The disadvantage for students who do not learn or understand evolution is twofold. From a conceptual point of view, it leaves students underprepared for taking higher-level life science courses, likely closing them off from the growing number of career pathways in health care and biotechnology. And from a practical point of view, students in many states need to understand evolution and natural selection to pass the high-stakes graduation exams in order to compete in the education and career market.

There are small but significant differences in science achievement between male and female students in US middle schools and high schools. While females tend to get higher grades in science classes (Voyer & Voyer, 2014), males score higher on national and international achievement tests. Over the past three test cycles, in both eighth and twelfth grades, males consistently scored significantly higher overall in science and in all content specific areas—physical science, life science and earth and space sciences on the Nation's Report Card (National Assessment of Educational Progress (NAEP), 2009, 2011, 2015 Science Assessments). While achievement gaps are narrowing for both grade levels, the most recent test (2015) indicates that the gender gap still exists. International assessments show similar results. Fourteen- and fifteen-year-old male students in the USA also scored significantly higher in science literacy than did females (Trends in International Mathematics and Science Study (TIMSS), 2015, and Program for International Student Assessment (PISA), 2015). With that as background, we assessed the understanding of the standard-based evolutionary concepts of students who have completed middle- and high school life science courses, independent of their environment.

## 2 Research Questions

Is there educational equality, so that students in a wide range of schools, regions, and SES settings understand the theory of evolution? Is there gender equality in understanding evolution as it is set out in the national standards?

## 3 Methodology

This research is part of Project MOSART-LS (Misconceptions-Oriented Standards-based Assessment Resources for Teachers in Life Science) at the Science Education Department of the Harvard–Smithsonian Center for Astrophysics. Data for this study were provided from two mixed-methods studies funded by the National Science Foundation (DRL#0830922, DRL #1316645). The main instruments developed during each of these projects were student assessments designed to measure scientific knowledge and misconceptions. The items on the assessments were designed by two MOSART-LS teams (consisting of educators and educational researchers) in alignment with the National Science Standards for the middle school assessment and the Next Generation Science Standards for the high school assessment. To develop the items, hundreds of studies in the science education research literature on student misconceptions were reviewed to determine the most common student misconceptions for each content strand (structures and processes, ecosystems, heredity, and biological evolution). The teams constructed hundreds of unique multiple choice items for middle school and high school that measure the degree to which test takers hold either a misconception or an accepted scientific view of a concept.

For example, this middle school evolution question developed includes the correct answer (a) and a strongly held misconception by students (b).

According to scientists, which of the following organisms have become extinct?
(a) many plants and animal species.
(b) many animal species, but only a few plant species.
(c) a few plant and animal species.
(d) no plant or animal species.
(e) many plant species, but only a few animal species.

All questions were reviewed by experts (biologists from universities, national laboratories, and industry) for clarity and accuracy. Items from particular subtopics were sent to those considered experts in the related fields. The reading-level characteristic was addressed through analyses of multiple readability criteria done by a recognized expert. For a more complete description of the item development process including analysis of validity and reliability for middle school items, see

Sadler et al. (2013). The analysis of high school data is currently in process and will be published within the year.

Both grade bands of assessment items were pilot tested to establish both validity and reliability. After the pilot test, a field test was conducted for each data set. Data for this study were generated from the field study tests for both instruments. For middle school students, a 35-item assessment was field tested. Due to a large number of items generated by the team, and their subsequent validity and reliability established on pilot tests, 13 paper-and-pencil forms were generated in order to field test all of the items in the item bank. However, the forms were not unique: Each form contained eight items in common with all other forms, which we called "anchor items." These items were meant to link all of the forms, as well as to create a basis for comparison between forms. In addition to the content questions, students were asked a variety of demographic questions. During the high school field test, 30 content items were tested. Twenty-two paper-and-pencil forms were needed to test all of the items developed by the team. For this assessment, each form contained six anchor items.

Teacher and department chairs were solicited to participate in each study with an invitation via postal mail; these solicitations were sent to a randomized list of middle and high school science educators/life science educators generated by Market Data Retrieval (schooldata.com). Each respondent who agreed to participate in the study distributed the assessment to their class (or classes) and returned both the assessment forms and the answer sheets. Additionally, teachers were asked to answer the survey questions to determine their knowledge of life science concepts and their knowledge of student misconceptions.

No student was reported to have refused to take the assessment. Students in both age bands were tested in April–June as near to the end of their life sciences course as was possible. Therefore, it can be assumed that the results measured student understanding after nearly two semesters of study of the relevant grade-band concepts. These data were then scanned, combined into one database, and cleaned. The test scores were calculated on a scale of 0–1, as were the average scores for anchor items and the items within the evolution standard.

In order to assess which variables were predictors for doing well on the evolution standard items, the first step was to run a correlation analysis for each age group, between average score on evolution items and the anchor average score. Then, multiple regressions models were constructed to assess the importance of a variety of variables, all entered as groups, on the same outcome variable (average score on the evolution standard), while controlling for the impact of anchor score: race/ethnicity (White, Black, Asian, or Hispanic), school type (public, non-sectarian private, Christian, or Catholic), or region (northeast, south, Midwest, or west). Previous ANOVA analysis results for middle school data implied that there was not a significant relationship between gender and evolution standard score ($F(1.15692) = 0.098$, $p = 0.754$); thus, gender was not included in the groups of possible predictor variables. For the high school data, there did appear to be a statistically significant difference in evolution standard average scores by gender. ANOVA results ($F(1.9738) = 43.014$, $p < 0.001$) showed that means were higher for females than for males; gender was therefore entered into the model for high school students.

## 4 Sample

In total, 16,383 students completed the middle school (grades 6 through 8) assessment. Of the respondents, 51.4% were male and 48.6% female. In regard to race/ethnicity, 52.7% of the respondents identified themselves as White, 9.6% as Black, 7.4% as Asian, 16.9% as Hispanic, and 10.8% as other race (with 2.6% choosing not to answer). A total of 95.6% of the students attend public school, while 1.1% attend non-sectarian private school, 1.6% attend non-Catholic Christian school, 1.5% attend Catholic school, and 0.2% attend other kinds of schools (for example, an overseas school on a military base). In regard to region, 15.5% of the students were in the northeast, 31.6% in the south, 21.8% in the Midwest, and 31.1% in the west (including Alaska and Hawaii). The sample closely approximates the national averages for middle school students and can be considered a nationally representative sample for this population.

In total, 9740 students completed the high school (grades 9 through 12) assessment. Of these respondents, 48.3% were male and 51.7% female. A total of 64.7% of the respondents identified as White, 8.6% as Black, 9.4% as Asian, 2.6% as Native American/Pacific Islander, and 14.7% as Other/Mixed Race; 16.2% of the total population self-identified as Hispanic. The majority of students in this cohort were in the ninth or tenth grades (82.4%, on average between 14 and 16 years old). A total of 90.9% of the population attend public school, 5.2% attend Catholic school, 1.6% attend non-Catholic Christian school, 1.5% attend non-sectarian private school, and 0.8% attend Jewish school. A total of 19.8% of the students attend schools in the northeast, 23.1% in the south, 32.4% in the Midwest, and 24.7% in the west (including Alaska and Hawaii). The sample closely approximates the national averages for high school students and can be considered a nationally representative sample for this population.

## 5 Results

### *5.1 Middle School*

The relationship between anchor item average and evolution standard average was investigated using Pearson product-moment correlation coefficient. There was a medium positive correlation between the two variables, $r = 0.47$, $n = 16{,}328$, $p < 0.001$. Hierarchical multiple regressions were then used to assess the ability of different variable groups to predict evolution standard average scores while controlling for anchor score average. Anchor score average was entered at Step 1, explaining 47% of the variance in evolution item average score. Neither race/ethnicity nor school type, while controlling for anchor item average, changed the total variance explained by the model. Region, however, raised the total variance explained by the model to 48.3%, $F(4.16321) = 1240.46$, $p < 0.001$. The region group accounted for an additional

1.3% of the variance in evolution item average score, $R$ squared change = 0.013, $F$ change (3.16321) = 89.479, $p < 0.001$. In the final model, all included variables were statistically significant contributors to predicting evolution standard average score, although region had far less of an impact than anchor score average, as implied by beta values (*beta*(anchor score average) = 0.474, $p < 0.001$, *beta*(south) = −0.138, $p < 0.001$, *beta*(Midwest) = −0.138, $p < 0.001$, *beta*(west) = −0.141, $p < 0.001$).

## 5.2 High School

Similar statistical analyses were applied to the data for this group as for the middle school data in order to be able to compare results as a pseudo-longitudinal study. The relationship between anchor item average and evolution standard average was again investigated via Pearson product-moment correlation coefficient. There was a medium positive correlation between the two variables, $r = 0.37$, $n = 9740$, $p < 0.001$. Hierarchical multiple regressions were then used to determine which, if any, variables were the best predictors of a high average score for the evolution items. For this cohort, test score was entered at Step 1, explaining 64.7% of the variance in evolution standard average score. Anchor item average score explained a further 1.6% of the total variance, $F$ change (2.9737) = 460.714, $p < 0.001$. Region, race, gender, and school type were then added to the model, and variables that were not statistically significant were removed. In the final model, six variables were statistically significant contributors to changes in variance, although beta values show that region, race, gender, and school type had significant, but negligible impact on evolution standard average score (*beta*(test score average) = 0.890, $p < 0.001$, *beta*(anchor score average) = −0.155, $p < 0.001$, *beta*(Midwest) = −0.042, $p < 0.001$, *beta*(Asian) = 0.030, $p < 0.001$, *beta*(Female) = 0.025, $p < 0.001$, *beta*(non-Catholic Christian school) = −0.016, $p < 0.001$).

# 6 Conclusions and Implications

The results of this study indicate that gender is not a statistically significant predictor of scoring higher on evolution items at the middle school level. At the high school level, however, there is a small but statistically significant advantage to being female. These results may imply that the stereotypically gendered distribution of hard sciences becomes a salient concept in high school, whereas in middle school (i.e., prepuberty), academic achievement may be less influenced by gender norms and expectations. Because this result is not necessarily specific to evolution, but more to life science as a discipline, it may be seen as evidence for the beginning of the gendered dichotomy in science that relegates physical science to male domination and life science and the "soft" sciences (social sciences) to female domination. Ideally, no subject should be the domain of one gender or the other, but have equal representation. More research needs to be done to discover when and how gendered associations with life science begin and how they may be counteracted.

In a similar vein, race and ethnicity were not statistically significant contributors to the model for middle school, but one race, Asian, was a negligible but statistically significant positive variable for the high school cohort. While there may be many reasons why this was a statistically significant result for the older students, the best proxy explanation may be that, much like gender, socially acceptable stereotype behaviors may become more pertinent notions in high school than they are in middle school. Stereotype pressure/threat that would stress the idea that Asian students excel in science may compel these students to try harder on a low-stakes assessment (as the assessment in this study was) than their peers.

In both the middle school and high school data, geographical region was a statistically significant factor in evolution average score. For the high school cohort, only the Midwest was a statistically significant factor (negative predictor), whereas the Midwest, West, and South were all negative predictors for middle school evolution average score. If the data are viewed as a pseudo-longitudinal study, it would be reasonable to say that the disadvantages of being from the West or South in middle school disappear by the end of a life science course in high school. The disadvantage of being from the Midwest may be due to many conflicting and/or conflated factors, not the least of which may be choice of text, the order in which units are taught, the curriculum being taught, emphases on concepts in the curriculum, and how state standards have been modified from the NGSS standards. Any of these, or a combination, would likely produce the results seen in this study.

It is perhaps most important to note that school type was not a statistically significant predictor to doing well on the evolution standard in middle school. This suggests that students in all schools are being taught evolution equally, and concerns about students in religious schools may be exaggerated. In fact, students attending non-Catholic Christian schools had a negligible (in terms of effect size) but statistically significant advantage on the high school assessment. The items on this assessment were phrased so that students would answer in a way that scientists would answer. For example, in the high school assessment, an item on adaptation does not provide the opportunity for students to answer in a way that is non-scientific (Table 2). Students who answer correctly are then indicating that they understand the science of the theory of evolution; we did not ask them to answer in a way that correlates with their personal or religious beliefs.

For example, in the high school science test, on a question about common ancestry, an item asks what conclusions a scientist might draw using data:

Frogs, lizards, and birds all have a similar arrangement of bones in their limbs. What might a scientist conclude about these animals?

(a) They move in the same way.
(b) ***They have a common ancestor.***
(c) They live in the same habitat.

**Table 2** Example of a high school adaptation item with response statistics. Some species of monkeys instinctively make noises to signal danger to each other. This behavior evolved because:

| Possible answer | Number of students who chose this answer | Percent who chose this answer (%) | Correct answer/most common misconception |
|---|---|---|---|
| (a) These monkeys had poor eyesight | 22 | 6 | |
| (b) Monkeys that learned to do this were liked better by other monkeys | 24 | 6 | |
| (c) Monkeys that did this survived and reproduced | 173 | 45 | Correct answer |
| (d) Individual monkeys adapted their behavior to best fit their environment | 141 | 37 | Most common misconception |
| (e) Individual monkeys got smarter with every generation | 25 | 6 | |

(d) They evolved at the same time.
(e) They are the same size as adults.

Or what evidence a scientist would use to draw a conclusion:

Three different species of lizard live in the same habitat. What evidence would a scientist use to conclude that these species share the same ancestor?
(a) They have similar body shapes.
(b) They have a similar diet.
(c) They have a similar population size.
(d) ***They have similar DNA***.
(e) They have similar behaviors.

The results reveal that the students about whom many have been concerned with regard to learning evolution, i.e., those students attending non-Catholic Christian schools, are (very slightly) better able to answer evolution items in a scientific manner than public school students. These results imply that general concern about learning, accepting, and believing in evolution should be focused on public school students (or all school students) rather than the minority of students at religiously affiliated schools.

We should feel some sense of relief from the results of this study: There *is* equality in evolution learning and understanding in the USA, especially for younger students, and there is some evidence that teachers are successfully navigating the difficulties of teaching a subject in which strong pre-existing cultural beliefs and loyalties are potentially in conflict with orthodox scientific understanding. However, the results seen in this study among American students remains somewhat worrisome. While we expected that the average test score would be lower than the average test score for standardized exams (Sadler et al., 2013), the average test score for this assessment was 0.45 for middle school (0.29 for the evolution standards) and 0.48 for high school (0.42 for the evolution standards). These scores indicate that, nationwide, American students are still not meeting the standards for life sciences in general and evolution in particular. While no specific group of students is at a disadvantage relative to others, there is still a great deal of work to be done to ensure that every student has a reasonable understanding of evolution.

## References

American Association for the Advancement of Science. (2012). *Evolution on the front line: An abbreviated guide for teaching evolution, from Project 2061 at AAAS*. August 14, 2012. http://www.aaas.org/news/press_room/evolution.

Blancke, S., Boudry, M., Braeckman, J., De Smedt, J., & Cruz, H. (2011). Dealing with creationist challenges. What European biology teachers might expect in the classroom. *Journal of Biological Education, 45*(4), 176–182.

Borczyk, B. (2010). Creationism and the teaching of evolution in Poland. *Evolution: Education and Outreach, 3*, 614–620.

Berkman, M., & Plutzer, E. (2011). Defeating creationism in the courtroom. But not in the classroom. *Science, 331*(6016), 404–405.

Carlson, D. (2005). America weighs in on evolution vs creationism in schools. Gallop Poll. http://www.gallup.com/poll/16462/americans-weigh-evolution-vs-creationism-schools.aspx.

Clement, P. (2015). Creationism, science and religion: A survey of teachers' conceptions in 30 countries. *Procedia—Social and Behavioral Sciences, 167*, 279–287.

Curry, A. (2009). Creationist beliefs persist in Europe. *Science, 323,* 1159.

Dobzhansky, T. (1973). Nothing in biology makes sense except in the light of evolution. *American Biology Teacher, 35*(3), 125–129.

Eglin, P. G. (1983). *Creationism versus evolution. A study of the opinions of Georgia science teachers* (Unpublished doctoral dissertation). Georgia State University, Atlanta, GA.

Ellis, W. E. (1983). Biology teachers and border state beliefs. *Society, 20*(2), 26–30.

Futuyama, D. J. (Ed.). (2000). *Evolution, science, and society: Evolutionary biology and the national research agenda*. Piscataway, NJ: Office of University Publications, Rutgers, The State University of New Jersey.

Hill, J. (2014). Rejecting evolution: The role of religion, education, and social networks. *Journal for the Scientific Study of Religion (2014), 53*(3), 575–594.

Kutschera, U. (2008). Creationism is Germany and its possible cause. *Evolution: Education and Outreach, 1,* 84–86.

Lac, A., Hemovich, V., & Himelfarb, I. (2010). Predicting position on teaching creationism (instead of evolution) in public schools. *The Journal of Educational Research*. Taylor & Francis, Ltd. http://doi.org/10.2307/20721588.

Long, D. E. (2012). The politics of teaching evolution, science education standards, and being a creationist. *Journal of Research in Science Teaching, 49*(1), 122–139. https://doi.org/10.1002/tea.20445.

Mervis, J. (2015). Why many U.S. biology teachers are 'wishy washy.' *Science, 347*, 1054.

Miller, J., Scott, E., & Okamoto, S. (2006). Public acceptance of evolution. *Science, 313*, 765–766.

Moore, R., & Kraemer, K. (2005). The teaching of evolution and creationism in Minnesota. *The American Biology Teacher, 67*(8), 457–466.

National Academy of Sciences. (2004). *Teaching about evolution and the nature of science*. Washington, DC: National Academy Press.

National Association of Biology Teachers. (2011). *NABT position statement on teaching evolution*. http://www.nabt.org/websites/institution/?p=92.

National Academies of Science Engineering Medicine. (2012). *Thinking evolutionarily: Evolution education across the life sciences: Summary of a convocation*. Washington, DC: The National Academies Press.

National Center for Science Education. (2008). *Evolution education: Understanding and teaching the science of evolution*. August 10, 2012. http://ncse.com/evolution.

NGSS Lead States. (2013). *Next generation science standards: For states, by states*. Washington, DC: The National Academies Press.

National Research Council. (1996). *National science education standards*. Washington, D.C.: National Academy Press.

National Research Council. (1998). *Teaching about evolution and the nature of science*. Washington, DC: The National Academies Press.

National Science Teachers Association. (1997). An NSTA position statement on the teaching of evolution. *Journal of College Science Teaching, 27*(1), 7–8.

National Science Teachers Association. (2013). *NSTA position statement: The teaching of evolution*. http://www.nsta.org/about/positions/evolution.aspx.

President's Council of Advisors on Science and Technology (PCAST). (2012, February). *Report to the president, engage to excel: Producing one million additional college graduates with degrees in science, technology, engineering and mathematics*. February, 2012. http://www.whitehouse.gov/administration/eop/ostp/pcast/docsreports.

Riddle, O., Fitzpatrick, F. L., Glass, H. B., Gruenberg, B. C., Miller, D. F., & Sinnott, E. W. (1942). *The teaching of biology in secondary schools of the united states: A report of results from a questionnaire*. Union of American Biological Societies, Committee on the Teaching of Biology.

Rutledge, M. L., & Mitchell, M. A. (2002). High school biology teachers' knowledge structure, acceptance & teaching of evolution. *The American Biology Teacher, 64*(1), 21–28.

Sadler, P., et al. (2013). Assessing the life science knowledge of students and teachers represented by the K–8 national science standards. *CBE Life Science Education, 12*(3), 553–575.

Sager, C. (Ed.). (2008). *Voices for evolution* (3rd ed.). Berkeley, CA: National Center for Science Education.

Shankar, G., & Skoog, G. D. (1993). Emphasis given evolution and creationism by Texas high school biology teachers. *Science Education, 77*(2), 221–233.

Troost, C. J. (1967). Teaching of evolution in the secondary schools of Indiana. *Journal for Research in Science Teaching, 5*(1), 37–39.

U.S. Department of Education, Institute of Education Sciences, National Center for Education Statistics. National Assessment of Educational Progress (NAEP). https://nces.ed.gov/nationsreportcard/.

U.S. Department of Education, Institute of Education Sciences, National Center for Education Statistics. Program for International Student Assessment (PISA). https://nces.ed.gov/surveys/pisa/.

U.S. Department of Education, Institute of Education Sciences, National Center for Education Statistics. Trends in International Mathematics and Science Study (TIMSS). https://nces.ed.gov/timss/.

Voyer, D., & Voyer, S. D. (2014). Gender differences in scholastic achievement: A meta analysis. *Psychological Bulletin, 140*(4), 1174–1204. https://doi.org/10.1037/a0036620.

Wiles, J. R. (2010). Overwhelming scientific confidence in evolution and its centrality in science education—And the public disconnect. *The Science Education Review, 9*(1), 18–27.

Wiles, J. R., & Alters, B. (2011). Effects of an educational experience incorporating an inventory of factors potentially influencing student acceptance of biological evolution. *International Journal of Science Education, 33*(18), 2559–2585.

Yasri, P., & Mancy, R. (2014). Understanding student approaches to learning evolution in the context of their perceptions of the relationship between science and religion. *International Journal of Science Education, 36*(1), 24–45. https://doi.org/10.1080/09500693.2012.715315.

Zimmerman, M. (1987). The evolution-creation controversy: Opinions of Ohio school biology teachers. *Ohio Journal of Science, 87*(4), 115–125.

**Jaimie L. Miller-Friedmann** is currently a doctoral candidate in the Department of Education at the University of Oxford. Prior to returning to studies, she was an Educational Researcher in the Harvard College Observatory Science Education Department, where she designed curricula, participated in projects researching formal education and assessment, informal learning, identity formation, career intention and educational sociology, and professional development. Her doctoral thesis is concerned with the gender gap in specific science disciplines in the UK.

**Susan E. Sunbury, EdD** is an Educational Researcher and Project Manager in the Smithsonian Astrophysical Observatory Science Education Department with more than twenty years of experience in the development and facilitation of formal and informal research and evaluation projects. She has also taught for seven years in elementary and middle school science classrooms. Most recently, Dr. Sunbury developed survey items for Project MOSART-HSLS (Misconception Oriented Standards-based Assessment Resource for Teachers in High School Life Science).

**Philip Sadler** is currently the Director of the Science Education Department at the Harvard Smithsonian Center for Astrophysics and F.W. Wright Senior Lecturer in Astronomy. He has co-authored pre-college curricula in astronomy and engineering, college calculus, and Apple Computer's first product training course. He has taught middle school mathematics and science, undergraduate astronomy, and graduate STEM teaching courses. His research includes assessment of student understanding in life, physical, and earth sciences, the effect of professional development on science teachers' knowledge, and how advanced high school coursework impacts student persistence and performance after they transition to college.

# Examining Teaching Assistants' (TA) Experiences Facilitating Traditional Versus Active-Learning-Based Tree-Thinking Curricula: TA Perceptions, Student Outcomes, and Implications for Teaching and Learning About Evolution

**Yi Kong, Nancy Pelaez, Trevor R. Anderson and Jeffrey T. Olimpo**

## 1 Introduction and Motivation for the Study

In order to help students grasp the full gamut of evolutionary patterns and processes, both microevolution and macroevolution should be taught in the classroom (Catley & Novick, 2009; Gibson & Hoefnagels, 2015; Novick, Schreiber, & Catley, 2014). However, given the strong instructional emphasis on microevolutionary mechanisms (such as natural selection) in post-secondary classroom contexts, traditional instruction in macroevolution has not always promoted students' conceptual development of macro processes such as evolutionary relatedness, speciation, and extinction (Catley, 2006). Importantly, reasoning about evolution with tree-thinking abilities is one way to understand a chronology for the history of life on Earth (Kong, Thawani, Anderson, & Pelaez, 2017). Even though the importance of understanding and interpreting phylogenetic trees as diagrams that depict the evolutionary relationship of taxa is widely recognized, various studies have reported that college students often lack tree-thinking abilities (Halverson, Pires, & Abell, 2011; Meir, Perry, Herron,

Y. Kong
College of Education, Fujian Normal University, Renwenlou 704, Fuzhou 350108, Fujian, China
e-mail: ykong@fjnu.edu.cn

N. Pelaez · T. R. Anderson
Purdue University, West Lafayette, USA
e-mail: npelaez@purdue.edu

T. R. Anderson
e-mail: ander333@purdue.edu

J. T. Olimpo (✉)
The University of Texas at El Paso, El Paso, USA
e-mail: jtolimpo@utep.edu



U. Harms and M. J. Reiss (eds.), *Evolution Education Re-considered*,
https://doi.org/10.1007/978-3-030-14698-6_7

& Kingsolver, 2007). Such evidence verifies the need to better address students' development of tree-thinking skills within undergraduate biology classrooms.

To more deeply understand phylogenetic trees, biology educators and students need to clearly understand the various reasons why current professional biologists use evolutionary trees (Kong, Anderson, & Pelaez, 2016). Toward this end, a Model of the Use of Evolutionary Trees (MUET) was developed to represent four components found in scientific research reports that made use of evolutionary trees (Kong et al., 2017). First, the data used as a basis for constructing a tree detail some variable properties of the research subjects organized on the tree. Second, different types of evolutionary tree shapes typically have nodes, tips, and branches, and each node has branches that connect to a set of tips. These branches and tips include descendants of an ancestral group at the node. Furthermore, a tree will often indicate a root for the most ancient ancestor shared by all groups on the tree. Third, an evolutionary tree is constructed based on analytical methods that may include the principle of parsimony, a problem-solving approach that assumes that the fewest number of evolutionary transitions will give the most likely genealogical history. Finally, scientists reason about genealogical history with trees to understand clades, to track chronology, and to read and interpret patterns of characteristics to distinguish cases of homology from homoplasy. In the biological sciences, homology refers to a characteristic shared by members of a clade (feathers on birds, for example) that is similar because it was inherited from a common ancestor. In contrast, homoplasy refers to a shared trait that has not been inherited from a common ancestor. Such traits indicate convergent evolution. Wings of birds and wings of insects allow both groups to fly, but not because they are closely related through a recent ancestor. There may be a need to introduce undergraduate students to ideas like these (that professional biologists depict with trees) as a foundation for pedagogical content knowledge to improve teaching with evolutionary trees in biology classrooms. In this study, we use the MUET to inform the design of a novel curriculum, and that curriculum was then compared to traditional instruction on tree-thinking with respect to TAs' perceptions about their teaching of both curricula and their perceptions of students' tree-thinking abilities in their classrooms.

## 2 Using the MUET to Design a Curriculum

The MUET was used to develop various classroom activities, including explicitly engaging students in exercises to assist them in developing an understanding of the chronology depicted with evolutionary trees. The science of ordering events by occurrence in time is chronology (O'Hara, 1988). Tree-thinking attempts to show a chronological sequence in the appearance of features identified in the targeted data sources. With guided practice, students were asked to explain the relationships depicted in a rooted tree that are missing in an unrooted tree. They also identified sets of taxa based on character information provided, constructed evolutionary trees using data sources collected from organisms to understand different methods used to build trees, and learned to build an alignment up stepwise, starting with the most similar traits and progressively adding the most dissimilar (divergent) ones, with the assumption that similar traits could be homologous, that is, have evolved from the same common ancestor. Students also practiced "reasoning with trees," which refers to the reasons why an evolutionary tree is used to represent research findings. Reasoning based on reading and interpreting evolutionary trees often raises questions about the chronology, cladistics, and homology or homoplasy that could be answered with additional data. In contrast, the traditional curriculum had students use tree-thinking to study and learn relationships, such as plant systematics, according to published textbook material (Walsh, Hotchkin, Dash, & Watts, 2013).

## 3 Research Design

Two tree-thinking curricular approaches for an introductory biology course were compared at a mid-sized Hispanic-serving Institution in the USA. One curriculum is the traditional approach that has been offered for many years in the introductory biology course. As indicated in the preceding section, this traditional curriculum requires students to use tree-thinking to learn plant systematics. The other curriculum was the above-mentioned MUET curriculum. Since Graduate Teaching Assistants (TAs) are responsible for a relatively high proportion of the undergraduate-level teaching in many American institutions (Bomotti, 1994), including at the institution at which this research occurred, we considered it important to investigate TAs' relative perceptions of their implementation of these two tree-thinking curricula. In order to accomplish this goal, we addressed the following research questions:

(1) Within the context of an introductory biology classroom, what perceptions did TAs possess about implementation of the MUET *versus* traditional curricula as they related to facilitation of tree-thinking instruction?
(2) What benefits did TAs believe they experienced as a result of facilitating the MUET curriculum?
(3) What perceptions did TAs possess about the design elements of the MUET curriculum?
(4) What perceptions did TAs possess about the effectiveness of the MUET curriculum before and after they were made aware of student performance on warm-up and wrap-up tree-thinking quizzes in their classrooms?

For the purpose of this study, we conceptualize TAs' perceptions of effectiveness as their capacity to produce desired student learning outcomes, in other words, their perceptions of how well students enrolled in their laboratory sections improved their evolutionary tree-thinking knowledge.

### 3.1 *An Exploratory Study Examining Traditional* Versus *MUET Curricula*

The work described here was conducted at a mid-sized Hispanic-serving Institution in the USA. Participants represented a convenience sample consisting of all students enrolled in the BIOL 1108: Organismal Biology laboratory course in the Spring 2016 semester. Twenty sections of this laboratory course are offered each semester, with each section enrolling a maximum of 24 students. Eleven TAs were assigned to teach the 20 sections: nine of them taught two sections per person, and the other two TAs taught one section each. For this study, it was not feasible to use random assignment of students to treatment groups since students were already enrolled in a particular section of the laboratory course. However, the curriculum (traditional *versus* MUET) implemented within each section was randomized, with each TA facilitating one traditional section and one MUET section. Individuals assigned to teach only one section of the laboratory course were excluded from our analyses. The BIOL 1108 laboratory course was a whole-semester course (15 weeks in duration) in which students spent two hours in class with the TAs each week. TAs who taught this course participated in a mandatory one-hour weekly training meeting. This TA training was offered by the laboratory coordinator to help TAs better implement the laboratory instruction planned for each week of the course.

In Spring 2016, the BIOL 1108 laboratory course was taught with two different types of tree-thinking curricula. The traditional curriculum was supported by a manual associated with this laboratory course (Walsh et al., 2013), which requires students to use tree-thinking to explore, understand, and better appreciate organismal diversity and phylogenetic relationships across plants. The MUET curriculum was developed by the Biology Education Research Group (BERG) at the Hispanic-serving Institution for the purpose of improving students' tree-thinking abilities by systemically teaching tree-thinking as an independent topic. TAs were trained to implement these two types of tree-thinking curricula. More specifically, they received a one-hour training session on how to use the laboratory manual of the BIOL 1108 laboratory course to implement the traditional curriculum. Since the MUET curriculum was a novel curriculum that was new to all TAs, they received two hours training about this curriculum. The TAs were introduced to the models and examples used to frame the MUET curriculum in the first hour, and then the MUET curriculum designer (Y.K.) introduced how to implement this curriculum in the classroom in the second hour. In support of the research design described above, each TA received both forms of tree-thinking professional development in order to enhance the likelihood that they would implement both the traditional and MUET curricula with high fidelity.

### 3.2 *TA Participants*

Nine TAs who each taught two sections of BIOL 1108 in Spring 2016 were recruited to participate in this study. Since these TAs had fully participated in the TA training,

**Table 1** Teaching assistant (TA) demographic information ($n = 9$)

| TA | Prior evolutionary tree-thinking knowledge | Gender | Religion | Ethnicity | Years of TA experience |
|---|---|---|---|---|---|
| 1 KF | Yes | M | None | White | 6 |
| 2 GQ | Yes | M | None | Mixed | 3 |
| 3 BL | Yes | F | Muslim | Mediterranean | 2.5 |
| 4 WQ | Yes | M | Catholic | Hispanic | 1 |
| 5 QC | Yes | M | Catholic | White | 0.5 |
| 6 FT | No | F | None | Hispanic | 5 |
| 7 KC | No | M | Catholic | Mixed | 2.5 |
| 8 BD | No | F | Catholic | Hispanic | 2 |
| 9 BM | No | M | Catholic | Hispanic | 0.2 |

they were able to implement the traditional curriculum in one classroom and the MUET curriculum in their other classroom. Demographic information of the TAs is listed in Table 1. A total of 318 students in these TAs' sections were also recruited. Their students included 121 males (38.1%) and 197 females (61.9%) of whom 266 were Hispanic (83.6%), 20 White (6.3%), 9 Asian (2.9%), and 8 African American (2.5%).

## *3.3 Collection of Students' Warm-up/Wrap-up Quiz Data and Semi-structured TA Interviews*

The fact that each TA was assigned to teach two different classroom sections afforded us the opportunity to compare the traditional curriculum to the MUET curriculum. Each TA was randomly assigned one section in which to implement the MUET curriculum while the traditional curriculum was implemented in their other section. Random assignment of students to treatment groups was not possible. A pre-quiz/post-quiz design was adopted to partially eliminate a major limitation of a post-quiz-only design, since this approach allows for empirical assessment of any differences in the two groups before examining the influence of each TA's approaches to teaching on students' tree-thinking abilities.

In order to examine students' tree-thinking abilities prior to their engagement in either the MUET or traditional curriculum, a 13-item warm-up quiz was administered. Quiz items were informed by the literature (Baum, Smith, & Donovan, 2005; Naegle, 2009; Smith, Cheruvelil, & Auvenshine, 2013). Students were asked to complete a wrap-up quiz containing the same 13 items following their participation in the lesson. The TAs were recruited, and the students of participant TAs were invited with Informed Consent for Research Involving Human Subjects to grant permission for the anonymous use of their evolutionary tree-thinking scores for research purposes according to a protocol that was reviewed and approved by The University of Texas at El Paso (UTEP) Institutional Review Board (# 822142-2).

A semi-structured interview protocol was conducted to explore TAs' perceptions about the two types of tree-thinking curricula. This interview lasted between 20 and 60 minutes based on TAs' responses. Each TA obtained his/her students' warm-up and wrap-up quiz scores near the end of the interview, and he/she was invited to provide thoughts about these scores.

### *3.4 Student Learning and TA Interview Data Analysis*

Students' knowledge of evolutionary tree-thinking was assessed at the beginning and at the end of each TA's implementation of the MUET curriculum in one section and the traditional tree-thinking curriculum in the other section. Descriptive statistics were tabulated as a means to characterize student performance on the warm-up and wrap-up quizzes.

TAs were briefly interviewed about their prior teaching experience, the importance of teaching students about tree-thinking, how effective and satisfied they were with the MUET and traditional curricula, how their own tree-thinking ability was influenced by the MUET, and how or why they would change the MUET curriculum, to improve it, in the future. Interview data were transcribed and analyzed by applying open coding and axial coding strategies of inquiry (Strauss & Corbin, 1990). The TAs' knowledge of the subject matter, of pedagogy, and of the context were each examined separately and also for integration in the act of teaching. Thus, evidence was examined to determine if the TAs were able to transfer and integrate their different types of knowledge to reflect each TA's Pedagogical Content Knowledge (PCK) (Gess-Newsome, 1999). In the integrative model, PCK is viewed as the integration of the knowledge of subject matter, pedagogy, and context. In this study, subject matter refers to a TA's knowledge of evolutionary tree-thinking, pedagogical knowledge refers to a TA's ability to select and apply appropriate teaching strategies that specifically help students to develop their tree-thinking skills effectively in the classroom, and contextual knowledge refers to a TA's knowledge of the nature of undergraduate students' learning in a social context. These three domains of knowledge were applied to evaluate TAs' PCK based on their interview data.

## 4 Discussion of Findings

The MUET curriculum was compared with the traditional curriculum in terms of perceived impact on TAs' subject matter knowledge of evolutionary tree-thinking and also in terms of TAs' pedagogical knowledge and their understanding about the nature of students' learning processes. In some cases, the MUET curriculum influenced

TAs' perceived effectiveness of their teaching, as detailed below, according to their interpretation of their students' warm-up and wrap-up quiz scores in the introductory-level biology classrooms that they facilitated.

### *4.1 TAs' Perceptions About Their Implementation of the MUET Curriculum in an Undergraduate Biology Classroom Compared with the Traditional Curriculum for Teaching Evolutionary Tree-Thinking*

Overall, the TAs held a consensus opinion that the MUET curriculum was far more effective in teaching tree-thinking than the traditional approach, which, in their view, was mainly ineffective in teaching tree-thinking. To support this opinion, TAs provided two primary explanations. First, they considered it a great advantage that the MUET curriculum had been designed to improve students' tree-thinking abilities by teaching tree-thinking as an independent subject. For example, one TA said, "I like the MUET curriculum because it's looking at phylogenetics and tree-thinking as an independent subject. So not just attached to plants; … it's not … a subtopic of the plant chapter, but the MUET curriculum, I think, emphasizes more importance of tree-thinking in general. So, I would like to see a chapter just focusing on tree-building rather than a subtopic of another chapter." In contrast, they considered the traditional curriculum to be very limited in scope where a tree-thinking tool was only presented as applicable to aiding students in understanding plant systematics.

The MUET curriculum required each TA to use the same classroom activities for students to practice and apply tree-thinking abilities, so that TAs were consistent in terms of their teaching of tree-thinking. In support of this finding, one TA said, "I think that will help in knowing that all the students learned the same exact thing in the same exact way … [to] ensure that the students are actually learning it. I think that's a problem a lot of the time is the TA doing different things and not being on the same page and students learning differently." On the other hand, in the traditional curriculum, tree-thinking was not a required independent subject, and TAs' attention to tree-thinking from an instructional standpoint was therefore variable. Some TAs rarely used tree-thinking, and some TAs who were tree-thinking experts had applied extra teaching materials related to tree-thinking. In turn, this made the tree-thinking instruction in some of the traditional sections examined superior to that in other sections, which then resulted in disparate levels of student learning about evolutionary tree-thinking.

In addition to the consensus opinions reported above, some TAs compared these two types of tree-thinking curricula by monitoring student engagement with curricular activities in the two laboratory classrooms that were taught with the two different methods. They thought that the MUET curriculum was structured and designed better than the traditional curriculum, as students showed more enthusiasm and were engaged with classroom activities of the MUET curriculum in ways that were more animated than was the case for the traditional curriculum. For example, a TA said, "I think it is totally effective. … for the ones that I used the MUET in class, they were more enthusiastic. They felt more comfortable … [and] for the ones that I didn't use the MUET, they were …, they hesitated more." The classroom activities in the MUET curriculum not only engaged students' learning of tree-thinking but also enhanced TAs' perceived effectiveness of their own practice in the classroom, as some TAs mentioned that they felt better able to engage students in understanding evolutionary tree-thinking when they implemented the MUET classroom activities.

Although TAs tended to prefer the MUET curriculum to the traditional curriculum when considering student learning objectives for the introductory organismal biology laboratory course, several TAs reported that they would have liked to combine these two types of curricula. As one TA explained, "I think the information that they get in the traditional approach is important, but I definitely think we need to implement the MUET concepts to get the tree-thinking knowledge as well. The traditional … focus on systematics is talking about plant evolution. And in terms of organismal biology, plants are incredibly important. So, they need to understand the evolution of plants… if you kind of combine the two, then I think it would be most effective."

## 4.2 Benefits TAs Believe They Experienced as a Result of Implementation of the MUET Curriculum

As a result of implementing the MUET curriculum, TAs reported gains in both the cognitive domain and the social-emotional domain. In the cognitive domain, TAs who claimed not to know much about tree-thinking stated that their tree-thinking abilities were greatly improved. Furthermore, two TAs who were tree-thinking experts mentioned that the MUET curriculum slightly influenced their understanding of trees. In fact, one TA stated that the MUET curriculum helped him refresh his tree-thinking knowledge, and the other TA stated that he had gained knowledge about the multiple applications of tree-thinking in the real world. Additionally, many TAs mentioned that the strategy deployed in the MUET curriculum to simplify trees and break down the complex information into simple pieces was a good strategy because it made the knowledge easy to teach in the classroom.

In the social-emotional domain, some TAs realized that their attitude toward teaching and their perceived teaching behaviors had changed. One TA stated that she was encouraged by students' positive performance in the MUET curriculum, which made her enjoy teaching and gave her confidence in her future teaching. In support of this finding, a second TA said, "I really don't feel confident about my teaching. But I think I cannot say I was the best but I was good, especially the second week [of the MUET curriculum]." Yet another TA stated that she had noticed the differences in learning progress displayed by various students and, in the future, would like to care more about students' reactions in her classroom.

### 4.3 Some TAs' Perceptions About the Design of the MUET Curriculum Reflect Their Lack of PCK

TAs mentioned their own prior knowledge and experience when they provided their perceptions about the design of classroom activities and the content knowledge students were able to gain from the MUET curriculum. With regard to the design of classroom activities associated with the MUET curriculum, some TAs noted that they would like to add more exercises to help students practice their tree-thinking abilities, such as the ability to understand chronology and know how to use the principle of parsimony to construct trees.

For the content covered in the MUET curriculum, TAs believed that all of it was meaningful, and some TAs mentioned specific content that they believed to be of particular importance, providing reasons for why this was the case. Specifically, homology and homoplasy were mentioned by five TAs, as this knowledge was also taught in the introductory organismal biology lecture course. For example, a TA said, "I really like doing these [activities] with the students. I guess it's after the cladistics homology section, where we have practised looking at the different evolutionary relationships, and … also the most fun, but it was challenging, was the homoplasy like the convergent evolution section and then the homoplasy explanation. Sometimes it's not very obvious, like if you just read the definition. But having the activity with the comparison slide helped." Cladistics was mentioned by two TAs, as they thought this knowledge was foundational for students to understand tree-thinking. Introduction of tree components was mentioned by two TAs as they thought this knowledge was very basic for students to understand how information is represented with evolutionary trees. Tree shapes were mentioned by one TA, as it can help students if they come to understand different types of trees that they may encounter in textbooks and/or research papers in the future. One TA thought the introduction of the importance of trees at the beginning of class can motivate students in learning tree-thinking in the classroom. Another TA thought the activity of constructing trees would help students form a comprehensive understanding of trees.

As described above, most of the content in the MUET curriculum was identified as meaningful from the TAs' perspectives. However, some TAs also pointed out specific content that, from their perspective, they found to be less meaningful in the MUET curriculum. For instance, one TA thought it was not necessary to introduce the differences between bifurcated and multifurcated trees, as this content was not important and might distract students' attention from understanding other tree-thinking knowledge. Their rationale for this claim was that knowledge of bifurcated and multifurcated trees did not show up in textbooks or quizzes. However, some TAs' perceptions about meaningless content in the MUET curriculum reflected their lack of PCK, as evidenced by four particular cases from this study. These four cases are described below. Interpretation of problems with PCK in each case is also provided.

*Case One*. A TA thought it was meaningless to introduce molecular data as a potential source to construct trees, as this knowledge should be taught in a genetics course in the future. This reflects a lack of contextual knowledge, as some students might not learn genetics in the future, and many students had learned about DNA sequences in high school. This idea also reflects a lack of pedagogical knowledge, as even though students are going to learn more about molecular data in the future, it is necessary for students to become familiar with all the types of data sources that can be used to construct trees, so that students form a more generalized understanding in their mind.

*Case Two*. A TA thought it was redundant to teach homology twice in the MUET curriculum, as students can easily understand the knowledge of homology with one introduction. However, the knowledge of homology was introduced the second time in the MUET curriculum with the purpose of helping students distinguish homology from homoplasy. This TA's idea reflected a lack of pedagogical knowledge, as she did not realize that teaching homology twice was a strategy to help students understand and distinguish two concepts, which are sometimes confused by students who do not yet know much biology.

*Case Three*. A TA reported that it should not be necessary to ask students to identify if a family tree is an evolutionary tree or not because a family tree is a topic that is not related to tree-thinking. Family trees share similar features with evolutionary trees, so asking students to distinguish between family trees and evolutionary trees is a good strategy to help students better understand these two different types of figures (Kong, Anderson, & Pelaez, 2016). This TA's statement reflects a lack of integrating subject matter knowledge with other components of pedagogical content knowledge.

*Case Four*. A TA thought is might be better to avoid teaching chronology, as this knowledge is difficult for students to understand. This TA cared about students' learning by stating that chronology is challenging for students to learn. Since BIOL 1108 students are required to achieve the course learning objective of understanding the evolutionary history of organisms on Earth, and knowledge of chronology is the fundamental way of achieving this objective, the above statement reflects a lack of contextual knowledge.

## *4.4 Examining the Extent to Which the MUET Curriculum and Students' Quiz Performance Across Sections Influenced TAs' Perceived Effectiveness of the MUET Curriculum and of Their Teaching*

Each TA was presented (near the end of the interview) with his/her students' evolutionary tree-thinking scores (Table 2) on the warm-up and wrap-up quizzes after facilitating both the MUET curriculum and traditional curriculum and was asked to share his/her perceptions about those outcomes as part of the interview process. Overall, TAs stated that they were not surprised that students performed well following participation in the MUET curriculum, whereas students who had participated in the traditional curriculum had only improved slightly in their tree-thinking abilities. Interestingly, three of the nine TAs expected higher evolutionary tree-thinking

**Table 2** Students' warm-up and wrap-up quiz scores in MUET and traditional curriculum sections stratified by TA

| TA name | MUET curriculum | | | Traditional curriculum | | |
|---|---|---|---|---|---|---|
| | *n* | Warm-up (*M*; *SD*) | Wrap-up (*M*; *SD*) | *n* | Warm-up (*M*; *SD*) | Wrap-up (*M*; *SD*) |
| 1 KF | 13 | 5.08 (1.71) | 7.77 (2.05) | 19 | 6.32 (2.67) | 6.74 (2.16) |
| 2 GQ | 15 | 5.93 (2.22) | 7.13 (2.03) | 21 | 5.19 (1.57) | 6.71 (1.35) |
| 3 BL | 18 | 6.61 (1.72) | 7.44 (2.04) | 13 | 7.15 (2.15) | 6.38 (2.60) |
| 4 WQ | 22 | 5.95 (2.15) | 7.64 (1.92) | 23 | 6.57 (1.93) | 7.26 (2.12) |
| 5 QC | 19 | 6.37 (1.71) | 6.89 (1.79) | 12 | 5.75 (1.36) | 6.42 (1.51) |
| 6 FT | 18 | 6.39 (1.82) | 8.22 (2.10) | 18 | 6.17 (1.10) | 6.11 (1.64) |
| 7 KC | 18 | 6.56 (1.54) | 8.33 (2.20) | 22 | 6.36 (2.08) | 6.45 (2.36) |
| 8 BD | 15 | 6.67 (1.99) | 7.13 (3.04) | 16 | 6.06 (1.91) | 6.38 (1.89) |
| 9 BM | 20 | 6.20 (2.09) | 7.00 (1.84) | 16 | 6.44 (2.28) | 6.38 (1.89) |

scores for the students who engaged in the MUET curriculum, as they thought that the MUET curriculum was very effective in teaching tree-thinking.

Two TAs noted that they believed that their students had improved in their tree-thinking abilities as a result of participation in the traditional curriculum. According to the demographic information of these two TAs (2GQ and 4WQ), both had prior teaching experience as well as an educational background in biology and evolutionary tree-thinking. These two TAs explained that they had provided extra teaching materials about tree-thinking in the traditional curriculum because tree-thinking is their focal area of interest and expertise.

As shown in Table 2, although most TAs expected that students would exhibit an increase in performance as a result of participation in the MUET curriculum, three TAs indicated that such increases were less than they had expected. These TAs provided some reasons about this unexpected result, which included students' missing attendance and lesser engagement in class as well as students' high performance on the evolutionary tree-thinking warm-up quiz. It is important to note for these TAs, however, that, according to their demographic information, one TA (9BM) lacked both teaching experience and biology background, one TA (5QC) was a tree-thinking expert but lacked teaching experience, and one TA (8BD) had rich teaching experience but lacked an educational background in biology and evolutionary tree-thinking. This demographic information suggests that TAs' lack of teaching experience, and their lack of training in the biological sciences (particularly in the area of evolutionary tree-thinking), might be mediating factors that influence students' understanding of tree-thinking in the classroom.

To further address this concern, TAs who had insufficient teaching experience and/or biology background were invited to share their perceptions about how these attributes potentially influenced their teaching of evolutionary tree-thinking within the classroom. With the exception of one TA, who stated that he was not used to teaching, the other TAs did not think that their lack of rich teaching experience was a challenge for them. To overcome the barrier of poor biology background, TAs sought professional help from the tree-thinking curriculum developer and reviewed pertinent content prior to implementing the MUET curriculum in their class.

## 5 Conclusion and Implications

In this chapter, we demonstrate that the types of tree-thinking curricula implemented in a classroom can clearly impact TAs' perceived effectiveness of their own teaching practice, especially for those lacking in pedagogical experience and/or knowledge of evolutionary tree-thinking. In interviews, TAs compared the MUET curriculum and the traditional curriculum, evaluated the content of the MUET curriculum and their students' learning from their teaching of this innovative curriculum within the classroom, and shared their thoughts about similarities and differences in students' tree-thinking abilities in the two sections that they facilitated with and without the MUET. In summary, responses to the first three research questions (discussed below) show that the TAs thought that the MUET curriculum was effective in teaching tree-thinking, and, in some cases, led the TAs to expect higher student performance than what they observed for individuals who had participated in the MUET curriculum intervention. Conversely, TAs were not required to teach tree-thinking as an independent subject as part of the traditional curriculum, which gave the TAs more flexibility in how they approached teaching this content.

Some insights from the semi-structured interviews help to explain the variable student outcomes observed with each approach (and especially with the traditional approach) based on TAs' perceptions about their implementation of the MUET curriculum compared with the traditional curriculum for teaching evolutionary tree-thinking in an undergraduate biology classroom (RQ1/RQ4). Their implementation of the two types of tree-thinking curricula provides valuable implications for the teaching of tree-thinking and the future training of graduate TAs. First, most of the TAs did not explicitly teach tree-thinking in the traditional curriculum, which resulted in various levels of improvement or decline in student performance on the wrap-up quiz relative to the warm-up. Only a few highly-trained individuals who were also experienced TAs found the MUET curriculum approach to not be necessary. Also, some TAs reported that their students' lack of tree-thinking abilities in the traditional curriculum might negatively impact their learning of plant systematics in the classroom. This highlights a need for more explicit instruction on evolutionary tree-thinking within the context described. However, some TAs would like to merge both of these two tree-thinking curricula in the future to help students form tree-thinking abilities and then apply those abilities to learn plant systematics. As an implication from this finding, the study suggests that the implementation of a designed curriculum (i.e., the MUET curriculum) can help TAs gain confidence and satisfaction from their teaching experience, which may help them form positive attitudes toward teaching about evolution in the future.

TAs reported experiencing several benefits as a result of implementing the MUET curriculum (RQ2). This included an increase in TAs' satisfaction with their teaching experience when implementing the MUET curriculum in their classroom. Specifically, they were more satisfied with their teaching, as students' positive performance in the classroom encouraged TAs to be more confident about teaching tree-thinking concepts in the future. An important finding is that some TAs gained knowledge about how to engage students as well as confidence in their ability to care about students' learning. Most of the TAs thought that the design and the content of the MUET curriculum were meaningful. However, some TAs' perceptions about the MUET curriculum reflected their lack of knowledge in all the three domains of PCK (RQ3) (i.e., contextual knowledge, pedagogical knowledge, and subject matter knowledge). By linking TAs' tree-thinking learning experience and teaching experience to their students' performance in both the traditional and the MUET curricula, it became apparent that TAs' lack of teaching experience and lack of biology background might negatively impact students' tree-thinking abilities in the classroom. However, TAs offered neither of these reasons as a potential explanation for observed student outcomes within their sections. On the one hand, they put effort into overcoming the barrier of poor tree-thinking knowledge and were satisfied with their teaching in the classroom. On the other hand, they thought students' absences and poor engagement in class might be a reason that negatively impacted students' tree-thinking abilities within the sections that they facilitated. Given these dichotomous viewpoints, we propose that future research continues to examine how TAs, including those who are pedagogically experienced but who may lack specific evolutionary biology content knowledge, engage with novel curricula in the field.

In detail, this exploratory study revealed that the MUET curriculum influenced TAs' perceptions of the effectiveness of their own praxis, especially for TAs with variable amounts of prior teaching experience and evolutionary tree knowledge (RQ3). The study also provides suggestions for how instructors can better promote TAs' effective teaching of tree-thinking in the future. Instructors might take into account TAs' prior teaching experiences and tree-thinking knowledge when decisions are made about who should be assigned to teach tree-thinking in the classroom. The MUET curriculum helped increase TAs' confidence, but the intervention was not enough for inexperienced teachers who were lacking in knowledge. The TAs who only lacked tree-thinking learning experience but had rich teaching experience were able to implement the MUET curriculum effectively in their classrooms. However, TAs who lacked teaching experience may need extra training to overcome this deficit before they implement either the MUET or the traditional curriculum in the classroom.

In addition to the need to help TAs gain teaching experience in their TA training program, this study also revealed that the current TA training was not sufficient in helping all TAs form and integrate the three domains of knowledge of PCK. Thus, more effort could be focused on augmenting the TA training. The MUET curriculum was an example that successfully improved TAs' subject matter knowledge of tree-thinking, but extra training is needed for TAs who lack pedagogical knowledge and their understanding about the nature of students' learning process in the context of an introductory-level organismal biology laboratory course. Since this was an exploratory study, further research is required to examine whether the findings from the small sample in the context of this study are more broadly applicable. However, the work presented here suggests that a designed curriculum and TA training could potentially be fruitful areas for future work to improve tree-thinking, in particular, and perhaps even evolution education, in general.

**Acknowledgements** We wish to thank the TAs and undergraduate biology students who agreed to participate in this study. We are also grateful to Michael Reiss, Ute Harms, and the authors of chapters for *Evolution Education Re-considered: Understanding what works* who met at Kiel, Germany, in September 2017 to provide feedback on this work as well as to Mr. Daniel Quijas for contributing to the transcription of the interview data associated with this study. Support for this work was provided by the BUILDing SCHOLARS core at The University of Texas at El Paso (NIGMS/NIH Award Numbers RL5GM118969, TL4GM118971, and UL1GM118970).

## References


Baum, D., Smith, S. D., & Donovan, S. (2005). The tree-thinking challenge. *Science, 310,* 979.

Bomotti, S. S. (1994). Teaching assistant attitudes toward college teaching. *The Review of Higher Education, 17*(4), 371.

Catley, K. M. (2006). Darwin's missing link—A novel paradigm for evolution education. *Science Education, 90*(5), 767–783. https://doi.org/10.1002/sce.20152.

Catley, K. M., & Novick, L. R. (2009). Digging deep: Exploring college students' knowledge of macroevolutionary time. *Journal of Research in Science Teaching, 46*(3), 311–332. https://doi.org/10.1002/tea.20273.

Gess-Newsome, J. (1999). Pedagogical content knowledge: An introduction and orientation. *Examining pedagogical content knowledge* (pp. 3–17). Netherlands: Springer.

Gibson, J. P., & Hoefnagels, M. H. (2015). Correlations between tree-thinking and acceptance of evolution in introductory biology students. *Evolution: Education and Outreach, 8,* 1–17.

Halverson, K. L., Pires, C. J., & Abell, S. K. (2011). Exploring the complexity of tree-thinking expertise in an undergraduate systematics course. *Science Education, 95,* 794–823.

Kong, Y., Anderson, T. R., & Pelaez, N. (2016). How to identify and interpret evolutionary tree diagrams. *Journal of Biological Education, 50*(4), 395–406.

Kong, Y., Thawani, A., Anderson, T. R., & Pelaez, N. (2017). A model of the use of evolutionary trees (MUET) to inform K-14 biology education. *The American Biology Teacher, 79*(2), 79–88.

Meir, E., Perry, J., Herron, J., & Kingsolver, J. (2007). College students' misconceptions about evolutionary trees. *The American Biology Teacher, 69,* 71–76.

Naegle, E. (2009). *Patterns of thinking about phylogenetic trees: A study of student learning and the potential of tree-thinking to improve comprehension of biological concepts* (Doctoral dissertation). Retrieved from ProQuest Dissertations and Theses database. (UMI3357304).

Novick, L. R., Schreiber, E. G., & Catley, K. M. (2014). Deconstructing evolution education: The relationship between micro-and macroevolution. *Journal of Research in Science Teaching, 51*(6), 759–788.

O'Hara, R. J. (1988). Homage to Clio, or, toward an historical philosophy for evolutionary biology. *Systematic Biology, 37,* 142–155.

Smith, J. J., Cheruvelil, K. S., & Auvenshine, S. (2013). Assessment of student learning associated with tree-thinking in an undergraduate introductory organismal biology course. *CBE-Life Sciences Education, 12,* 542–552.

Strauss, A., & Corbin, J. (1990). *Basics of qualitative research: Grounded theory procedures and techniques*. Newbury Park, CA: Sage Publications.

Walsh, E., Hotchkin, P., Dash, S., & Watts, S. (2013). *Organismal biology 1108 laboratory manual* (2nd ed.). Dubuque, IA: Kendall Hunt Publishing Company.

**Yi Kong, Ph.D.** is an Associate Professor at Fujian Normal University in China and previous postdoctoral fellow in the Department of Biological Sciences at The University of Texas at El Paso. She received her Ph.D. in Biology Education from the Department of Curriculum & Instruction at Purdue University. Her research related to evolutionary education includes building a conceptual framework and developing new strategies to help undergraduate students form expert-level tree-thinking abilities.

**Nancy Pelaez, Ph.D.** Associate Professor of Biological Sciences at Purdue University, is a physiologist and biology educator who serves as Convener of the Biology Education Area for Scholarship and Teaching (BEAST) in Purdue's Department of Biological Sciences. As a former school science teacher with a K-12 California single subject teaching credential in both Life Science and Physical Science, Pelaez, who is a Fellow of the Royal Society of Biology (FSRB), received a Fulbright Award for a semester in Vienna, Austria, in 2015, and the American Physiological Society 2016 Guyton Educator of the Year Award. Pelaez is an Associate Editor of the journal *CBE-Life Science Education* and PI on the NSF-funded Advancing Competencies in Experimentation—Biology (ACE Bio) Network.

**Trevor R. Anderson, Ph.D.** is a biochemist and biochemistry education researcher who serves on the Editorial Board of *Biochemistry and Molecular Biology Education* (*BAMBEd*). His Visualization in Biochemistry Education (VIBE) research group focuses on visual literacy and representational competence, reasoning with core concepts and representations, science inquiry and reasoning about experiments including problem-solving in biochemistry laboratories, student assessment design and validation, and curriculum and faculty development. Central to his faculty development activities has been the *BAMBEd* Bridging-the-Gap series aimed at encouraging scientists to apply educational research to teaching practice.

**Jeffrey T. Olimpo, Ph.D.,** is an Assistant Professor of Biological Sciences at The University of Texas at El Paso. His research is in the area of bioeducation and focuses primarily on cognitive and non-cognitive outcomes resultant from novices' participation in course-based undergraduate research experiences as well as the impact of professional development opportunities on the academic and career growth of undergraduate teaching assistants (UTAs). Olimpo currently teaches a large-enrollment, introductory cell and molecular biology course. He furthermore serves as PI on the NSF-funded Multi-Institutional *Tigriopus* CURE initiative and is a Tips & Tools Section Editor for the *Journal of Microbiology and Biology Education.*

# Utility of Context-Based Learning to Influence Teacher Understanding of Evolution and Genetics Concepts Related to Food Security Issues in East Africa

**Timothy A. Goodale**

## 1 Context

A recent visit to Kenya's 'Dandora Dump' by Pope Francis underscores the region's longstanding troubles with poverty, disease and long-term stability. Direct and indirect impacts of poverty have caused many to suffer from a wide range of diseases including lung cancer, skin diseases and lead poisoning that have caused stunted growth and triggered mental disabilities in children. Issues surrounding food security and hunger cause many East African citizens to scour wastelands like Dandora and compromise their health and well-being in order to survive.

The agricultural crop *Cassava* is a crucial food source and has the potential to help alleviate many of the issues in East Africa surrounding food access and security that drive impoverished people to risk their long-term health. Cassava has the potential to increase farm incomes, reduce rural and urban poverty and help alleviate many food security issues. Cassava can be produced with family labour and with minimal tools, making it an attractive and low-risk crop for poor farmers. In addition, cassava is available to low-income rural households in the form of simple food products and can be cultivated in many extreme environments. Cassava is an important subsistence food crop in Kenya and Tanzania, especially in the semi-arid areas, and is an important crop within the region's famine reserve. Cassava diseases such as brown streak and mosaic virus are transmitted through the vectors (white fly), and subsequent infected plant materials cause crop yield losses of up to 70%. In addition, poor crop management and use of popular seed varieties add to these losses, hence exacerbating an already vulnerable food supply situation.

This backdrop establishes an intriguing context within which to teach and learn about genetics and evolution. Context-Based Learning (CBL) is the use of real-life and or fictitious examples in teaching environments in order to learn through the

T. A. Goodale (✉)
Elizabeth City State University, Elizabeth City, NC, USA
e-mail: Tagoodale@ecsu.edu



U. Harms and M. J. Reiss (eds.), *Evolution Education Re-considered*,
https://doi.org/10.1007/978-3-030-14698-6_8

actual, practical experience with a subject rather than just its theoretical parts. Food security is an ongoing problem to which most people can relate. A vast majority probably have not personally experienced food security issues, but people can generally relate to the potential detriment of a food shortage and its implications to health. Therefore, within this project the issue of food security in East Africa serves as the central problem that teachers and students investigate and for which they propose possible solutions. These outcomes are accomplished in parallel with actual scientists who represent diverse demographic backgrounds and are conducting similar work that is represented within the curricular intervention. The variables associated with this particular issue offer a unique manner in which to teach genetics and evolution that differs from conventional approaches.

## 2 Introduction

Among the major national and international science organizations (AAAS, 2001; NSTA, 1997, etc.), a broad consensus exists that supports the teaching of the theory of evolution and related concepts in genetics as a unifying theme in biological science. However, within the long history of science education in the USA, this topic has continued to be a focus of conflict for teachers, students and local communities. Beginning with the Scopes Monkey Trial (1925) and *Epperson v. Arkansas* (1968) through recent cases such as *Kitzmiller v. Dover Area School District* (2005), religious and community perspectives have often been at odds with the scientific community regarding the teaching and learning of evolution in public schools (Lee, 2006; Linder, 2007). Unfortunately, science teachers are at the forefront of this conflict. In addition, many science educators, especially those in biological fields, experience personal conflict between their religious beliefs and scientific perspectives in regard to evolutionary theory (Brickhouse, Dagher, Letts, & Shipman, 2000; Aguillard, 1999; Eve & Dunn, 1990; Shankar & Skoog, 1993; Rutledge & Warden, 2000). Research findings reveal that many teachers in the USA hesitate to teach evolution in a thorough fashion, often choosing to include creationism or omit evolution from the curriculum (Moore, 2002; Rutledge & Warden, 2000; Weld & McNew, 1999). These social and personal conflicts can create consequences for impressionable students who may seek to pursue an academic degree or career in the sciences. Research has revealed that teachers' attitudes and views about subject matter can impact their curricular and instructional decisions (Carlesen, 1991). Likewise, a teacher's conception and knowledge structure of evolution impacts student understanding and achievement within this important and unifying concept (Diekoff, 1983). This situation exacerbates the shortage of STEM academic and career pursuits by secondary students as teacher effectiveness in science is one the bigger predictors of future college success in STEM (Adelman, 1999). The quandary presented within teacher apprehension and content limitations in evolution and its subsequent impact on student understanding and achievement provide the key focus of this study. In summation, this study seeks to determine if an academic intervention and subsequent teacher training focused on evolution and genetics and grounded in 'context-based learning' can influence teacher understanding and/or acceptance of evolution and subsequently improve student knowledge and achievement.

## 3 Review of Relevant Literature

### *3.1 Context-Based Learning (CBL)*

In theory, utilization of context-based learning is supposed to address challenges in science education like a lack of clear purpose within instruction, breadth of content, an absence of relevance to students and minimal higher-order cognitive understanding. Overarching goals of CBL seek to foster positive attitudes towards science while, at the same time, providing a sound basis of scientific understanding for further study to improve student motivation, problem-solving and achievement within specific scientific disciplines. In study, implementation of CBL has had significant positive influence on student achievement, attitude and motivation in science (Magwilang, 2016) and has led to deeper understanding of content in chemistry and biology compared to traditional instruction (Ulusoy & Onen, 2014). A related study found that CBL implementation led to improvement in attitudes towards science and a deeper understanding of scientific ideas while specifically establishing more positive attitudes towards science in both girls and boys and reducing gender differences in attitude (Bennett, Lubben, & Hogarth, 2007). Studies that examined teachers and CBL utilization found that teachers believe a context-based approach makes science more relevant, more motivating to teach, and that their students were more interested in science, in terms of both their immediate responses in lessons and their increased likelihood of deciding to pursue science in tertiary study (Bennett & Lubben, 2006). These findings provide a solid foundation from which to deduce that a properly vetted and implemented instructional intervention focused on genetics and evolution and grounded in CBL could positively influence both teachers and students in regard to understanding and interest in the subject matter.

### *3.2 Teaching and Learning Interventions in Evolution and Genetics*

Many factors have the potential to influence the effective teaching and learning of evolution and genetics. These can range from demographics, attitudes and beliefs of either the teacher or student to pedagogical approaches in the learning environment and cognitive dispositions that learners bring to the classroom (Smith, 2010). Instructional interventions typically aim to address one or more of these variables and in the past have shown mixed results. A foundational approach to evolution education centres on conceptual change theory grounded in prior work of Piaget (1964) and Posner, Strike, Hewson, and Gertzog (1982). In conceptual change classroom approaches, teachers explicitly address student prior conceptions and provide opportunities for students to employ those conceptions in contexts that lead to learning opportunities from which teachers can help students generate understandings that are more generalized (Smith, 2010). In a holistic sense, approaches related to 'Conceptual Change Theory' require that students or teachers alter their worldview or go against previously held beliefs and therefore it is difficult to determine significant influences or outcomes related to these variables.

Intervention studies that have shown impact include Banet and Ayuso (2003) that utilized an approach grounded within 'learning based on understanding through action'. Upon completion of an instructional intervention, student views that were consistent with a modern evolutionary synthesis increased by 44% from pre- to post-assessment. Likewise, similar studies that focused on acceptance of evolution included the use of situated learning through interactive software (Crawford, Zembal-Saul, Munford, & Friedrichsen, 2005) and college coursework that focused on the acceptance of evolution as science (Ingram & Nelson, 2006) both demonstrated strong gains in participant acceptance of evolution. Comparatively, studies that have focused on gains in content knowledge with respect to evolution and genetics demonstrated similar growth. Nehm and Schonfeld (2007) investigated the impact of a 14-week course that utilized collaborative learning, concept mapping and field excursions on pre-service teacher understanding of evolution and determined that over 75% of participants demonstrated growth but overall levels on understanding were still quite low. In addition, Kampourakis and Zogza (2009) reported on the effects of a course that employed lecture/discussion with a 'constructivist perspective' on 14–15 year-old students and found strong content (from 33 to 62% passing) gains as measured from testing prior and after the intervention, though again the resulting knowledge was low. Overall, Smith (2010) identified 11 intervention studies in which gains in evolution understanding or acceptance have occurred since the year 2000. Each of these studies has their individual strengths and weaknesses, but they do provide a foundation from which to expand and eventually improve.

These works form the foundation of the aim and related research questions of this particular study. Utilization of context-based learning in evolution and genetics serves as a novel instructional intervention from which to assess impact on teacher content knowledge, acceptance and efficacy. Based on prior research, it would be reasonable to surmise that current science educators would have low levels of content knowledge and varying levels of acceptance with respect to evolution. Training and professional development in the utilization of context-based learning with a unique focus in international food security could have a positive impact on content knowledge, acceptance and/or efficacy in teaching the subject matter. The following three research questions drove the methods and outcomes of this study:

I. What are current levels of content knowledge and acceptance of principles associated with evolution and genetics in beginning science educators?

II. Does training and exposure to a context-based learning curricular intervention focused on international food security issues positively influence content knowledge and or acceptance evolution and genetics?
III. Does training and exposure to a context-based learning curricular intervention focused on international food security issues positively influence beginning teacher efficacy with respect to teaching evolution in the classroom?

## 4 Methodology

The participants of this study were 20 beginning science teachers representative of two urban areas in the south-east region of the USA. All of the educators were recent graduates at the bachelor's degree level from two universities, and all came from a traditional teacher preparation programme that consisted of 40 credit hours of core classes, 50 credit hours of science content and 30 credit hours of education/teacher preparation content. All participants were between the ages of 20–25 years old, approximately 15 were female, and 5 were male. Each participant held full certification to teach science and specifically biology in each of his or her respective states.

Each participant underwent the same recruitment, assessment and training procedures. Involvement in the training and related measures of evaluation was voluntary. An email advertisement was sent to all the prospective science teacher alumni at each of the universities. This call sought voluntary participation in a free professional development training session for science teachers geared towards teaching genetics and evolution. Time commitment was one instructional day or approximately six hours. Teachers confirmed their participation though email correspondence and responses provided the time and location of the training. In total, eleven teachers participated in a workshop at the first university and nine teachers participated in an identical workshop at the second university.

In each training session, participants were pretested on several measures, which included content knowledge and understanding of evolution, knowledge and attitude towards genetics and acceptance of evolutionary principles. The pretest was inclusive of previously utilized instruments from Rutledge and Warden (1999) 'Acceptance of the Theory of Evolution', Perez et al. (2013) 'Inventory of Students' Understanding of Evolutionary Developmental Biology', Smith, Wood, and Knight (2008) 'Genetics Concept Assessment', and Haga et al. (2013) 'Attitudes Toward Genetics'. Session participants were allotted thirty minutes to complete the combined questionnaire. Separately, most of these instruments underwent various validity and reliability vetting. The evolution content knowledge tool by Perez et al. (2013) was validated by experts and administered to 1191 students in field trials whose responses were used to evaluate the readability, difficulty, discriminability, validity and reliability. Initial analysis found that items ranged in difficulty from 0.22 to 0.55 and in discriminability from 0.19 to 0.38. Findings suggest the tool is effective for assessing student understanding of concepts in evolution among undergraduate biology majors. The 'Inventory of Students' Understanding of Evolutionary Developmental Biology'

(Smith et al., 2008) was vetted for face and content validity through expert review and student pilot testing, and reliability was determined via the test–retest method to calculate an r value of above 0.7 which is an acceptable coefficient of stability. Likewise, the Genetics Concept Assessment instrument was reviewed by genetics experts, validated by student interviews, and taken by 600 students; results showed that differences in the item difficulty and item discrimination index values can be used to distinguish between concepts that are retained poorly or effectively. The Attitudes Toward Genetics instrument was validated by 25 genetics professionals and educators and evaluated using 400 students; reliability estimates were 0.995 for the pretest and 0.997 for the post-test iterations. In summation, these instruments were utilized to provide a solid indicator of participant knowledge, attitude and acceptance of genetics and evolutionary principles. By utilizing this baseline information, the impact of the intervention can be judged and trends and relationships among variables explored.

Upon completion of the pretesting, participants took part in an active teacher training that spanned four hours. In these workshops, participants learned how to mimic the genetic mapping of the cassava plant by utilizing classroom exercises that demonstrate how scientists isolate DNA, determine sequencing of codons and determine the eventual mapping of genes and their expression of traits. The other three units focused on identifying mutation types associated with vector transmission and subsequent evolutionary changes that lead to cassava diseases, utilizing gel electrophoresis to identify infected plants and utilizing argumentation to propose long-term solutions involving gene therapy, GMOs and large-scale vector control. Beginning teachers were guided through each of the four activities, as they would be implemented in a classroom setting. In addition, each teacher took part in each of the activities including DNA extraction, worksheets associated with coding, sequencing and trait identification, mutation types and vector transmissions. Participants also utilized a gel electrophoresis laboratory and identified an infected/sick plant sample compared to a healthy sample. Lastly, using guiding principles of scientific argumentation, participants had to propose a single-page solution to the food security issue surrounding cassava mosaic disease and discuss the relevant pros and cons.

At the conclusion of the teacher training and introduction to the instructional intervention, participants were post-tested via a focus group and structured interview. This structure was utilized due to time constraints and guided through prior research that suggests pre- and post-tests should be three to six weeks apart (Brown, Irving and Keegan, 2008). Given the immediacy of the workshop, researchers believed it was important to capture any changes that might occur during the specific time frame and capture participants' perspectives as to what guides their attitudes and beliefs. Grouping consisted of five members and lasted approximately thirty minutes. The focus group questioning served two purposes: (1) report and record any changes in acceptance and or content responses from the pretest; (2) probe for foundations of belief systems of current and future practices in evolution education. Utilizing these methodologies and measures provides unique insight to address the related research questions and generate information about current levels of knowledge and respective attitudes towards genetics and evolution.

## 5 Findings

In investigating participant levels of content knowledge in evolution related to Perez et al.'s (2013) instrument Inventory of Students' Understanding of Evolutionary Developmental Biology, the overall group of twenty teachers demonstrated a weak grasp of fundamental principals. Performance on this 10-item multiple-choice assessment ranged from 10% (1/10) to a high score of 90% (9/10), which shows high variability in content knowledge with respect to evolution. Figure 1 depicts performance on each of the specific items and overall average score on the assessment in total. Overall, the group average was 37%.

Conversely, participant content knowledge in genetics as measured in part by the Smith et al. (2008) Genetics Concept Assessment was much higher as scores on this multiple-choice assessment ranged from 68 to 100%. Overall performance for the group was at 87%. Likewise, participant attitudes towards genetics were generally positive as measured by Haga et al. (2013) Attitudes Toward Genetics. Table 1 depicts the 13-item assessment and average rate of agreement with each statement. A score of 3 is neutral, anything above is agreement, and anything below is disagreement.

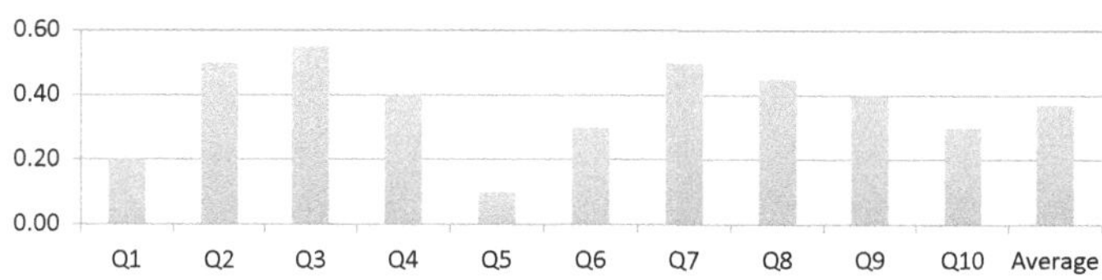


**Fig. 1** Participant content knowledge in evolution

**Table 1** Attitude toward genetics

| Item | Mean |
|---|---|
| 1. I think the development of DNA research is hopeful for the treatment of diseases | 4.60 |
| 2. I think that the development of DNA research is a positive medical progress | 4.53 |
| 3. I approve of using DNA testing for early detection of diseases | 3.55 |
| 4. I would inform my children about the results of a DNA test for a specific disease | 3.80 |
| 5. I want to know whether my disease is hereditary | 4.35 |
| 6. I would inform my siblings about the results of a DNA test for a specific disease | 4.30 |
| 7. I worry about the penalties of DNA testing for being able to affect health insurance | 2.45 |
| 8. The possibility of a DNA test will change one's future | 2.90 |
| 9. As long as a disease cannot be treated, I don't want a DNA test | 4.35 |
| 10. If I had a DNA test done, my family does not need to know about the result | 4.30 |
| 11. I don't want a DNA test to tell me that I am at risk for a certain disease | 2.50 |
| 12. I worry about the consequences of DNA testing for the chances of finding a job | 3.00 |
| 13. The idea of a DNA test frightens me | 2.50 |

**Table 2** Acceptance of evolution in science teachers

| Item | Mean |
|---|---|
| The evidence used to support evolution is weak or inconclusive | 1.7 |
| The theory of evolution is the product of good science | 3.7 |
| Evolutionary biology is not really science | 1.95 |
| Evolutionary theory is well supported by scientific data, research and study | 4.05 |
| The current theory of evolution is the best scientific explanation for the origin of species | 3.85 |
| Evolutionary theory explains why chimpanzees and humans share characteristics | 4 |
| Evolution can be used to develop sound explanations about living things of today | 3.9 |
| Humans do not evolve | 1.8 |
| Evolution is happening right now | 4.3 |
| Species exist today in the same form they always have | 1.6 |
| Any species could be evolving right now | 4.25 |
| Humans have evolved from previously existing species | 4.05 |
| New species arise from previously existing species | 3.95 |
| Everyone should understand evolution | 4.2 |
| It is important to let people know how strong the evidence is that supports evolution | 3.9 |
| People who plan to become biologists need to understand evolution | 4.45 |
| I would be willing to argue in favour of evolution in a public or church setting | 3.45 |
| Understanding evolution helps me understand other parts of biology | 3.85 |
| Evolution is a good explanation of how humans first emerged on the earth | 3.65 |

In investigating teacher acceptance and understanding of evolution with the Rutledge and Warden (1999) and Smith, Snyder and Devereaux (2016) Acceptance of the Theory Evolution measurements, there were no major trends that arose with regard to non-acceptance of major principles within the theory of evolution. Table 2 depicts the measurement item and average rate of agreement with each statement. A score of 3 is neutral, anything above is agreement, and anything below is disagreement.

Lastly, the focus group post-test evaluation consisted of four open-ended questions. Responses to each of these prompts were recorded, compiled and evaluated to determine themes or group consensus. Each question is depicted below, major categories of responses are shown, and frequencies are tabulated.

1. After participating in the context-based learning curriculum would you change any of your answers from the initial questionnaires? Explain
   a. Yes responses $N = 2$
      i. Review responses on evolution content knowledge instrument, perceived poor performance
   b. No responses $N = 18$
      i. Confident in initial choices

   ii. Position on issues hard to change in short time frame

2. What do you feel is the strongest indicator or foundational element of either your belief/acceptance or disbelief/non-acceptance of the concept of evolution? (Participant should state if they believe or not and why) examples conflict with religious beliefs, scientific understanding/knowledge, family perspectives/influence?

   a. Strongly Accept $N = 2$
      i. Scientific knowledge shapes beliefs
   b. Strongly Non-accept $N = 3$
      i. Evolution conflicts with religious view and family values
   c. Conflicted $N = 15$
      i. Believe many elements of the theory of evolution
      ii. Disagree with human origins
      iii. Have strong religious views that do not coincide

3. In your experience, what is the strongest or most prevalent indicator of someone that holds an opposing view of yours on evolution? (if there are divergent perspectives, be sure to ask/prompt both)

   a. Strongly Accept $N = 2$
      i. Religious beliefs
   b. Strongly Non-accept $N = 3$
      i. Perceived understanding of science
      ii. Atheism
   c. Conflicted $N = 15$
      i. People do possess a 'grey area', or partisan takes on the subject
      ii. Those with concrete agendas and that are not open for dialogue

4. (A) Do you feel that a teacher that has a strong disbelief in the concept of evolution should teach that subject matter? (B) Do you think students of those teachers are underserved/cheated? (C) Can a teacher overcome their personal bias and teach the concepts effectively as a believer? (Prompts asked one at a time)

   A. Yes responses $N = 0$
      No responses $N = 20$
      i. Bias would be palpable to students (10)
      ii. Knowledge would be compromised (6)
      iii. Effort and effectiveness would differ vastly (4)
   B. Yes responses $N = 20$
      i. Weakness in content parlays to ineffective teaching and thus learning (14)
      ii. Bias will influence kids (6)
      No responses $N = 0$
   C. Yes responses $N = 4$
      i. Lots of training on effective pedagogy required
      No responses $N = 16$

i. Religious and personal beliefs are too long held to be overcome with interventions
ii. Schools better off hiring specialists to cover content like they do with sex education, too controversial for many

Lastly, the Pearson correlation analysis was conducted between participant performance and responses on content knowledge and acceptance of evolution. The correlation was essentially neutral at (−0.025), albeit slightly negative. Thus, a relationship between knowledge and acceptance of evolution could not be established.

## 6 Discussion

Findings from this work provide for three interesting points of discussion. Major themes that arose include the relatively low level of content knowledge in evolution of beginning science educators, general acceptance of the theory of evolution but with some level of apprehension, and a widely held belief that teachers should be knowledgeable and have a strong level of acceptance of the theory of evolution. Overall, these results exhibit various limitations; however, they provide some unique findings to explore in future research involving evolution education especially within teacher training and preparation and school policy. Improving student learning outcomes within the teaching and learning of evolution is of great importance. The unifying elements of the theory are critical for fundamental understanding of most of the essential concepts in biological sciences. Participants in this study all have higher education in biology and evolution, yet still struggle with both rote recall of concepts associated with evolution and critical understanding. These participants are the future teachers that K-12 students will learn under. When students are underprepared in secondary education, it could lead to attrition in post-secondary science programmes and thus limit the future pipeline of STEM-based professionals, which is a major economic concern in many countries.

One finding that stood out was the low level of content knowledge with respect to principles of evolutionary biology. The group pretest average of 37% represents broad limitations in content knowledge with respect to evolution. Since all participants were certified science educators in their respective states and completed requisite degree coursework in biology and related sciences and passed eligibility examinations, the rate of achievement is surprising. The small number of participants ($N = 20$) and assessment items ($N = 10$) is a contributing factor to this initial finding as outliers have a more profound impact with a smaller sample size and reliability of findings is compromised when fewer items are utilized. Nonetheless, initial findings do trigger some concern in regard to the ability of educators to teach concepts of evolution. It is a long-held stance that content knowledge is a dispositional pillar in regard to teaching effectiveness. From foundational studies such as Shulman (1987) to Hill, Rowan and Ball (2005) and a plethora of others, findings have shown repeatedly that when a teacher possesses high/higher levels of content knowledge. It positively influences student achievement. Foundational knowledge of the subject matter also leads to higher levels of teaching efficacy (Lee & Tsai, 2010) and pedagogical content knowledge (Park & Oliver, 2008), each of which can again positively affect student achievement in various scientific disciplines. The importance of a strong background in content knowledge is undeniable. Influences on content knowledge, especially within biological sciences, are broad and numerous. In specific reference to evolution, the primary factor is most likely academic preparation. Biology is already a broad field and in traditional teacher preparation (in the USA), approximately 30 credit hours of content-related (biology) coursework are replaced with courses in education. Specific courses in evolution are often upper-level electives and are avoidable by students. So more often than not, the only exposure to evolution for many traditionally trained educators in the USA is a few weeks of focus within an introductory life science course, often during their first year of study. Lack of initial training coupled with the deficiency of access to quality professional development in the teaching and learning of evolution leads to a less than ideal situation for both teachers and students alike. In working with teachers, future endeavours in evolution education probably need to focus on attaining a baseline benchmark of content knowledge and spending time on remediating deficiencies and misconceptions so that classroom interventions have an increased likelihood to succeed. Content knowledge is vital to the efficacy and effectiveness of an educator and subsequently the achievement, understanding and future success of their students.

A second finding of note is the wide acceptance of both genetics and evolution in regard to controversial sociocultural beliefs. In general, it would be easy to presume that science educators, specifically those trained in biology or life science, would have a strong belief or acceptance of the theory of evolution. However, research exists that contradicts this finding. As previously cited, personal conflict between a teacher's religious beliefs and scientific perspectives in regard to evolutionary theory are common and can influence instructional approaches and subsequently student learning (Brickhouse et al. 2000; Aguillard, 1999; Eve & Dunn, 1990; Shankar & Skoog, 1993; Rutledge & Warden, 2000). Findings from this work support these earlier studies in that many teachers believe in several tenets of evolutionary theory but also experience a personal conflict with concepts related to the origin of humans. Again, the small sample size in this study can hide some of the data presented as averages or mean scores. In addition, many of the average scores were either a high three or low four, demonstrating agreement but not strongly held agreement. A noteworthy outcome is that there were three science educators who strongly opposed the theory of evolution. This is a small number, but in reality if 15% (3/20) of educators teaching biology have strong opposition to the theory, over time and considering a broader population, the impact on the field could be significant. Overall, a large number of teacher participants (18/20) were either conflicted or had strong opposition to the theory of evolution. Much like content knowledge, a working understanding of the general concepts of evolution is necessary to effectively teach it. Having conflicting conceptions can negatively impact teaching or cause teachers to ignore or superficially cover important material. Future educational research or interventions should again focus on identifying pre-held misconceptions and work to correct these errors with respect to content knowledge. Overcoming personal beliefs is a much more daunting task. For those who are conflicted as to their acceptance, interventions that require more facilitation and allow students personal exploration of the content may be a strong approach.

Lastly, the utilization of a focus group provided a venue for novel findings and the ability to probe beyond a simple agreement scale. Based on the findings, contradicting views came about based on qualitative inputs and quantitative results from content and attitudinal questionnaires. The fact that most of the participants did not feel they needed to change their answers on the content pretest on attitudinal scales did not come as a surprise. The intervention was short or limited in time frame, and time to reflect is often necessary to change perspectives. Moreover, the responses as to the foundation of participant thoughts on evolution were also of little surprise. The main points of interest arose from further probing as to reasons of conflict among those who identified as conflicted. Many cited the fact that many popular medical doctors and scientists to not adhere wholeheartedly to the theory of evolution. Examples of Dr. Ben Carson and Dr. Phillip Skell were cited as credible scientists who are sceptical of evolution and provide a level of credence to personal beliefs. A majority of students cited that they believe a common ground between religion and science, and specifically evolution, can exist and should be further explored. Most do not deny that evolution and natural selection occur; it is the cause or starting point that is debatable in their minds. Obviously, there is a large field of research that explores the intersection of religion and science and there are many points of agreement. However, this research is more in the social science realm of attitudinal commonalities than that of traditional science approaches. In the end, it is a tricky situation to approach. The value of validating religious perspectives could be a growth in more people accepting basic tenants of evolutionary theory. The thought is if you discredit a personal stance from the start, people could tune themselves out of the content related to the lesson or activity. This outcome simply exacerbates the issue of non-acceptance. On the other hand, much like the intelligent design movement, it is dangerous to allow a concept equal footing in a classroom. At the same time, scientists need to remember that very little of the field is viewed as law or fact and much of the content is labelled as 'until proven otherwise'. Utilization of scientific argumentation techniques within evolution educational interventions could be useful in validating multiple perspectives while requiring evidence to substantiate stances. In conclusion, it may be better to allow teachers to work in some level of religious inclusion that can be vetted through scientific argumentation rather that have a flat-out denial or the exclusion of the topic.

A final element of the focus group that was a noteworthy outcome was the consensus that a teacher who holds strong beliefs against the theory of evolution would be ineffective in teaching the content. The ironic point here was that three of the teachers viewed themselves as strong non-acceptors and would fall into this scenario. In addition, most believed that a teacher in this category would have a hard time overcoming personal bias and that most training initiatives would not be impactful. However, the idea of an 'Evolution Specialist' in schools was an interesting concept. In the USA, many school districts tackle delivery of content associated with sex education through the use of contracted specialists or specially trained personnel. This is typically a two-week unit of instruction on varying concepts of anatomy, physiology and health related to human sexuality. Comparatively, evolution is taught in a similar time frame and is equally controversially and dreaded by many teachers. The question that needs to be explored and discussed related to this possibility is: Does opting out of teaching and 'really knowing' evolution hinder other realms of biological science? In essence, does being weak in evolutionary biology make you weak in many of the other fields? One last point to consider is that some of the endeavours to utilize a specialist in sex education have not shown significant academic benefit (Wight et al., 2002). This study demonstrated weakness in the implementation of the intervention, but nonetheless in relation to evolution education significant differences in intervention implementation would probably be rare in most cases.

In closing, the intervention proposed in this study has the potential to ignite research agendas to address content weakness and teacher perspectives in regard to evolution. The context-based learning intervention that focuses on food security issues in Africa is in its beginning stages and needs further refinement and a broader implementation that includes student outcomes. These are planned over the next three years. Findings from this work will improve implementation of teacher training. One crucial element will be to better refine an assessment of evolution content knowledge and to have participants take that assessment prior to arriving to workshops. Results can be shared, and weaknesses and misconceptions addressed within the workshop. A post-test could help determine levels of improvement in content knowledge attributable to the teacher training. In addition, the use of CBL and scenario-based learning helps teachers become more of a facilitator of knowledge and students direct more of their learning experiences. In addition, with the use of the curriculum, teachers can demonstrate mutation, genetic change and species evolution within one unit while teaching all the requisite standards. Much of the controversy surrounding human origins and religious conflict can be avoided through active problem solving compared to conversing on theories and personal viewpoints. The intervention has promise to help struggling teachers while adding to the body of knowledge with respect to instructional interventions within evolution education. This sociocultural-scientific issue is longstanding and also has a long way to go with respect to any type of meaningful resolution.

## 7 Future Research Considerations

Several interesting foci for future research could be explored from findings from this and similar work in evolution education. First, a deeper exploration of the spectrum of content knowledge of practising teachers with respect to the primary facets of evolutionary theory needs to be benchmarked. From here, impact of varying content knowledge levels on teaching pedagogy, efficacy, effectiveness and impact on student learning/interest could be investigated. This would provide insight on training and preparation along with identifying variables that most impact the effective teaching of evolution. Second, the concept of an 'evolution content specialist' would be interesting to explore and would provide for rigorous framework for social science study of its impact. Referring back to content, knowledge and efficacy, it is possible that a uniquely trained individual in content, communication and teaching could have a discerning impact on student learning outcomes. A third possibility that could be explored is the longitudinal impact of teachers on students' academic pursuits and/or success in the science fields. Often, introductory science courses are 'gatekeeper' courses, in that large numbers of students drop out of these STEM disciplines and pursue others due to academic struggles. It would be interesting to conduct follow-up investigations with students to see if secondary preparation/current understanding of principles had any impact on those who drop out and those who persist/succeed. Lastly, it would be interesting to investigate the impact of sharing pretest results (presuming poor performance) with participants and gauge the impact of academic standing on efficacy to teach content related to evolution and its influence of perspectives of necessary teacher qualities.

## References

AAAS. (2011). *Vision and change in undergraduate biology education: A call to action*. Washington, D.C: AAAS.

Adelman, C. (1999). *Answers in the tool box: Academic intensity, attendance patterns, and bachelor's degree attainment*. U.S. Department of Education. Washington, DC: Office of Educational Research and Improvement.

Aguillard, D. (1999). *Evolution education in Louisiana public schools: A decade following: Edwards v Aguillard*. The American Biology Teacher (pp. 182–188).

Banet, E., & Ayuso, G. E. (2003). Teaching of biological inheritance and evolution of living beings in secondary school. *International Journal of Science Education, 25*(3), 373–407.

Bennett, J., & Lubben, F. (2006). Context-based chemistry: The Salters approach. *International Journal of Science Education, 28*(9), 999–1015.

Bennett, J., Lubben, F., & Hogarth, S. (2007). Bringing science to life: A synthesis of the research evidence on the effects of context-based and STS approaches to science teaching. *Science Education, 91*(3), 347–370.

Brickhouse, N. W., Dagher, Z. R., Letts, I. V., William, J., & Shipman, H. L. (2000). Diversity of students' views about evidence, theory, and the interface between science and religion in an astronomy course. *Journal of Research in Science Teaching, 37*(4), 340–362.

Brown, G. T. L., Irving, S. E., & Keegan, P. J. (2008). *An introduction to educational assessment, measurement, and evaluation: Improving the quality of teacher-based assessment* (2nd ed). Auckland, NZ: Pearson Education NZ.

Carlesen, S. W. (1991). Effects of new biology teachers' subject-matter knowledge on curricular planning. *Science Education, 75*, 6631–6647.

Crawford, B. A., Zembal-Saul, C., Munford, D., & Friedrichsen, P. (2005). Confronting prospective teachers' ideas of evolution and scientific inquiry using technology and inquiry-based tasks. *Journal of Research in Science Teaching, 42*(6), 613–637.

Diekhoff, G. M. (1983). Testing through relationship judgements. *Journal of Educational Psychology, 75*, 227–233.

Epperson v. Arkansas. (1968). 393 US, 97. *S. Ct., November, 12*, 1968.

Eve, R. A., & Dunn, D. (1990). Psychic powers, astrology & creationism in the classroom? Evidence of pseudoscientific beliefs among high school biology and life science teachers. *The American Biology Teacher, 52*(1), 10–21.

Haga, S. B., Barry, W. T., Mills, R., Ginsburg, G. S., Svetkey, L., Sullivan, J., et al. (2013). Public knowledge of and attitudes toward genetics and genetic testing. *Genetic testing and molecular biomarkers, 17*(4), 327–335.

Hill, H. C., Rowan, B., & Ball, D. L. (2005). Effects of teachers' mathematical knowledge for teaching on student achievement. *American Educational Research Journal, 42*(2), 371–406.

Ingram, E. L., & Nelson, C. E. (2006). Relationship between achievement and students' acceptance of evolution or creation in an upper-level evolution course. *Journal of research in science teaching, 43*(1), 7–24.

Kampourakis, K., & Zogza, V. (2009). Preliminary evolutionary explanations: A basic framework for conceptual change and explanatory coherence in evolution. *Science & Education, 18*(10), 1313–1340.

Lee, B. (2006). Kitzmiller v Dover area school district: Teaching intelligent design in public schools. *Harv CR-CLL Rev, 41,* 581.

Lee, M. H., & Tsai, C. C. (2010). Exploring teachers' perceived self efficacy and technological pedagogical content knowledge with respect to educational use of the World Wide Web. *Instructional Science, 38*(1), 1–21.

Linder, D. (2007). *State v. John Scopes* (The Monkey Trial).

Magwilang, E. B. (2016). Teaching chemistry in context: Its effects on students' motivation, attitudes and achievement in chemistry. *International Journal of Learning, Teaching and Educational Research, 15*(4).

Moore, R. (2002). Teaching evolution: Do state standards matter? *BioScience, 52*(4), 378–381.

National Science Teachers Association. (1997). *Position statement on the teaching of evolution.* Available on NSTA.org.

Nehm, R. H., & Schonfeld, I. S. (2007). Does increasing biology teacher knowledge of evolution and the nature of science lead to greater preference for the teaching of evolution in schools? *Journal of Science Teacher Education, 18*(5), 699–723.

Park, S., & Oliver, J. S. (2008). Revisiting the conceptualisation of pedagogical content knowledge (PCK): PCK as a conceptual tool to understand teachers as professionals. *Research in Science Education, 38*(3), 261–284.

Perez, K. E., Hiatt, A., Davis, G. K., Trujillo, C., French, D. P., Terry, M., et al. (2013). The EvoDevoCI: A concept inventory for gauging students' understanding of evolutionary developmental biology. *CBE-Life Sciences Education, 12*(4), 665–675.

Piaget, J. (1964). Part I: Cognitive development in children: Piaget development and learning. *Journal of Research in Science Teaching, 2*(3), 176–186.

Posner, G. J., Strike, K. A., Hewson, P. W., & Gertzog, W. A. (1982). Accommodation of a scientific conception: Toward a theory of conceptual change. *Science Education, 66*(2), 211–227.

Rutledge, M. L., & Warden, M. A. (1999). The development and validation of the measure of acceptance of the theory of evolution instrument. *School Science and Mathematics, 99*(1), 13–18.

Rutledge, M. L., & Warden, M. A. (2000). Evolutionary theory, the nature of science and high school biology teachers: Critical relationships. *The American Biology Teacher, 62*(1), 23–31.

Shankar, G., & Skoog, G. D. (1993). Emphasis given evolution and creationism by Texas high school biology teachers. *Science Education, 77*(2), 221–233.

Shulman, L. (1987). Knowledge and teaching: Foundations of the new reform. *Harvard educational review, 57*(1), 1–23.

Smith, M. U. (2010). Current status of research in teaching and learning evolution: II. Pedagogical issues. *Science & Education, 19*(6–8), 539–571.

Smith, M. U., Snyder, S. W., & Devereaux, R. S. (2016). The GAENE—generalized acceptance of evolution evaluation: development of a new measure of evolution acceptance. *Journal of Research in Science Teaching, 53*(9), 1289–1315.

Smith, M. K., Wood, W. B., & Knight, J. K. (2008). The genetics concept assessment: A new concept inventory for gauging student understanding of genetics. *CBE-life sciences Education, 7*(4), 422–430.

Ulusoy, F. M., & Onen, A. S. (2014). A research on the generative learning model supported by context-based learning. *Eurasia Journal of Mathematics, Science and Technology Education, 10*(6), 537–546.

Weld, J. & McNews, J. C. (1999). *Attitudes toward religion*. The Science Teacher, December 1999.

Wight, D., Raab, G. M., Henderson, M., Abraham, C., Buston, K., Hart, G., et al. (2002). Limits of teacher delivered sex education: interim behavioural outcomes from randomised trial. *British Medical Journal, 324*(7351), 1430.

**Tim Goodale** is Associate Professor of Education at Elizabeth City State University. He earned his BS, MS and Ph.D. from Old Dominion University in Norfolk, VA. His areas of expertise/research interests include K-12 STEM professional development for teachers, integrating emerging sciences in the K-12 curriculum, programme evaluation and educational assessment. Findings from contributed work to this volume stem from his role as a lead researcher in an NSF PIRE (Partnerships in International Research and Education) Grant (Award # 1545553), involving ongoing investigations surrounding the cassava plant in East Africa.

# Bridging the Gap Towards Flying: *Archaeopteryx* as a Unique Evolutionary Tool to Inquiry-Based Learning

**Alexandra Buck, Sofoklis Sotiriou and Franz X. Bogner**

## 1 Introduction

As evolution plays a crucial role in understanding life on earth and the deeper connections of organisms, it is considered one of the ten most important central issues in science (Harlen, 2010). School-level teaching of evolution thus is viewed as laying important foundations to formally introduce scientific understanding (Mead, Hejmadi, & Hurst, 2017). Yet many students tend to experience problems in grasping basic principles of evolution, although from a scientific point of view they are regarded as obvious (Mayr, 1997; To, Tenenbaum, & Hogh, 2017). This apparent gap seems to foil young students' understanding of scientific phenomena which often may build upon many intuitive (or alternative) conceptions (Evans & Lane, 2011). The principle itself is a well-discussed issue in many research fields, highlighting the processes as to how concepts originate or change over the course of an individual person's lifetime by integrating different fields such as cognitive psychology and cognitive development, leading to psychology and, of course, science education (e.g. Sinatra, Kienhues, & Hofer, 2014; Goodale, 2017). Consequently, by converging all involved fields, interdisciplinarity was an essential impetus combining and integrating all sub-fields into commonly agreed 'conceptual change' research (e.g. Driver & Easley, 1978; Posner, Strike, Hewson, & Gertzog, 1982; Vosniadou & Brewer, 1992).

Various studies report low levels of understanding of evolution even among university freshmen (Jakobi, 2010; Cunningham & Wescott, 2009) or pre-service teachers (Goodale, 2017). Learning initiatives about evolutionary topics may require interac-

A. Buck · F. X. Bogner (✉)
Department of Educational Biology, Centre of Math and Science Education, University Bayreuth, Z-MNU, University Campus, NW-1, 95447 Bayreuth, Germany
e-mail: franz.bogner@uni-bayreuth.de

S. Sotiriou
Ellinogermaniki Agogi, R&D, Pallini, Athens, Greece
e-mail: sotiriou@ea.gr



U. Harms and M. J. Reiss (eds.), *Evolution Education Re-considered*,
https://doi.org/10.1007/978-3-030-14698-6_9

tive, experimental, self-motivational and creative thinking contexts where problems based on everyday examples demand individual solutions (Blancke, Boudry, Braeckman, deSmedt, & deCruz, 2011). Although the theory of evolution doubtless provides the essential basic conception in biology, in secondary school syllabuses, curricula are often restricted to human evolution (Gropengießer, Harms, & Kattmann, 2013); what is more, teaching Darwin's theory is usually not introduced before grade 8. Although the natural history of birds (and vertebrates) is commonly taught in grade 5 and grade 6, species' evolution in general is not explained as an isolated genealogical tree but as constant co-evolution in competition, cooperation and correlation of species (Hammann & Aschoff, 2013). It is often recommended to assign to evolution theory a central and constant role in biology lessons, by for instance following Dobzhansky (1973) who promoted this view as a general vision: 'Nothing in biology makes sense except in the light of evolution'.

Introducing evolution within the concept of the origin of species (e.g. birds descending from extinct reptiles by using the authentic *Archaeopteryx* fossil as a bridging links) certainly may support understanding theoretical concepts of phylogenetic relationships (Kattmann, 2017). 'Why birds fly' may help students to understand the evolutionary advantage to occupy new biotopes (Nachtigall, 1985). To encourage young children with no pre-knowledge of evolution to explore evolutionary topics of dinosaurs and ancient birds, sixth graders may grant good candidates to discover the mystical world of *Archaeopteryx* as well as to enjoy learning with authentic hands-on experience on fossils. Studying birds may lead to enhanced interest in learning science (Peter, 1994; Weber & Kattmann, 1991). The question of the last conjoint ancestor of a monophyletic group like birds is optimally answered by the existence of *Archaeopteryx*, which as a two-legged reptile with feathers is closely related to dinosaurs (Wellnhofer, 2008). Proving the evidence of evolutionary linkage between birds and dinosaurs, *Archaeopteryx* represents the 'missing link' of Darwin's theory (Catley, 2006). Moving from Jurassic times to modern times, students understand the morphology and physical body features of flight adaptions on bird skeletons and discover bird species in today's environments via wildlife documentation media.

Focusing on a classroom as an effective and authentic learning environment, a diversity of children with various backgrounds needs attendance (Cavallo & McCall, 2008). Science classes with different abilities and interests may work together to discover, experiment, engage or even enjoy being creative, integrating both highly skilled children and those with special educational needs.

As inspiring science learning plays a major role in tomorrow's classroom (Sotiriou & Bogner, 2011), motivating and encouraging students plays an essential role in boosting long-term interest and simultaneously establishing a solid knowledge base (e.g. Randler & Bogner, 2006; Schumm & Bogner, 2016). Structured inquiry-based learning may be a suitable tool to bolster knowledge acquisition and to prompt awareness of learning outcomes (Schmid & Bogner, 2015), by specifically supporting self-determination and autonomous learning (Anderson, 2002). Introducing the combination of arts and science is regarded as an innovative way to enhance motivation and interest in learning any science. Creativity and the motivation potential of students may imply inquiry-based science education (IBSE) learning which builds upon several skills, for instance, question drafting, problem-solving, performing school experiments, awareness, science protocol writing and evaluating. Science education should be an essential component of a learning continuum for all, from pre-school to actively engaged citizenship (Sotiriou, Bybee, & Bogner, 2016). The European initiative *CREATIONS*, with the specific intent to 'develop and engage science classrooms', was intended to provide synergies between science, arts, creativity and innovation in schools at all levels. The *Archaeopteryx* and bird flight programme contributed to the European *responsible research and innovation* (RRI) initiative and offered an example of shifting from STEM to STEAM subjects (the 'A' stands for arts). As innovative ideas and creative solutions often emerge at the margins of disciplines (Mumford, 2002), linking 'science and arts' may build upon creative scientific concepts, leading to an engagement in multiple arts and science assignments, bringing a variety of variables into play, from connecting both cerebral hemispheres to different gender preferences towards science and arts. The use of digital media such as simulations as well as authentic artefacts for design may support individual inquiry (Sotiriou et al., 2016).

In view of different gender preferences in science learning, combining science with arts may encourage girls in particular to better learn and engage in science lessons, and to connect with technical and engineering tools. Especially for young children, gender aspects still seem to play a role in learning behaviour. Therefore, it is crucial for girls to be adequately integrated into science lessons. Linking arts and handicraft with natural science tools may enhance creative learning and activate a broader interest in learning science. Connecting arts and science may benefit girls and boys equally to choose their favourite learning approach and train different skills. Research shows that curiosity about the world around us, learning and understanding science and thinking like a scientist are important for students of all ages and provide a solid foundation for future success (Hampden-Thompson & Bennett, 2013; DeWitt, Archer, & Osborne, 2013). This combination is assumed to contribute to personal well-being, to support creativity and innovation as well to enable students of all ages, backgrounds and talents to be more autonomous and become active citizens. A key for reaching that goal may lie in the support of acquisition of key competences rather than simply learning numbers and facts (Sinatra et al., 2014). Being able to collaborate, listen to the ideas of others, think critically, take initiative, be creative, solve problems, assess risk and constructively manage emotions are considered essential for success in adult life and the basis for lifelong learning (OECD, 2012).

Learning at hands-on workstations, which are structured in small steps, is assumed to support effective and interactive learning. Authentic materials such as fossils, feathers and stuffed bird species in combination with multimedia tools are considered supportive tools to provide effective and motivational aspects of learning. Situational emotions and interest are regarded as leading to successful learning in classrooms (Schönfelder & Bogner, 2017). Basic knowledge about the life history and identification of animal species is known to be an important trigger for learning about biodiversity (Randler & Bogner, 2002). Original objects are also seen as a primary motivation of learning (Sturm & Bogner, 2008; Kossack & Bogner, 2012; Goldschmidt & Bogner, 2015), tapping into the favourite interests of children (Morgan, 1992). Hands-on study of birds and fossils will make theoretical knowledge more transferable and understandable. Students will come into the role of young science researchers with questions and answers to find out more about the life of *Archaeopteryx*: Was it capable of flight? Which body features made flight possible? What are fossils made of? What is thermal uplift and how does a stork use it to travel to Africa? Students may connect individual explanations and ideas with scientific knowledge and improve their answers by comparing them with information on cards. For a successful learning process, teachers offer tutoring only when needed. Instructional information guides students through the experimental procedure. After completing the workstations, students discuss their results and ideas with other group members and evaluate the learned topics in a protocol booklet. Reflection at school as well as feedback questionnaires completes a subsequent consolidation (Fremerey & Bogner, 2015).

Our study applied an educational module to sixth graders with no pre-knowledge of evolution who were given the opportunity to explore evolution in an unconstrained, playful and creative learning-by-doing way based on IBSE principles. We regard the topic of dinosaurs and ancient birds as ideal for sixth graders as they are expected to be fascinated to discover the world of *Archaeopteryx* during the Jurassic, and to enjoy learning via imaginary journeys and authentic hands-on experience of fossils. We focused on the application of inquiry-based learning methods incorporated into a single instructional classroom intervention to detect short-term effects on students' evolutionary and scientific knowledge. We had three research questions: (1) Do students show cognitive achievement after performing inquiry-based learning methods once during a classroom intervention? (2) Are the learning methods at workstations appropriate to knowledge gain? (3) Do gender differences play a role?

# 2 Methods

## 2.1 *Participants and Instructional Unit*

A total of 139 sixth graders (42.9% females; age M $\pm$ SD $=$ 11.19 $\pm$ 0.39 years) participated. Students were included when teachers were willing to participate and parents had given permission. Students were not informed in advance about any testing schedule, for instance regarding post-test application (c.f., Bogner, 1998). Missing data of just a few participants were substituted with mean scores of classmates.

Our aim was to examine the learning potential of an evolutionary unit with novice students. We sought to promote interest in evolutionary aspects of the *Archaeopteryx* fossil and the natural phenomenon of bird flight. We used a replica of the original specimen to make the *Archaeopteryx* directly accessible. Our unit consisted of six hands-on stations (165 min in total) split into 2 $\times$ 2 school hours. The programme was integrated into regular school schedules and may reflect a realistic everyday

**Fig. 1** Schedule of each test point for knowledge questionnaires T0 (1–2 week before), T1 (immediately after intervention), T2 (6–8 week later)

teaching unit. Structured into three interactive learning modules, the unit covered contents of biology, palaeontology and physics. Mainly evolutionary aspects of the *Archaeopteryx* fossil were discussed, including the evolutionary origin of birds and the phenomenon of bird flight.

Small groups of 3–4 participants (assembled by free choice) completed in teams the tasks at the stations, guided by a workbook with instructions for each task; each team completed a written protocol to record the answers. After a short introduction about the architecture of our learning programme, all working stations were completed autonomously by rotating each group within a 15–20-min schedule. Educational material and additional information were provided on an information board. A variety of learning methods was chosen following the IBSE learning model. Participants could acquire the content of a station in written information boards, short documentary movies, information material or hands-on experiments. All workstations were designed as student-centred and self-learning units (Fig. 1). Except for station 1 with the *Archaeopteryx* replica fossil, there were always two workstations per instructional unit to ensure efficient workflows. An additional station was provided for fast-working teams. To ensure self-directed and student-centred learning, teachers stayed in the background, ready to offer help on request with organizational, technical questions. One instructor and one class teacher were present as supervisors. A work booklet with solutions was accessible if requested.

Workstations were structured into three sections: (1) arts in science with the fossil, (2) multimedia and (3) hands-on experiments (see Fig. 1). Multimedia tools employed wildlife documentary videos and virtual flight simulation; hands-on stations included feathers, dinosaur bones and stuffed birds (seagull, hawk and blackbird) as demonstration objects (Thomas 2013; Bossert 1998; Peter 1994). For arts in science, collaborative handicraft artwork with natural fossils and paper flight modelling was applied. Students learned about the evolutionary link between birds and dinosaurs (reptiles) by discovering distinct skeleton features through direct hands-on experience with an *Archaeopteryx* fossil replica. For a detailed description of activities and the educational units, see Table 3. While experimenting with phenomena such as fossilization or gliding with thermal uplift, students slipped into the role of a 'science researcher' following the creative and multisensory approach of scientific thinking.

## 2.2 Test Design

We applied pre- and post-test questionnaires monitoring cognitive achievement before (T0) and immediately after participation (T1). The items covered relevant knowledge dimensions about the programme contents (Anderson & Kratwohl, 2001), each containing four potential multiple choice answers, of which one was correct (see Table 1). Distractors were chosen on the basis of clarity, relevance and plausibility, as well as, of course, the common misunderstandings regarding evolution as obtained from a pilot test run which was applied twice with two sixth-grade pilot classes. The same procedure applied for adjustment purposes of educational materials. To minimize test effects and avoid potential memorizing of answers, the order of items was randomly varied for each test time. Participants were never aware of testing cycles (Bogner, 1998). According to the literature (e.g. Geier & Bogner, 2011; Dieser & Bogner, 2015) no significant difference in knowledge achievement scores was expected in mere test–retest applications (i.e. without any intervention),

**Table 1** Module's integration of art in science, multimedia and hands-on

| Arts in science with fossils | Multimedia | Hands-on |
|---|---|---|
| 1. How fossils get formed? Collaborative handicraft fossil artwork made of natural material and cement | 2. Feather structure, learning quiz advantage of flying?<br>3. Flight simulation: How does uplift works?<br>4. Bird observation with video about different flight types | 5. Discovering bird and dinosaur skeleton features with *Archaeopteryx* fossil<br>6. Migrant bird experiment: How does a stork use thermal uplift for gliding?<br>7. Paper flight rally with origami arts |

The numbers 1–7 represent the working stations' themes

**Table 2** Examples of knowledge test items; (X) indicates the correct answer

| Item | Sample questions |
|---|---|
| N_KN_2 | Which body features do *Archaeopteryx* and birds have in common?<br>(1) Teeth, (2) wings (X), (3) 3-digit-claws, (4) Tail spinal cord |
| N_KN_12 | In which geological area and environment did *Archaeopteryx* live?<br>(1) Stone age (Neanderthal), (2) Jurassic (dinosaurs) (X), (3) Modern era (humans), (4) Pre-Cambrian (bacteria, volcanoes) |
| N_KN_13 | Why can birds fly? Because of … (mark the wrong answer with a cross)<br>(1) Hollow bones, (2) Concave wing shape, (3) Streamline body form, (4) Egg laying (X) |
| N_KN_14 | Who was the first inventor of a paragliding aircraft?<br>(1) Daniel Bernoulli, (2) Otto von Lilienthal (X), (3) Otto von Bismarck, (4) Albert Einstein |
| N_KN_15 | Thermal uplift occurs when the air on land …?<br>(1) Cools and sinks, (2) Heats up and rises (X), (3) Heats up with thundery clouds, (4) Heats up and streams north-west |

**Table 3** Description of the individual working stations

| Duration | Workstation | Content | Student activity |
|---|---|---|---|
| 10 | Introduction of the learning programme | Introduction by instructor: general information about the workstations, group work, distribution of workbooks, general instructions | Paying attention, working with the introductory pages and instructions as guides for the workstations with information boards |
| 15 | Demonstration How fossils are formed? | Collaborative handicraft fossil artwork made of natural material and cement | Working with original objects, understanding the process of fossilization linked with artwork |
| 20 | Workstation 1 *Archaeopteryx*—missing link | Discovering bird and dinosaur skeleton features with the *Archaeopteryx* fossil as real object | Experiencing the *Archaeopteryx* original replica, drawing distinct skeleton features |
| 20 | Workstation 2 *Archaeopteryx* | Learning about the mystical life of the ancient bird in Jurassic times | Answering questions in workbook from information boards; card quiz game with fossils |
| 20 | Workstation 3 How birds can fly? | Flight simulation shows uplift with airplane wing profile, body adaptions for bird flight | Exploring the physical and physiological conditions for creating uplift, wing structure |
| 20 | Workstation 4 How storks use gliding forces? | School experiment to show thermal uplift with feather floating in warm air for bird gliding | Experimenting with natural phenomena of thermal uplift, learning about migratory birds |
| 20 | Workstation 5 bird characteristics, flight types | Ecology of bird species in different environments and various flight types; Observation on real padded birds for demonstration | Bird observation with video to learn different flight adaptions in different environments; Bird Quiz with real objects |
| 15 | Workstation 6 Paper flight rally | Constructing different paper flight models, game with paper flight rally | Creating paper flight models with origami handicraft technique, hands-on with various wing shapes |
| 20 | Questionnaires and feedback | Post-test and résumé as well as feedback questionnaires complete the intervention for consolidation | Students reflect on the inquiry process and their learning |

so we did not include disclaimed test–retest groups at all. The questionnaire was used immediately after completion of all workstations at (T1) to ensure its direct relationship to programme participation. The questionnaire's completion required about 20–25 min (example, see Table 2).

## 2.3 *Statistical Analysis*

For statistical analysis, we used IBM SPSS (version 22.0). A total of 160 complete data sets of the knowledge questionnaire (T0 pre-test, T1 post-test, T2 delayed post-test) were obtained (Field, 2009). For each student, a mean score for each of the 23 items was calculated. After considering both the item difficulty and corrected item-total correlation for item selection, two items with a score <0.2 were excluded (c.f., Scharfenberg, Bogner, & Klautke, 2006). We tested normal distribution of our data using Wilcoxon's test ($p < 0.001$) and based all further analyses on nonparametric tests with one independent sample. Responses of the knowledge questions were mean-scored; a correct answer scored with 1, an incorrect with 0. Individual test scores for each test point were calculated as the number of correct answers. High mean in knowledge test scores indicates good comprehension. A reliability analysis of the knowledge questionnaire yielded a Cronbach's alpha of 0.61 for T0 and 0.86 for T1 for all knowledge items, and thus can be accepted as reliable (Lienert, 1969). Item difficulty showed in pre-test a mean of 0.36 as well as a mean item variance of 0.19; in post-test a mean of 0.63, again with a mean item variance of 0.19. Post-test sum scores showed a significantly higher score after the intervention. To avoid bias in the experiment, we have included all students irrespective of their performance. Additionally, we tested for potential gender differences in the achieved learning performance. We used the same questionnaire and tested the knowledge level for pre- and post-test in comparing boys' and girls' knowledge achievement. Data were compared with the scores obtained after the implementation and subjected to pairwise nonparametric Mann–Whitney U test for independent samples (Table 3).

# 3 Results

A comparison of the obtained frequencies just by eye already signals a substantial shift towards higher achievement sum scores (the maximum of correct answers was 26). While the mean of T0 scored 10.91 (±0.32 SE), T1 shifted to 16.43 (±0.59 SE). While just a small portion of students seems to resist any learning (scoring below 10), a considerable shift beyond 15 (out of 26) after participation is apparent.

A Wilcoxon's test showed a significant difference in knowledge level between all three testing cycles: cognitive achievement scores significantly increased due to our learning unit independently of gender differences. The implementation increased the

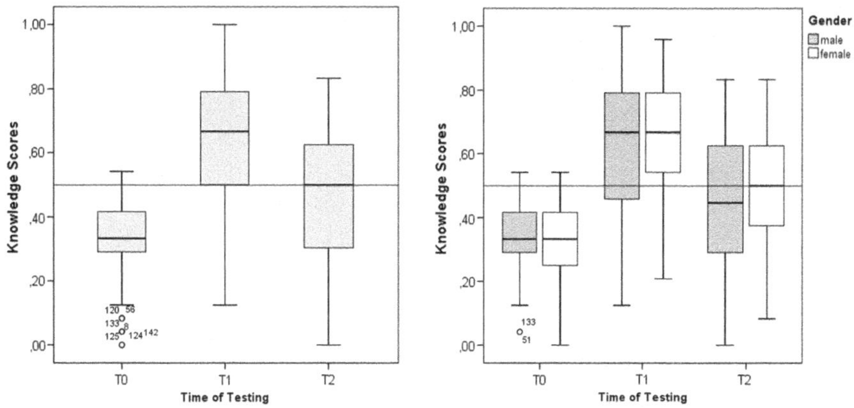


**Fig. 2** Left side: Boxplots of knowledge mean scores levels at the pre-test T0 (left), the post-test schedule T1 (middle) and the retention-test schedule (T2). Right side: Boxplots of knowledge mean scores levels separated by gender. T0 = 1–2 weeks before intervention; T1 = immediately after; T2 = 6–8 weeks after intervention

**Table 4** Correlation matrix of knowledge score at T0, T1 and T2 as well as regarding gender

| | | Gender | T0 KN | T1 KN | T2 KN |
|---|---|---|---|---|---|
| T0 KN | Correlation coefficient | −0.090 | 1.000 | | |
| | Sig. (2-tailed) | n.s. | . | | |
| | *N* | 160 | 160 | | |
| T1 KN | Correlation coefficient | 0.041 | 0.254** | 1.000 | |
| | Sig. (2-tailed) | n.s. | 0.001 | . | |
| | *N* | 160 | 160 | 160 | |
| T2 KN | Correlation coefficient | 0.102 | 0.190* | 0.362** | 1.000 |
| | Sig. (2-tailed) | n.s. | 0.025 | 0.000 | . |
| | *N* | 139 | 139 | 139 | 139 |

n.s. = Nonsignificance
*<0.05
**<0.01

knowledge level of our participants substantially and sustained it for 6–8 weeks at least (Fig. 2) (Table 4).

As expected, students have learnt better the better their pre-knowledge scores: the latter scores (T0) predicted the short-term learning success (T1). The same is true for the long-term effect (T2), for both the pre-knowledge (T0) and the post-knowledge (T1) as well: the more they had learned, the more they still knew after 6–8 week of delay. Gender equality analysis yielded no significant differences between male and female students (Mann–Whitney U test = 0.60, $p = 0.05$); there were differences, but they did not reach statistical significance. Both groups rate the knowledge questionnaire similarly, resulting in similar mean scores in pre-test and post-test for knowledge increase. Female and male students both improved knowledge levels compared to pre-test T0 and post-test conditions T1 (Fig. 2, right side) equally.

## 4 Discussion

Our first result clearly points to a substantial knowledge gain after participation in our educational unit ($p < 0.001$); thus, our first research question is confirmed. One reason for the high mean scores of 0.8 (at T1) might be the low pre-knowledge level and the student-centred inquiry approach. The latter may include the use of authentic tools, multimedia and most importantly student-centred, self-exploring activities. That student-centred compared to teacher-centred approaches regularly produce higher achievement scores has been repeatedly reported (e.g. Minner, Levy, & Century, 2002; Sturm & Bogner, 2008; Fremerey & Bogner, 2015). Similarly, a group size of three to four students has often been shown to be most effective (Randler & Bogner, 2002). Related studies may further underpin our result with similar knowledge shifts, as occurs with eLearning in classroom (Bissinger & Bogner, 2017) and zoological garden field trips to teach evolutionary adaptions within marine ecosystems (Sattler & Bogner, 2017). We can assume that the method of inquiry-based on student-centred learning at hands-on workstations may lead to successful learning, given the large effects, at least more than just a mere rote memorization of 'factual knowledge'. Individual interest in animals like birds or the fascination with the ancient fossil may certainly have further contributed. Earlier studies had assured a tenacity effect as cognitive knowledge scores were still detectable after longer periods, such as a half-year or even one-year duration (Schmid & Bogner, 2015; Marth & Bogner, 2018). In both studies, the delayed post-text (of about 6–8 weeks after implementation) produced scores which did not drop any further. Consequently, a delayed post-test measure was shown to be sufficient in unveiling the sustained knowledge level (reached by an intervention).

Another reason could lie in our chosen age group of sixth graders who were novices regarding any pre-knowledge in evolution. This may have contributed to the apparent effectiveness. Participants had no lesson instruction about evolution in general and *Archaeopteryx* or bird flight in particular; that is why we regard our participants as novices. In searching for the best way to teach evolution, Mead et al. (2017) tested whether the order of teaching may improve an understanding of evolution: when students first learn fundamental concepts of genetics, understanding of evolution was reported to be slightly better, with a shift of 7%. As teaching evolutionary issues are scheduled mainly in grade 9, any earlier school contact to evolution theory may improve the subsequent understanding of evolution.

We decided to avoid control group designs or even test–retest conditions with no intervention. First, some regard it as unethical that some students are not taught (Mead et al., 2017). Second, the considerable number of earlier studies including test–retest may also justify this decision as no effect on knowledge scores was reported: a simple repeated completion of knowledge questionnaires has never produced any learning effect (summaries see, e.g., Fremerey & Bogner, 2014; Dieser & Bogner, 2017). Third, our results reached a significantly high score of 0.8, although novices neither had pre-knowledge nor were aware about any testing cycles, so test effects can be excluded. Fourth, methods of defining equivalent control groups are controversial (Schumm & Bogner, 2016).

Our study design was not focused on teaching evolutionary theory, as this is required in curricula of a later grade, but more importantly on offering a first access to learning how evolution, birds and natural phenomena are connected in a playful, unrestricted way where students have space to experiment and interactively learn basics on which they can later build (Deadman & Kelly, 1978). The inquiry process includes learning tasks, assessments and teaching strategies that support learning by exploration and discovery (White & Frederiksen, 1998) and provides authentic and even complex learning experiences by giving students opportunities to flourish and participate in scientific practice (Sotiriou & Bogner, 2011). As an alternative to classroom intervention, the unit of *Archaeopteryx* and bird flight could equally be applied during an out-of-school visit to a Natural History Museum (Sturm & Bogner, 2010). Nevertheless, given the fact that original fossils of *Archaeopteryx* are available at only six natural history and palaeontological museums around the world (for instance, London or Berlin), a replica of *Archaeopteryx* shows great potential for classroom interventions anywhere.

The origin of our study was an initiative labelled *Natural Europe* that had provided an educational tool such as *Archaeopteryx* as outreach-learning settings together with the engagement of teacher communities to establish competences in scientific inquiry learning (Sotiriou et al., 2016). Contributing to the *RRI* goals '*responsible research and innovation*', our educational project showed the potential that strengthening students' and teachers' involvement by interacting with researchers may enhance participation, playfulness and creativity in everyday science education practice. As a follow-up project, in *CREATIONS,* a creative approach is used, e.g. paper flight modelling and collaborative handicraft fossil art. Science and arts can complement each other for inquiry and promote creative thinking. Another advantage is interdisciplinary learning: Although science subjects such as biology or physics are taught separately in school, they relate to each other. The method of crosslinking interdisciplinary knowledge is becoming popular, as science subjects use similar concepts and were practically applicable for *Archaeopteryx* and bird flight teaching tools. *Archaeopteryx* has many advantages for teaching evolution for novices in school: as a labelled missing link in the evolution of flight, a feathered reptile may have started to fly by flapping its wings and changing its diet to pursue insects and conquer airspace (Ostrom, 1976; Dodson, 1985). It is commonly accepted and proven in research that *Archaeopteryx* was able to fly (Wellnhofer, 2008; Erickson et al., 2009; Foth, Tischlinger, & Rauhut, 2014). Introducing *Archaeopteryx* in combination with bird flight and linking it to modern birds by using interdisciplinary tools and explaining physical aspects of flight like aerodynamics, thermal uplift and gliding makes science learning interesting. It is a new approach in crosslinking evolutionary aspects with other science subjects. Evolution can be presented as a story, which is not limited to explaining narratively how species may have evolved (Zabel & Gropengießer, 2011).

Our hands-on tools also included two multimedia stations, one with a virtual flight simulation and the other presenting bird flying features. According to Mayer and Moreno (1998), testing students with computer-generated animations, multimedia learners can process words and pictures more easily when the words are presented in spoken rather than visual form. We used observational instructions on birds and self-regulatory tools with flight simulation in two workstations separately. In our view, multimedia can be an interesting supplementary instrument for scientific discovery learning (Girwidz, Bogner, Robitzko, & Schaal, 2006a; Girwidz, Robitzko, Schaal, & Bogner, 2006b) but it cannot replace hands-on experience with authentic natural objects like fossils and birds. However, virtual experience may offer additional value for hand-on classrooms when applied accordingly (Barrett, Stull, Hsu, & Hegarty, 2015).

Our study's frame was a European initiative with a variety of partners of different mother tongues. Studies concerning bilingual language usages in science class have mainly been conducted in settings where both the teacher and the students speak the same language. Only a handful of foreign languages are taught in bilingual/dual language programmes within mainstream classroom (Cummins, 2005). For instance, Bogner (1999) described a national conservation programme in German-speaking Switzerland, where alpine swifts were in focus which breed in Switzerland but spend most of the year in Senegal. Besides various activities (e.g. constructing nest-boxes, watching bird breeding, etc.), participants exchanged letters in French with pen-friends in Senegal where the migrating bird (*Tachymarptis melba*) spends its winter. A bilingual sub-module focused on sharing observations of the bird's wintering region. In the present study, the teaching language was German, although both languages for information boards and instructions were made available to make complex meaning of scientific expressions more understandable.

Gender did not contribute to any difference in knowledge scores. Earlier studies have not reported such differences either. Liefländer and Bogner (2014), for instance, reported for a week-long outreach intervention ('Life in Water') similar cognitive learning levels and explained the absence of gender differences with developmental characteristics of their chosen age group. In our case, perhaps the use of arts in combination with inquiry learning at workstations may have suited students' interests. In earlier studies, female students generally were assumed to show less interest in technical and natural sciences (e.g. Scharfenberg & Bogner, 2010). Asking teachers about subject preferences, girls on average tend to show more interest in language, arts and crafts issues, whereas boys tend to prefer technical and maths-based subjects in school. Efforts in educational institutions and schools globally have been made to promote girls' interest in STEAM subjects, e.g. via girls' days in physics, and much more. Finding a teaching approach with arts in science, by using various tools and subjects to address abilities by challenging them with new methods and at the same time creating enthusiasm, will lead to a higher interest and willingness to learn in depth, in our case about evolutionary and science topics. Both genders will profit equally from instruction in *Archaeopteryx* and bird flight in our inquiry-centred module.

In conclusion, the positive effect of our implemented educational unit is reflected in a significant knowledge increase and thus we see our intervention as a reasonable approach to teaching evolution to novice students (Hermann, 2008). Various modifications were implemented in this study to give students a broader access to science learning, including: (a) arts in science; (b) interdisciplinary learning (c) authentic tools. Interdisciplinary hands-on workstations with inquiry tools may represent an appropriate method of choice accompanied, if applicable, by arts in science. Teaching *Archaeopteryx* and bird flight prior to evolution teaching may help improve an understanding of evolutionary contexts and enhance fascination to learn more about science (Stamos, 2008). This method may help both genders to raise awareness about natural phenomena and get in touch with evolutionary issues, even for fifth–sixth graders. Nevertheless, further research is needed to disentangle the contribution of the different approaches of our applied module, especially whether they may contribute separately or synergistically.

**Acknowledgements** This work was supported by the European HORIZON-2020 framework labelled CREATIONS: Developing an Engaging Science Classroom (Grant Agreement No. 665917; http://creations-project.eu). We would like to thank all students and teachers who supported our study. We kindly thank M. Wiseman for constructive discussion and assistance in statistical analyses, H.-D. Haas at the Jura-Museum in Eichstaett for providing the *Archaeopteryx* fossil replica as well as him and K. Elsner-Mann for supporting our first pilot test runs.

## References

Anderson, R. D. (2002). Reforming science teaching: What research says about inquiry. *Journal of Science Teacher Education, 13*(1), 1–12.

Anderson, L. W., & Krathwohl, D. R. (Eds.). (2001). *A taxonomy for learning, teaching, and assessing: A revision of Bloom's taxonomy of educational objectives*. Boston: Allyn and Bacon.

Barrett, T. J., Stull, G. S., Hsu, T. M., & Hegarty, M. (2015). Constrained interactivity for relating multiple representations in Science. When virtual is better than real. *Computers & Education, 81,* 69–81.

Bissinger, K., & Bogner, F. X. (2017). Environmental literacy in practice: Education on tropical rainforests and climate change. *Journal Environment, Development and Sustainability*, 1–16.

Blancke, S., Boudry, M., Braeckman, J., deSmedt, J., & deCruz, H. (2011). Dealing with creationist challenges. What European Biology teachers might expect in the classroom. *Journal of Biology Education, 45*(4), 176–182.

Bogner, F. X. (1998). The influence of short-term outdoor ecology education on long-term variables of environmental perspective. *Journal of Environmental Education, 29*(4), 17–29.

Bogner, F. X. (1999). Empirical evaluation of an educational conservation programme introduced in Swiss Secondary Schools. *International Journal of Science Education, 21,* 1169–1185.

Bossert, U. (1998). Archaeopteryx—Untersuchung eines Fossilfundes. [A.—Investigating a fossil]. *Biologie in der Schule*, 47(6), 333–335.

Catley, K. M. (2006). Darwin's missing link—A novel paradigm for evolution education. *Science Education, 90*(5), 767–783. https://doi.org/10.1002/sce.20152.

Cavallo, A. M., & McCall, D. (2008). Seeing may not mean believing: Examining students' understandings & beliefs in evolution. *The American Biology Teacher, 70*(9), 522–530.
Cummins, J. (2005). A proposal for action: Strategies for recognizing heritage language competence as a learning resource within the mainstream classroom. *The Modern Language Journal, 89,* 585–592.
Cunningham, D. L., & Wescott D. J. (2009). Still more "fancy" and "myth" than "fact" in students' conceptions of evolution. *Evolution: Education and Outreach, 2*(3), 505–517.
Deadman, J. A., & Kelly, P. J. (1978). What do secondary school boys understand about evolution and heredity before they are taught the topics? *Journal of Biological Education, 12*(1), 7–15.
DeWitt, J., Archer, L., & Osborne, J. (2013). Nerdy, brainy and normal: Children's and parent's constructions of those who are highly engaged with science. *Research in Science Education, 43*(4), 1455–1476.
Dieser, O., & Bogner, F. X. (2015). Young people's cognitive achievement as fostered by hands-on-centred environmental education. *Environmental Education Research.* https://doi.org/10.1080/13504622.2015.1054265.
Dieser, O., & Bogner, F. X. (2017). How individual environmental values influence knowledge acquisition of adolescents within a week-long outreach biodiversity module. *Journal of Global Research in Education and Social Science, 9*(4), 213–224.
Dobzhansky, T. (1973). Nothing makes sense in biology except in the light of evolution. *The American Biology Teacher, 35*(3), 125–129.
Dodson, P. (1985). Review of the international archaeopteryx conference. *Journal of Vertebrate Palaeontology, 5*(2), 177–179. https://doi.org/10.1080/02724634.1985.10011856.
Driver, R., & Easley, J. (1978). Pupils and paradigms: a review of literature related to concept development in adolescent science students. *Studies in Science Education, 5,* 61–84.
Erickson, G. M., Rauhut, O. W. M., Zhou, Z., Turner, A. H., Inouye, B. D., Hu, D., & Norell, M. A. (2009). Was dinosaurian physiology inherited by birds? Reconciling slow growth in Archaeopteryx. *PLoS One, 4*(10), 1–9.
Evans, E. M., & Lane, J. D. (2011). Contradictory or complementary? Creationist and evolutionist explanations of the origin of species. *Human Development, 54,* 144–159. https://doi.org/10.1159/000329130.
Field, A. (2009). *Discovering statistics using SPSS* (3rd ed.). Thousand Oaks, CA: Sage.
Foth, C., Tischlinger, H., & Rauhut, O. W. M. (2014). New specimen of *Archaeopteryx* provides insights into the evolution of pennaceous feathers. *Nature, 511,* 79–82.
Fremerey, C., & Bogner, F. X. (2014). Learning about drinking water: How important are the three dimensions of knowledge that can change individual behaviour? *Education Sciences, 4,* 213–228.
Fremerey, C., & Bogner, F. X. (2015). Cognitive learning in authentic environments in relation to green attitude preferences. *Studies in Educational Evaluation, 44,* 9–15.
Geier, C. S., & Bogner, F. X. (2011). Learning at workstations. Students' satisfaction, attitudes towards cooperative learning and intrinsic motivation. *Journal for Educational Research Online, 3*(2), 3–14.
Goldschmidt, M., & Bogner, F.-X. (2015). Learning about genetic engineering in an outreach laboratory: Influence of motivation and gender on students' cognitive achievement. *International Journal of Science Education, Part B, 6*(2), 166–187.
Goodale, T. A. (2017). Utility of context-based learning to influence teacher understanding of evolution and genetics concepts related to food security issues in East Africa (in press).
Girwidz, R., Bogner, F. X., Robitzko, T., & Schaal, S. (2006a). Media assisted learning in science education: An interdisciplinary approach to hibernation and energy transfer. *Science Education International, 17*(2), 95–107.
Girwidz, R., Robitzko, T., Schaal, S., & Bogner, F. X. (2006b). Theoretical concepts for using multimedia in science education. *Science Education International, 17*(2), 77–93.
Gropengießer, H., Harms, U., & Kattmann, U. (2013). Auswahl und Verknüpfung der Lerninhalte [Selecting and connecting learning contents]. In *Fachdidaktik Biologie*. Donauwörth: Aulis.

Hammann, M., & Aschoff, R. (2013). *Schülervorstellungen im Biologieunterricht: Ursachen für Lernschwierigkeiten [Student conceptions in Biology lessons: Causes of learning difficulties]*. Berlin: Springer.
Hampden-Thompson, G., & Bennett, J. (2013). Science teaching and learning activities and student's engagement in science. *International Journal of Science Education, 35*(8), 1325–1343.
Harlen, W. (2010). Principles and big ideas of science education. Hatfield: Association for Science Education. http://www.ase.org.uk/documents/principles-and-big-ideas-of-science-education.
Hermann, R. S. (2008). Evolution as a controversial issue: A review of instructional approaches. *Science & Education, 17,* 1011–1032.
Jakobi, S. R. (2010). "Little Monkeys on the Grass …" How people for and against evolution fail to understand the theory of evolution. *Evolution: Education & Outreach, 3*(3), 416–419.
Kattmann, U. (2017). Geschichte und Verwandtschaft der Lebewesen. Ein Basiskonzept der Evolutionsbiologie [History and relationship of animals. A basic concept of evolution]. *Unterricht Biologie, 421*, 39.
Kossack, A., & Bogner, F. X. (2012). How does a one-day environmental education programme support individual connectedness with nature? *Journal of Biological Education, 46*(3), 180–187.
Liefländer, A., & Bogner, F. X. (2014). The effects of children's age and sex on acquiring pro-environmental attitudes through environmental education. *Journal of Environmental Education, 45*(2), 105–117.
Lienert, G. A. (1969). *Testaufbau und Testanalyse [Test construction and Analysis]* (3rd ed.). Weinheim: Julius Beltz.
Marth, M., & Bogner, F. X. (2018). Does the issue of bionics within a student-centered module generate longterm knowledge. *Studies in Educational Evaluation, 55*, 117–124.
Mayer, R. E., & Moreno, R. (1998). A split-attention effect in multimedia learning: Evidence for dual processing systems in working memory. *Journal of Educational Psychology, 90*(2), 312–320.
Mayr, E. (1997). *This is biology*. Cambridge, MA: Harvard University Press.
Mead, R., Hejmadi, M., & Hurst, L. D. (2017). Teaching genetics prior to teaching evolution improves evolution understanding but not acceptance. *PLoS Biology, 15*(5), e2002255.
Minner, D. D., Levy, A. J., & Century, J. (2002). Inquiry-based science instruction—What is it and does it matter? Results from a research synthesis years 1984 to 2002. *Journal of Research in Science Teaching*. https://doi.org/10.10002/tea.20347.
Morgan, J. M. (1992). A theoretical basis for evaluating wildlife-related education programs. *The American Biology Teacher, 54,* 153–157.
Mumford, M. D. (2002). Social innovations: Ten cases from Benjamin Franklin. *Creativity Research Journal, 14*(2), 253–266.
Nachtigall, W. (1985). *Warum Vögel fliegen [Why birds fly]*. Hamburg: Rasch & Röhring.
OECD. (2012). *Programme for the international assessment of adult competencies (PIAAC)*. Paris.
Ostrom, J. H. (1976). Archaeopteryx and the origin of birds. *Biological Journal of the Linnean Society, 8,* 91–182.
Peter, D. S. (1994). Entwicklung des Vogelflugs [Development of the bird flight]. *Praxis der Naturwissenschaften, 43*(7), 10–14.
Posner, G. J., Strike, K. A., Hewson, P. W., & Gertzog, W. A. (1982). Accommodation of a scientific conception: Toward a theory of conceptual change. *Science Education, 66*(2), 211–227.
Randler, C., & Bogner, F. X. (2002). Comparing methods of instruction using bird species identification skills as indicators. *Journal of Biological Education, 36*(4), 2–9.
Randler, C., & Bogner, F. X. (2006). Cognitive achievement of group-based hands-on identification skill training. *Journal of Biological Education, 40*(4), 161–166.
Sattler, S., & Bogner, F. X. (2017). Short- and long-term outreach at the zoo: Cognitive learning about marine ecological and conservational issues. *Environmental Education Research, 23*(2), 252–268. https://doi.org/10.1080/13504622.2016.1144173.
Scharfenberg, F.-J., & Bogner, F. X. (2010). Instructional efficiency of changing cognitive load in an out-of-school laboratory. *International Journal of Science Education, 32*(6), 829–844.

Scharfenberg, F.-J., Bogner, F. X., & Klautke, S. (2006). The sustainability of external control groups for empirical control purposes: A cautionary story in science education research. *Electronic Journal of Science Education, 11*(1), 22–36.
Schmid, S., & Bogner, F. X. (2015). Effects of students' effort scores in a structured inquiry unit on long-term recall abilities of content knowledge. *Education Research International,* (Article ID 826734).
Schönfelder, M., & Bogner, F. X. (2017). How to sustainably increase students' willingness to protect pollinators. *Environmental Education Research*, 2–13.
Schumm, M., & Bogner, F. X. (2016). The impact of science motivation on cognitive achievement within a 3-lesson unit about renewable energies. *Studies in Educational Evaluation, 50,* 14–21.
Sinatra, G. M., Kienhues, D., & Hofer, B. K. (2014). Addressing challenges to public understanding of science: Epistemic cognition, motivated reasoning and conceptual change. *Educational Psychologist, 49*(2), 123–138.
Sotiriou, S., & Bogner, F. X. (2011). Inspiring science learning: Designing the science classroom of the future. *Advanced Science Letters, 4,* 3304–3309.
Sotiriou, S., Bybee, R., & Bogner, F. X. (2016). PATHWAYS: A case of large-scale implementation of evidence-based practice in science inquiry-based science education. *International Journal of Higher Education, 6*(2), 1–12.
Stamos, D. N. (2008). *Evolution and the big questions. Sex, race and other matters.* Malden: Blackwell Publishing.
Sturm, H., & Bogner, F. X. (2008). Student-oriented versus teacher-centred: The effect of learning at workstations about birds and bird flight on cognitive achievement and motivation. *International Journal of Science Education, 30*(7), 941–959.
Sturm, H., & Bogner, F. X. (2010). Learning at workstations in two different environments: A museum and a classroom. *Studies in Educational Evaluation, 36*(1–2), 14–19.
Thomas, B. (2013). Archaeopteryx. Kinder lernen den einzigartigen Urvogel kennen [Kids get to know the unique fossil bird]. *Grundschule Sachunterricht, 57,* 28–33.
To, C., Tenenbaum, H. R., & Hogh, H. (2017). Secondary school students' reasoning about evolution. *Journal of Research in Science Teaching, 54*(2), 247–273.
Vosniadou, S., & Brewer, W. F. (1992). Mental models of the earth: A study of conceptual change in childhood. *Cognitive Psychology, 24,* 535–585.
Weber, I., & Kattmann, U. (1991). Archaeopteryx—ein befiederter Dinosaurier? [A.—A feathered dinosaur]. *Unterricht Biologie, 15*, 41–43.
Wellnhofer, P. (2008). *Archaeopteryx. Der Urvogel von Solnhofen [The fossil bird of Solnhofen]* (pp. 205–216). München: Pfeil.
White, B. Y., & Frederiksen, J. R. (1998). Inquiry, modelling and metacognition: Making science accessible to all students. *Cognition and Instruction, 16,* 3–118.
Zabel, J., & Gropengießer, H. (2011). Learning progress in evolution theory: Climbing a ladder or roaming a landscape? *Journal of Biological Education, 45*(3), 143–149.

**MA Alexandra Buck** worked part-time for 18 months in the European Horizon-2020 project with the label: CREATIONS. Within that context, she was responsible for the lesson development and classroom implementation.

**Dr. Sofoklis Sotiriou** is the director of the R&D science education department of Ellinogermaniki Agogi (Greece). Within the CREATIONS project, he was leading the work package 'User Community Support' where all selected case studies (such as the presented one) were supervised and logistically coordinated.

**Franz X. Bogner** holds the Chair of Biology Education at the University of Bayreuth, leading the Centre of Math & Science Education (Z-MNU). With experience of 11 years as high school teacher, he simultaneously built up his career as scientist. After a Ph.D. in neurobiology (1986), US postdoctorate in chemical ecology (1989–1991) and habilitation in didactics (1996), he was appointed to a Full Professorship in 1997. His research focuses on inquiry-based science education and environmental literacy; his four-times independently confirmed 2-MEV model has been translated into 31 languages. The total citation of his research papers exceeds 5200 (h-index: 37). He was leading and coordinating the European HORIZON-2020 project with 16 partner institutions acronymed CREATIONS.

# Overcoming Motivational Barriers to Understanding and Accepting Evolution Through Gameful Learning

**David C. Owens**

## 1 Introduction

Though the vast majority of the scientific community recognizes the import and validity of evolutionary theory, many in the American (US) public struggle to understand or accept it (Miller, Scott, & Okamoto, 2006). Cognitive constraints, including essentialism and teleology, as well as an inadequate understanding of the nature of science and insufficient exposure to evolution education have been identified as contributors to this rejection of evolution and serve as barriers to conceptual change (Kampourakis, 2014). Yet, the biggest barrier to the understanding and acceptance of evolution may be the lack of motivation to do so, which may result from a conceptual commitment to worldviews that are perceived to conflict with evolutionary theory (Sinatra, 2005). In this chapter, I set out to describe and discuss a gameful, inquiry-based learning environment and its effects on students' motivation to learn about the evolutionary life history of plants—an instructional intervention with the potential to aid learners in overcoming motivational barriers to understanding and accepting evolution.

### 1.1 The Role of Motivation in Conceptual Change

One's prior to conceptual knowledge, or *conceptual ecology* (Toulmin, 1972), governs all aspects of information processing and knowledge acquisition. When new information is congruent with prior knowledge, it is *assimilated* rather effortlessly into a learner's existing paradigm of understanding (Piaget, 1970). However, when individuals encounter information that goes against previously constructed knowl-

D. C. Owens (✉)
Georgia Southern University, Armstrong Campus, 11935 Abercorn Street University Hall 205, Savannah, GA 31419, USA
e-mail: dcowens@georgiasouthern.edu



U. Harms and M. J. Reiss (eds.), *Evolution Education Re-considered*,
https://doi.org/10.1007/978-3-030-14698-6_10

edge, *accommodation*—a radical restructuring or replacement of those alternative frameworks—must occur, if that new knowledge is to be acquired (Vosniadou & Brewer, 1987). As alternative frameworks are highly resistant to change, the accommodation of new information that results in conceptual change is an effortful process. In the case of evolution, accommodating an understanding of evolution can be a difficult, effortful process for individuals who perceive it to contradict their worldview.

The first conceptual change model (CCM) was first advanced as a domain-specific framework of accommodation (Posner, Strike, Hewson, & Gertzog, 1982) to be used as a lens for understanding the conditions that must be met before conceptual change can occur in light of one's oppositional conceptual ecology. The model included four conditions:

1. An individual must become *dissatisfied* with his or her current conception; otherwise, the impetus to explore alternative concepts would not exist.
2. The new conception must be *intelligible* so as to enable a better explanation of an experience than the existing conception.
3. The new conception must be *plausible* in terms of its consistency with other knowledge and its potential to solve problems generated by previously held conceptions.
4. The new conception must appear *fruitful* in addressing new areas of inquiry.

However, the CCM did not account for any affective components of conceptual change. Rather, it was built on the assumption that conceptual change was controlled strictly by cognitive means and occurred for learners in the same rational and logical manner as it does for scientists (Pintrich, Marx, & Boyle, 1993)—idyllic even in formal science practice, but certainly not the case for the average learner whose cognitive engagement is necessarily influenced by affective factors, such as goals for learning, motivational beliefs, and a need for cognition (Pintrich & Schrauben, 1992), as well as situational classroom contexts that may or may not contribute to learning (Blumenfeld et al., 1991). Recognizing the limitations to their *cold* model of conceptual change, Strike and Posner (1992) suggested that "a wider range of factors needs to be taken into account in attempting to describe a learner's conceptual ecology. Motives and goals and the institutional and social sources of them need to be considered" (p. 162).

In the decades since, a *warming trend* has emerged to address the need for including affective components, such as motivation, when considering the potential for conceptual change (Sinatra, 2005), such as that which is often required for the understanding and acceptance of evolution to occur. For one, each individual's need for cognition varies in terms of his or her desire to find relevance in and make sense of the natural world (Cohen, Stotland, & Wolfe, 1955). Those naturally possessing a need for cognition are more motivated to engage in sense-making practices, such as analyzing problems and developing solutions, while others are satisfied with heuristic processing of ideas and lack motivation to engage in the active or effortful evaluation of ideas (Chaiken, 1987). However, interest and emotional involvement stemming from environmental conditions, including social contexts, have the potential to motivate engagement and comprehension (Hidi, 1990), even among those who lack an apparent need for cognition or have yet to become dissatisfied with their existing ideas. Furthermore, learners' active use of concepts that have experiential value in their life and expand their perception of the world around them can result in transformative experiences that lead to conceptual change (Pugh, 2002, 2011).

## *1.2 The Motivational Nature of Gameful, Inquiry-Based Learning*

The purpose of this chapter is to describe an intervention directed at addressing the affective components of conceptual change and discuss the effect of the intervention on students' motivation to learn biology in the context of plant evolutionary life history. The intervention hinged on inquiry-based learning in which students were tasked with developing relevant questions that required higher-order thinking to be answered and critiquing those questions and answers in the context of a game. Inquiry-based and gameful learning are introduced below.

### 1.2.1 Inquiry-Based Learning

Inquiry-based learning is an active, student-centered pedagogy by which the learner engages in open-ended exploration of some curiosity that he or she finds interesting and, in doing so, actively develops knowledge. Learning through inquiry typifies scientific thinking and has been championed in circles of science education as a means for enhancing students' ability to think critically. A classroom environment rooted in inquiry-based learning allots time for learners to make discoveries, understand new ideas, develop questions that require significant cognitive engagement to be answered, and critique those questions and answers using evidence—all seemingly conducive to conceptual change (Dole & Sinatra, 1998; Pugh et al., 2010). In this study, all participating students engaged in inquiry-based learning by tasking them with (a) asking questions that made the content relevant to their lives and required higher-order thinking to be answered, (b) developing thorough, accurate explanations for those questions prior to class, (c) communicating the value and relevance of those questions and answers to their peers during class, and (d) critiquing their peers' questions and answers in terms of their relevance, importance, and accuracy.

### 1.2.2 Gameful Learning

If the ever-increasing amount of time individuals are spending playing video games is any indication of their motivational potential, then the use of game design elements to structure more formal learning environments, or *gamification* (Deterding, Dixon, Khaled, & Nacke, 2011, p. 10), holds promise as a mechanism for enhancing motivation to learn in formal academic settings (Owens, 2017)—including those of evolution education. Two common elements of game design are the leaderboard and repeat-testing. *Leaderboards* advertise each player's points and rank, including badges that highlight achievements or accomplishments, and promote the demonstration of competence and enable comparisons among students. The *repeat-testing* element of games enables individuals to repeat levels with minimal risk until satisfied with the competence they developed. The gameful learning environment induced by the inclusion of leaderboards and repeat-testing were expected to enhance students' motivation to learn about plant evolutionary life histories—a component critical to the promotion of conceptual change (Sinatra & Pintrich, 2003).

### *1.3 Motivation to Learn Biology*

The components of motivation that enhance student enjoyment of science and recognition of its value can be distinguished by their intrinsic or extrinsic nature (Glynn, Brickman, Armstrong, & Taasoobshirazi, 2011). For example, intrinsic motivation is often characterized by an enjoyment of learning that comes from an inherent interest in the concept or activity (Deci & Ryan, 1985). Self-determination contributes to the maintenance of intrinsic motivation when a learner is optimally stimulated and their needs for competence and autonomy are being met (Deci & Ryan, 1985), as does self-efficacy, which describes one's belief in one's ability to execute a specific task (Bandura, 1977). Extrinsic motivation to learn often arises from reasons outside one's immediate interest or enjoyment, such as obtaining a desired grade or job. For the purpose of this study, motivation to learn biology was considered a multi-component construct (Glynn et al., 2011), consisting of the following attributes of motivation: intrinsic motivation; self-determination; self-efficacy; career motivation; and grade motivation.

The research was guided by the following question:

How does inquiry-based learning instruction designed with two gaming elements, the leaderboard and repeat-testing, affect students' motivation to learn biology in the context of plant evolutionary life history?

## 2 Methods

### *2.1 Research Context*

#### 2.1.1 Sample

Participants in the study were 140 undergraduates enrolled in one of eight sections of a second-semester introductory biology laboratory course for science majors at a large public university in the southeastern region of the USA. A demographic survey indicated that 64% were female. The ethnicity of participants was as follows: 61.4% Caucasian, 22.1% African American, 5.7% Hispanic, 2.9% Asian, 0.7% Native American, and 7.1% did not specify.

### 2.1.2 Course Description

The intervention took place in a second-semester biology laboratory course for science majors. The laboratory course met once weekly for two hours and 45 min and accompanied a lecture course that also met for two hours and 45 min each week. Together, the lecture and laboratory comprised of a four-credit-hour course. The focus of the course was on plant and animal evolutionary life history, including anatomy, physiology, and classification. The study examined the first four weeks of the course, which involved only the portion of the curriculum concerning plants. Content covered included the following phyla: Hepaticophyta, Bryophyta, Pterophyta, Lycophyta, Cycadophyta, Ginkgophyta, Coniferophyta, Gnetophyta, and Anthophyta.

The laboratory course was taught by graduate teaching assistants (GTAs) from within the biology department. GTA instructors met prior to the start of each week's laboratory sessions to ensure familiarity with inquiry-based learning, as well as to discuss alternative conceptions that might arise during laboratory sessions so that the GTAs could effectively recognize and address them during student-centered instruction. GTAs also worked together to develop an eight-item multiple-choice assessment that would be administered to all students at the end of the laboratory session. The quiz aligned with the content that students reviewed before class to support their creation of questions and the discussion of those questions during class.

## *2.2 Learning Conditions*

### 2.2.1 Inquiry-Based Learning

All students engaged in an inquiry-based learning environment developed from the National Research Council's (NRC, 2000) assertion that students should be "asking scientifically-oriented questions, giving priority to evidence in responding to questions, formulating explanations from evidence, connecting explanations to scientific knowledge, and communicating and justifying explanations" (p. 23), a declaration reiterated as *essential practices* in the *Framework for K-12 Science Education* (NRC, 2011). The class structure is described below.

**Questions**: Prior to each laboratory session, students were required to read the corresponding material from the laboratory manual (Vodopich & Moore, 2014), generally about six pages and covering one or two plant phyla. From this material, students were to develop two questions, one each from the first and second halves of the assigned reading that extended the concepts they read and made the concepts more relevant to their lives or satisfied a curiosity. Questions were to be developed using Bloom's taxonomy so as to require a higher level of thinking than understand or remember (i.e., apply, analyze, evaluate, or create), and students were to answer the questions using at least one piece of outside information, preferably from a reputable source, such as a science journal. Students were to submit their two questions with corresponding answers 12 h prior to class, so that their GTA instructor could formatively assess understanding and direct classroom instruction accordingly. The GTA was also tasked with grading each question and answer and providing feedback according to the level of Bloom's taxonomy it was structured with, its relation to the concepts in the reading, its relevance to real life, and any alternative conceptions the student might have been included in their answers.

**Negotiation of question quality**: Upon entry to class each week, students were randomly assigned to one of six laboratory tables in teams of three or four (24 student maximum per class). Students were provided with 30 min to share their questions about the first half of the assigned reading material and work toward developing a team best question and answer that they could present to their peers using the slides and specimens available to them. Afterward, teams were provided with an additional 30 min to repeat the process in developing a best question and answer concerning the second half of the material.

**Presentation**: Each of the six laboratory tables was labeled 1, 2, 3, 4, 5, or 6, and the seats at each laboratory table were marked 1, 2, 3, or 4 prior to the start of the laboratory. Once teams had developed a best question, a 1, 2, 3, or 4 was drawn from a deck of cards, and the person in that seat at each laboratory table would serve as the presenter for their team. At that point, each team had five minutes to prepare the individual in the seat that was chosen to present their team's question and explain its answer. Once the preparation time ended, one of the six teams was drawn to present. The presenter of that team would communicate their question and explain its answer, while the other team members served in a supportive role by managing the specimens and slides on the overhead, helping create diagrams on the board, etc.

**In-class feedback**: Students both created feedback for and received feedback from their peers and from the GTA as part of their learning experience. Upon completion of the presentation, each team composing the peer audience developed feedback for the presenting team by using a rubric to provide a score, as well as written feedback, for each of the following items:

1. Team correctly categorized question using Bloom's taxonomy.
2. All content material was covered, including slides and specimens.
3. The relevance and importance of the concept were related.
4. Presenter clearly understood concept.

Each team was also free to share feedback with the presenting team. The GTA then provided his or her own feedback to the team, identifying any alternative conceptions that may have been promulgated and pointing out any important concepts that may not have been addressed. The process of presenter selection and presentation, followed by feedback, was undertaken a second time for the question addressing the second half of the assigned reading. Any time remaining after the second presentation was allotted for student review of slides and specimens.

**Assessment**: Students in all laboratory sections were provided with 20 min at the end of each laboratory session to complete the eight-item multiple-choice quiz.

**After-class feedback**: Prior to the following week's laboratory session, the GTA was tasked with reading each team's feedback concerning the presentations and providing his or her own feedback as to the quality of each team's critique using a rubric that consisted of the following items:

1. Team identified any error in question construction using Bloom's taxonomy.
2. Team identified any misrepresented concepts by presenter.
3. Team offered a better/alternative question or enhanced the connection/relevance of the question to the content.
4. Team provided written feedback for all items on the rubric for each presentation.

The GTA's feedback was written directly onto the rubrics each team had filled out and posted the following class for students to review as a means for enhancing their ability to critique science ideas and effectively communicate constructive criticism.

### 2.2.2 Gameful Learning

While all laboratory sections were structured with the inquiry-based learning format described above, two of the eight laboratory sections were each randomly assigned to either control, leaderboard, repeat-testing, or leaderboard with repeat-testing conditions (Table 1). Each of the four GTAs was responsible for instructing two of the eight sections of the laboratory course. However, no GTA taught two laboratory sections assigned to the same condition. Furthermore, the researcher observed each class to ensure that the student-centered instruction that had been agreed upon by GTAs was implemented with fidelity.

**Table 1** Description of conditions, including the course sections, the number of students composing each, and the GTAs responsible for teaching them

| Condition | Description | GTA | Lab sections | N |
|---|---|---|---|---|
| Control (C) | Included neither leaderboard nor repeat-testing | A, B | 3, 5 | 34 |
| Leaderboard (LB) | Included leaderboard but no repeat-testing | A, C | 1, 6 | 36 |
| Repeat-Testing (RT) | Included repeat-testing but not the leaderboard | C, D | 4, 7 | 37 |
| Leaderboard with Repeat-Testing (LBRT) | Included both the leaderboard and repeat-testing | B, D | 2, 8 | 33 |

**Repeat-testing**: Students in repeat-testing and leaderboard with repeat-testing conditions used laptops to log into the university's online grading system to take their quiz. Incorrect responses were accompanied by automated feedback specific to the misconception that likely led to the incorrect answer choice. Students assigned to a condition that included repeat-testing were allowed to repeat the test until satisfied or as was allowed by the 20-minute time limit. Students in control and leaderboard conditions took the same quiz and had 20 min to complete it, but were only allowed one attempt.

**Leaderboard**: Students assigned to a condition that included the leaderboard arrived at the start of each class to a leaderboard (e.g., Fig. 1) that was projected on the front wall by way of a document camera. Each student was anonymously represented on the leaderboard by a pseudonym—one of many genus names (e.g., *Quercus*) from a list of plant species that would be observed in the laboratory over the course of the study. The pseudonyms on the leaderboard were ordered by the total number of points each individual had accrued in previous weeks and accompanied by any badges each individual had earned up to that point. Weekly score totals were calculated by adding each individual's team score with their individual score. Also included on the leaderboard was each individual's rank for the current week, as well as the number of positions up or down the leaderboard that individual had moved from the previous week.

| Genus | Evo Status | Grp Ave | Quiz3 | wk 3 score | Total | wk3 rank | rank^ |
|---|---|---|---|---|---|---|---|
| *Quercus* | | 5 | 23 | 28 | 78.5 | 1 | 2 |
| *Pinus* | | 4 | 22 | 26 | 77 | 2 | 0 |
| *Salvinia* | | 5.5 | 13 | 18.5 | 72.5 | 3 | -2 |
| *Selaginella* | | 5.5 | 21 | 26.5 | 72 | 4 | 1 |
| *Gnetum* | | 6 | 20 | 26 | 71 | 5 | 1 |
| *Lycopodium* | | 5.5 | 24 | 29.5 | 70.5 | 6 | 3 |
| *Eucalyptus* | | 3 | 19 | 22 | 64 | 7 | 1 |
| *Equisetum* | | 5.5 | 12 | 17.5 | 58 | 8 | 2 |
| *Polytrichum* | | 6 | 11 | 17 | 55.5 | 9 | 2 |
| *Isoetes* | | 3 | 6 | 9 | 55.5 | 10 | -6 |
| *Zea* | | 3 | 14 | 17 | 50.5 | 11 | 3 |
| *Azolla* | | 4 | 19 | 23 | 49.5 | 12 | 4 |
| *Sphagnum* | | 5 | 16 | 21 | 46 | 13 | 7 |
| *Capsella* | | 4 | 16 | 20 | 46 | 14 | 4 |
| *Zamia* | | 5.5 | 6 | 11.5 | 45.5 | 15 | -2 |
| *Ginkgo* | | 5.5 | 5 | 10.5 | 45 | 16 | -4 |
| *Marchantia* | | 0 | 0 | 0 | 44.5 | 17 | -10 |
| *Ranunculus* | | 5 | 8 | 13 | 42.5 | 18 | -3 |
| *Lilium* | | 4 | 10 | 14 | 39.5 | 19 | 0 |
| *Helianthus* | | 6 | 7 | 13 | 39 | 20 | -3 |
| *Psilotum* | | 0 | 0 | 0 | 20 | 21 | 0 |

**Fig. 1** Example of the leaderboard posted in a laboratory section assigned to the leaderboard with repeat-testing condition

*Team score* In leaderboard conditions (i.e., leaderboard and leaderboard with repeat-testing), the GTA instructor ranked the quality of each team's presentation feedback so that each individual received a team score ranging from 6, being the highest quality feedback, to 1, being the lowest. This served as the team score portion of each individual's weekly score on the leaderboard.

*Individual score* Individuals in the leaderboard with repeat-testing condition earned points based on the outcome of their quiz relative to their peers. The number of points each individual received was determined first by the number of correct responses on the last quiz he or she attempted, then by the number of attempts it took to earn that score, and finally by the amount of time it took for that individual to complete their last attempt (quiz submission via laptop provided this information for each individual). Thus, in a class of 24, the individual with a perfect score, who completed the quiz with the fewest attempts and in the least amount of time, earned 24 points. Because individuals in the leaderboard-only condition took their quizzes via Scantron and were only allowed one attempt, their quiz scores could not be ranked by time or number of attempts. Scores of individuals in the leaderboard condition ranged from 0 to 8 depending on the number of correct responses on their single quiz attempt.

*Badges* At the end of each week's learning session but prior to the end of class quiz, each individual anonymously voted for their most valuable team member (MVTM) based on their contribution to the development of questions, presentations, discussions, and critique. Each week, MVTM badges appeared as an image next to each recipient's pseudonym on the leaderboard. The images represented adaptations that were significant in the evolution of plants (e.g., vascular system, height, pollination).

*Presentation of the leaderboard* At the start of each class, the researcher presented the leaderboard to the class, congratulating the three highest ranking individuals, those individuals who were selected to receive badges by their teammates, and those with the highest increase in rank from the previous week (for more on the gameful learning design, see Owens, Smith-Walters, & Barlow, in press).

## 2.3 Study Design

A convergent parallel mixed methods approach was used to understand the effects of gameful learning on students' motivation to learn biology. The inclusion of both quantitative and qualitative data enabled triangulation and provided the clearest understanding of the motivational effects of gameful learning. Quantitative data provided generalizable results and enabled the calculation of change in motivation to learn biology from pre- to post-gameful intervention, as well as the parsing out of differential effects of the elements of gamification on students' motivation to learn biology. Thematic analysis of qualitative data enabled elucidation of the quantitative results by providing insight into participants' perspectives concerning the motivational effects of gameful learning that included leaderboards and repeat-testing.

### 2.3.1 Quantitative

Quantitative data were collected by way of the Biology Motivation Questionnaire II (BMQ; Glynn et al., 2011), a 25-item survey that measures motivation to learn science; however, the word *science* was changed to *biology* for each of the survey items so that they were biology-specific. Five items address each: intrinsic motivation, self-efficacy, self-determination, career motivation, and grade motivation—the sum of which was considered to be an individual's motivation to learn biology in the context of plant evolutionary life history. Participants completed pre- and post-BMQs prior to the start of the week 1 laboratory and prior to the start of the week 4 laboratories, respectively. Split-half reliability coefficients with a Spearman–Brown correction for pre- and post-motivation to learn biology were calculated to be 0.75 and 0.78, respectively. Levene's test of homogeneity of variance was violated concerning the pre-intervention data, so nonparametric tests were used.

Quantitative data were analyzed in three ways:

1. Kruskal Wallis tests were conducted to ensure that the four conditions did not significantly differ in terms of ethnic and gender makeup, as well as in pre-intervention motivation to learn biology, where effect size was indicated by Cohen's $w$ $\left(\sqrt{\frac{\chi^2}{N}}\right)$.
2. Change in motivation to learn biology for each condition was determined using Wilcoxon signed-rank tests, where effect size was indicated by $r$ $\left(\frac{Z}{\sqrt{N}}\right)$. In reporting these results, the term *significant* was used to refer to $p$ values $< 0.05$ and *effect size* to indicate practical significance of the change in motivation for each condition using Cohen's (1988) benchmarks of effect for Pearson's $r$ and Cohen's $w$: small ($d = 0.1$), medium ($d = 0.3$), and large ($d = 0.5$).
3. Differential treatment effects were determined using Kruskal Wallis tests on gain scores ($G$ = postscore%—prescore%), where effect size was indicated by Cohen's $w$ and significance was set at $p < 0.05$. Six Mann–Whitney U tests (MWU) were conducted as pairwise post hoc analyses. For these MWU analyses only, Bonferroni adjustments reduced the significance level to $p < 0.008$ to lessen the potential for Type I error.

### 2.3.2 Qualitative

Qualitative data were collected using open-ended questionnaires (OEQs) with the purpose of elucidating the quantitative results of the BMQ. Plant genera were used as pseudonyms for each participant to ensure anonymity. Students whose responses from OEQs were included in the results were represented by their pseudonym, condition, and laboratory section number (e.g., *Azolla*G1).

**Open-ended questionnaires**: An open-ended questionnaire (OEQ) was administered to all participants prior to the start of the week 4 laboratory to solicit their perspectives of the gameful intervention in which they partook (i.e., leaderboard, repeat-testing, or leaderboard with repeat-testing). Participants assigned to the control condition did not experience either element of gameful learning and thus, did not respond to OEQ.

# 3 Results

## 3.1 Quantitative

Kruskal–Wallis tests showed that there were no significant differences among the four conditions in terms of gender $\chi^2(3) = 3.340$, $p = 0.342$, $w = 0.15$, ethnicity $\chi^2(3) = 2.982$, $p = 0.394$, $w = 0.15$, or pre-test motivation $\chi^2(3) = 5.543$, $p = 0.136$, $w = 0.20$.

Wilcoxon signed-rank tests were used to indicate significant pre- to post-change in motivation to learn biology for each condition. Motivation to learn biology significantly decreased in the control ($Z = -3.883$, $p < 0.001$, $r = 0.67$), leaderboard ($Z = -3.630$, $p < 0.001$, $r = 0.61$), and repeat-testing ($Z = -2.421$, $p = 0.015$, $r = 0.40$) conditions. The leaderboard with repeat-testing condition did not significantly change in terms of motivation to learn biology in the context of plant evolutionary life history ($Z = -0.260$, $p = 0.795$, $r = 0.05$) (Table 2 and Fig. 2).

Kruskal–Wallis tests indicated significant main effects for condition $\chi^2(3) = 11.587$, $p = 0.009$, $w = 0.29$. MWU follow-up tests indicated that gain in motivation to learn biology was significantly higher for individuals in the leaderboard with repeat-testing condition that for those in the control ($U = 318.000$, $p = 0.002$, $r = 0.37$) or leaderboard ($U = 350.500$, $p = 0.003$, $r = 0.35$) conditions. Individuals in the leaderboard with repeat-testing condition also had greater gains in motivation

**Table 2** Descriptive statistics, reliability, and change in motivation to learn biology

| | Pre | | | Post | | | | | | |
|---|---|---|---|---|---|---|---|---|---|---|
| | *M* | *SD* | $\alpha$ | *M* | *SD* | $\alpha$ | $M_{gain}$ | *r* | *Z* | *p* |
| Total Motivation | | | 0.75 | | | 0.78 | | | | |
| C | 77.3 | 11.9 | | 70.6 | 13.8 | | −6.7 | 0.67 | −3.883 | <0.001 |
| LB | 79.0 | 13.9 | | 70.3 | 20.0 | | −8.7 | 0.61 | −3.630 | <0.001 |
| RT | 79.9 | 13.0 | | 76.0 | 14.8 | | −3.9 | 0.40 | −2.421 | 0.015 |
| LBRT | 84.5 | 7.2 | | 84.5 | 8.3 | | 0.0 | 0.05 | −0.260 | 0.795 |

*Note* C = Control, LB = Leaderboard, RT = Repeat-Testing, LBRT = Leaderboard with Repeat-Testing

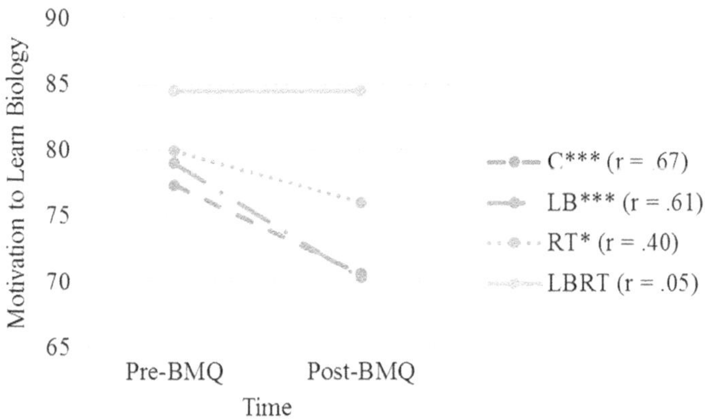


**Fig. 2** Change in motivation to learn biology across conditions according to the Biology Motivation Questionnaire (BMQ, 100 possible points). Effect size for each condition is indicated by *r*. *Note.* C = Control, LB = Leaderboard, RT = Repeat-Testing, LBRT = Leaderboard with Repeat-Testing, BMQ = Biology Motivation Questionnaire. $^{*}p < 0.05$, $^{**}p < 0.01$, $^{***}p < 0.001$

to learn biology than did individuals in the repeat-testing condition, though that difference was only marginally significant ($U = 457.500$, $p = 0.071$, $r = 0.22$). No other pairwise comparisons between conditions were significant.

These results suggest that the inclusion of gaming elements (in this case, the leaderboard with repeat-testing) has the potential to enhance motivation to learn in settings of evolution education—motivation necessary for conceptual change to occur (Dole & Sinatra, 1998; Pugh et al., 2010).

## 3.2 *Qualitative*

When asked how the gameful learning environment affected her motivation and ability to learn about plants, *Anthoceros*GRT2 responded, "It has made me a little more open-minded … I never thought the biology of plants would be so interesting." *Ranunculus*GRT8 agreed. "There has definitely been an increase of interest" (OEQ). Others noted the compelling nature of the gameful learning environment, which "motivated [*Salvinia*GRT8] to do better in this class" (OEQ) and "motivated [*Azolla*GRT2] tremendously and helped [him] actually enjoy learning about the material" (OEQ). *Psilotum*GRT8, too, indicated that gameful learning "definitely made me motivated to be on top of my studying so that I may be able to answer questions in class" (OEQ). Thus, increased motivation to learn about plants was a particularly important outcome of gameful learning considering the nature of the subject under study: "I read through what we were learning before the semester started. Plants was my least favorite, but now I realize I enjoyed the plant portion" (*Zea*GRT8, OEQ). These responses suggest that the gameful learning environment contributed to students' interest, motivation, and engagement in learning about plant evolutionary life history—a subject they might have otherwise found to be disinteresting. The following sections include student perspectives specific to the leaderboard and repeat-testing elements of gamification, respectively.

### 3.2.1 Leaderboard

The leaderboard served to motivate learning in a variety of ways. *Quercus*GRT8 called it "A fun tool to keep the class' interest and learning incentive" (OEQ) while *Helianthus*GRT8 recognized it as a "way to get everyone involved more" (OEQ). Some of that interest and engagement resulted from the leaderboard's facilitation of competition and comparison with others. "The competitive aspect was fun and engaging. I believe it will encourage others to be more prepared by knowing they are in an active environment competing with peers" (*Zea*GRT8, OEQ). *Psilotum*GRT2 agreed. "This part I did like a lot because it brought in competition into this class and made a great course even better" (OEQ).

The leaderboard not only facilitated motivation through individual competition, it also provided encouragement for students to support their peers' learning. For example, *Azolla*G1 felt that the leaderboard "motivated [her] to be prepared for [her] team's sake" by arriving with good questions, as well as an understanding of the concepts so that she could support her team in presenting and critiquing peers (OEQ). *Selaginella*GRT8 also indicated that the leaderboard "keeps you on your toes and wanting to strive to do better as well as help others" (OEQ), recognizing that contributing to one's teammates' mastery of science ideas and the ability to communicate them reflected positively on both the individual and the team in terms of the advancement of science understanding, as well as enhancing one's own position on the leaderboard.

### 3.2.2 Repeat-Testing

Students offered a variety of reasons for why they felt repeat-testing contributed positively to their learning experience. For example, *Eucalyptus*GRT2 indicated that "Retak[ing] [his] quiz as many times as [he] wanted made the material easier to learn" (OEQ). *Equisetem*GRT2 elaborated on how that could be. "Retaking the quiz … gives great feedback on my mistakes and allows me to fix them" (OEQ). Others appreciated that repeat-testing reduced their test anxiety. "It has increased my ability to learn about plants … When I get very nervous over taking a test I tend to forget information I would otherwise be able to recall" (*Azolla*RT4, OEQ). While all individuals loved the result of repeating-tests—increased likelihood of perfect score—many noted appreciation for being able to focus on the material rather than having to memorize and regurgitate just to make an A, such as *Pinus*RT7. "I was able to focus more on understanding patterns and concepts versus memorizing in order to ace one chance at a grade" (OEQ), suggesting the potential of repeat-testing to facilitate deeper processing—known to contribute to conceptual change (Sinatra, 2005).

It should be noted that not all individuals appreciated the leaderboard or repeat-testing. Some individuals did not like the competitive nature brought on by the leaderboard, such as *Isoetes*G1 "Biology is not a competition" (OEQ), though evolutionary fitness could certainly be framed as such. Others were intimidated by the prospect of being ranked lower than their peers were, or, alternatively, felt bad for those whom they ranked higher than. Concerning repeat-testing, while all individuals appreciated the opportunity to enhance their score, some suggested that because they could just guess until they got a perfect score, they did not feel the need to prepare for the quiz.

## 4 Discussion

"Rather than the learner being controlled solely by external factors (i.e., the nature of content or instruction), the leaner plays a significant role in choosing whether to consider alternative points of view" (Sinatra, Southerland, McConaughy, & Demastes, 2003, p. 511). That being the case, putting learners in an environment that motivates their interest in and engagement with content that they might find disinteresting or even contradictory to their current perceptions is crucial in promoting conceptual change (Dole & Sinatra, 1998; Hidi, 1990). In this study, we found that the inclusion of a leaderboard and repeat-testing could significantly contribute to students' motivation to learn about plants—a subject many participants had perceived to be boring and irrelevant, with some even indicating a hatred for plants (see Owens et al., in press). This disdain for plants likely contributed to a general negative trend in motivation to learn across all participants. However, we found that gameful learning can enhance students' motivation to learn about biology in the context of plant evolutionary life history, as well as to their interest in plants and their perspective concerning relevance of plants in their lives—especially when both repeat-testing and leaderboard elements were present. This newfound understanding should aid the ability of teachers and researchers to enhance the motivational nature of evolution instruction that is more conducive to conceptual change.

Considering that the goals one sets provide the purpose and direction behind their motivation to learn (Pintrich & Schunk, 2002), future studies of gameful learning in the context of evolution education and conceptual change would be well served by considering motivation through the lens of achievement goal theory, where gaming elements, such as the leaderboard and repeat-testing, are characterized by the goal messages they send to students. For example, students generally recognize the leaderboard as a performance classroom goal structure in that it sends a message to the learner that outperforming others is important, whereas repeat-testing is often recognized as a mastery classroom goal structure, as it emphasizes the importance of developing competence out of intrinsic interest and a personal desire to enhance one's own understanding (Owens, Smith-Walters, Oslund, & Barlow, 2018). Classroom goal structures, such as those resulting from the inclusion of leaderboard and repeat-testing, can affect the goals behind one's pursuit of competence and the intensity with which he or she pursues it, and thus, their motivation to learn, as well as other important learner characteristics, such as persistence and depth of processing, that may result (Hulleman, Schrager, Bodmann, & Harackiewicz, 2010). To this point, researchers of motivation and conceptual change have suggested that promoting mastery goals in the classroom with goal structures, such as repeat-testing, might be more likely to aid students' inclination to "disregard prior beliefs in order to reach their goal of understanding" (Linnenbrink & Pintrich, 2002, p. 119). However, the results of this study suggest the potential for including performance goal structures, such as the leaderboard, alongside mastery to maximize the motivation to learn that is necessary for conceptual change to occur.

## 5 Limitations and Implications

Participants in this study were enrolled in one of eight sections of a one-credit-hour laboratory course, as well as one of three sections of a corresponding three-credit-hour lecture course—each taught by a different professor. As a result, any assessment of content understanding to accompany the motivation assessment would have been confounded by differential reinforcement of content, both within and among learning conditions, by whichever professor was teaching the lecture section in which each individual was enrolled. Future studies focused on testing the potential for interventions to aid individuals in overcoming motivational barriers to understanding and accepting of evolution would be well served by including measures of motivation and understanding. Additionally, students assigned to one laboratory learning condition may have shared a lecture portion of the course with students from a different laboratory learning condition. It is not clear how knowledge of the different learning conditions may have affected students' motivation.

In reality, the impediment keeping many individuals from understanding and accepting evolution often has little to do with the evolutionary life history of plants. Rather, the idea that humankind evolved from a common ancestor often contradicts closely held values and serves as a barrier for some individuals to consider alternative evidence-based conceptions (Pew Research Center, 2016). While the gameful, inquiry-based learning environment that resulted from the intervention described above was not tested in terms of its ability to motivate conceptual change in the broader context of evolution, it certainly holds potential for doing so.

**Acknowledgements** I am grateful to Angela Barlow and Cindi Smith-Walters for their help in planning and conducting the study, to Dennis Mullen and Eric Oslund for valuable feedback concerning the study design and statistical analyses, to Steve Howard for logistical and moral support, and to the graduate teaching assistants who were responsible for facilitating the inquiry-based instruction.

## References


Bandura, A. (1977). Self-efficacy: Toward a unifying theory of behavioral change. *Psychological Review, 84,* 191–215.

Blumenfeld, P. C., Soloway, E., Marx, R. W., Krajcik, J. S., Guzdial, M., & Palincsar, A. (1991). Motivating project-based learning: Sustaining the doing, supporting the learning. *Educational Psychologist, 26,* 369–398.

Chaiken, S. (1987). The heuristic model of persuasion. In M. P. Zanna, J. M. Olson, & C. P. Herman (Eds.), *Social influence: The Ontario symposium* (Vol. 5, pp. 3–39). Hillsdale: Lawrence Erlbaum Associates.

Cohen, J. (1988). *Statistical power analysis for the behavioral sciences.* New York: Routledge Academic.

Cohen, A. R., Stotland, E., & Wolfe, D. M. (1955). An experimental investigation of need for cognition. *The Journal of Abnormal and Social Psychology, 51,* 291.

Deci, E. L., & Ryan, R. M. (1985). *Intrinsic motivation and self-determination in human behavior.* New York: Plenum.

Deterding, S., Dixon, D., Khaled, R., & Nacke, L. (2011). From game design elements to gamefulness: Defining "gamification." In: *Proceedings of the 15th International Academic MindTrek Conference: Envisioning Future Media Environments.* New York: Association for Computing Machinery.

Dole, J. A., & Sinatra, G. M. (1998). Reconceptualizing change in the cognitive construction of knowledge. *Educational Psychologist, 33,* 109–128.

Elliot, A. J., & Murayama, K. (2008). On the measurement of achievement goals: Critique, illustration, and application. *Journal of Educational Psychology, 100,* 613.

Gallup, G. (2017). American beliefs: Evolution vs. Bible's explanation of human origins. http://www.gallup.com/poll/21811/americanbeliefs-evolution-vs-bibles-explanation-human-origins.aspx. Accessed December 17, 2017.

Glynn, S. M., Brickman, P., Armstrong, N., & Taasoobshirazi, G. (2011). Science motivation questionnaire II: Validation with science majors and non-science majors. *Journal of Research in Science Teaching, 48,* 1159–1176.

Hidi, S. (1990). Interest and its contribution as a mental resource for learning. *Review of Educational Research, 60,* 549–571.

Hulleman, C. S., Schrager, S. M., Bodmann, S. M., & Harackiewicz, J. M. (2010). A meta-analytic review of achievement goal measures: Different labels for the same constructs or different constructs with similar labels? *Psychological Bulletin, 136,* 422–449.

Kampourakis, D. (2014). *Understanding evolution.* Cambridge: Cambridge University Press.

Linnenbrink, E. A., & Pintrich, P. R. (2002). The role of motivational beliefs in conceptual change. In M. Limon & L. Mason (Eds.), *Reconsidering conceptual change: Issues in theory and practice* (pp. 115–135). Dordrecht: Kluwer Academic.

Miller, J. D., Scott, E. C., & Okamoto, S. (2006). Public acceptance of evolution. *Science, 313,* 765–766.

NRC. (2000). *Inquiry and the national science education standards.* Washington, D.C.: National Academies Press.

NRC. (2011). *A framework for K–12 science education: Practices, crosscutting concepts, and core ideas.* Washington, D.C.: National Academies Press.

Owens, D. C. (2017). Issues with tissues: A tale of gameful learning in an introductory undergraduate biology laboratory course. *Journal of College Science Teaching, 47,* 38–42.

Owens, D. C., Smith-Walters, C., & Barlow, A. T. (In Press). Enhancing motivation to learn in a biology laboratory course through gaming: A design case. *International Journal of Designs for Learning: Special Issue on Games for Learning.*

Owens, D. C., Sadler, T. D., Barlow, A. T., Smith-Walters, C. (In Press). Student motivation from and resistance to active learning rooted in essential science practices. *Research in Science Education*.
Owens, D. C., Smith-Walters, C., Oslund, E. L., Barlow, A. T. (2018). *The motivational nature of gameful learning environments*. Paper presented at the American Educational Research Association Annual Conference, New York, April 13–17, 2018.
Pew Research Center. (2016). *Israel's Religiously Divided Society*. http://www.pewforum.org/files/2016/03/Israel-Survey-Full-Report.pdf. Accessed January 10, 2018.
Piaget, J. (1970). Piaget's theory. In Mussen, P. H. (Ed.), *Carmichael's manual of child psychology* (Vol. 1, 3rd edn., pp 703–732). New York: Wiley.
Pintrich, P. R., Marx, R. W., & Boyle, R. A. (1993). Beyond cold conceptual change: The role of motivational beliefs and classroom contextual factors in the process of conceptual change. *Review of Educational Research, 63,* 167–199.
Pintrich, P. R., & Schrauben, B. (1992). Students' motivational beliefs and their cognitive engagement in classroom academic tasks. In D. Schunk & J. Meese (Eds.), *Student perceptions in the classroom* (pp. 149–183). Hillsdale: Lawrence Erlbaum Associates.
Pintrich, P., & Schunk, D. (2002). *Motivation in education: Theory, research, and applications* (2nd ed.). Upper Saddle River: Merrill/Prentice Hall.
Posner, G. J., Strike, K. A., Hewson, P. W., & Gertzog, W. A. (1982). Accommodation of a scientific conception: Toward a theory of conceptual change. *Science Education, 66,* 211–227.
Pugh, K. J. (2002). Teaching for idea-based, transformative experiences in science: An investigation of the effectiveness of two instructional elements. *Teachers College Record, 104,* 1101–1137.
Pugh, K. J. (2011). Transformative experience: An integrative construct in the spirit of Deweyan pragmatism. *Educational Psychologist, 46,* 107–121.
Pugh, K. J., Linnenbrink-Garcia, L., Koskey, K. L., Stewart, V. C., & Manzey, C. (2010). Motivation, learning, and transformative experience: A study of deep engagement in science. *Science Education, 94,* 1–28.
Resnick, L. B., Levine, J. M., & Teasley, S. D. (Eds.). (1991). *Perspectives on socially shared cognition*. Washington, D.C.: American Psychological Association.
Sinatra, G. M. (2005). The "warming trend" in conceptual change research: The legacy of Paul R. Pintrich. *Educational Psychologist, 40,* 107–115.
Sinatra, G. M., & Pintrich, P. R. (2003). *Intentional conceptual change*. Mahwah: Lawrence Erlbaum Associates.
Sinatra, G. M., Southerland, S. A., McConaughy, F., & Demastes, J. (2003). Intentions and beliefs in students' understanding and acceptance of biological evolution. *Journal of Research in Science Teaching, 40,* 510–528.
Strike, K. A., & Posner, G. J. (1992). A revisionist theory of conceptual change. In R. A. Duschl & R. Hamilton (Eds.), *Philosophy of science, cognitive psychology, and educational theory and practice* (pp. 142–176). Albany: State University of New York Press.
Toulmin, S. (1972). *Human understanding*. Princeton: Princeton University Press.
Vodopich, D. S., & Moore, R. (2014). *Biology laboratory manual* (10th ed.). New York: McGraw Hill Education.
Vosniadou, S., & Brewer, W. F. (1987). Theories of knowledge restructuring in development. *Review of Educational Research, 57,* 51–67.

**David C. Owens** is Assistant Professor of Science Education at Georgia Southern University. After serving as a high-school biology teacher for five years with Memphis City Schools, he earned a master's degree in aquatic ecology and a PhD in mathematics and science education with a concentration in biology. His research interests include understanding the characteristics of learning environments that make science relevant, foster conceptual change, and enhance learners' motivation and ability to recognize and evaluate science and non-science considerations when making decisions in their everyday lives. He also serves as an instructor of outdoor education and character development with Outward Bound in the Pacific Northwest, Appalachia, and the Everglades.

# Using Human Examples to Teach Evolution to High School Students: Increasing Understanding and Decreasing Cognitive Biases and Misconceptions

**Briana Pobiner, William A. Watson, Paul M. Beardsley and Constance M. Bertka**

## 1 Introduction

The science education community broadly accepts that understanding evolution is a critical aspect of scientific literacy, as is evident by the prominence of evolution in the US education system as a 'big idea' or 'core concept' in the Next Generation Science Standards (NGSS, NGSS Lead States, 2013), the Advanced Placement (AP) Biology Curriculum Framework (The College Board, 2011), and AAAS's Vision and Change for Undergraduate Biology (AAAS, 2011) as well as the national curricula in other English-speaking countries (e.g., England (Department of Education, 2014), Australia (Australian Curriculum, Assessment and Reporting Authority [ACARA], 2014). Researchers and educators also broadly recognize that there are many barriers to students learning about evolution, including many teachers avoiding teaching about evolution altogether (e.g., Berkman, Pacheco, & Plutzer, 2008; Pobinerm, 2016). When teachers do teach the content, students often have cognitive biases and misconceptions, especially in the realm of a mechanism of evolution (natural

B. Pobiner (✉)
Human Origins Program, National Museum of Natural History, Smithsonian Institution, Washington, DC, USA
e-mail: pobinerb@si.edu

W. A. Watson
Learning Research Across the Divide, Haddon Heights, NJ, USA
e-mail: billwatson3@gmail.com

P. M. Beardsley
Department of Biological Sciences, Cal Poly Pomona, Pomona, CA, USA
e-mail: pmbeardsley@cpp.edu

C. M. Bertka
Science and Society Resources, LLC, Potomac, MD, USA
e-mail: cbertka@scienceandsocietyresources.com



U. Harms and M. J. Reiss (eds.), *Evolution Education Re-considered*,
https://doi.org/10.1007/978-3-030-14698-6_11

selection), common ancestry, deep time, and 'tree-thinking.' The goals of many life science educators are both to bring student understanding more in line with scientific ideas about evolution and to reduce the frequency of student misconceptions and the use of biased ways of thinking about change over time and shared ancestry.

At present, there is no consensus in the evolution education community about the most promising curricular or pedagogical strategies to use to achieve these two goals, especially at the pre-collegiate level. Reviews of the evolution education literature (e.g., Beardsley, Bloom, & Wise, 2012; Pobiner, 2016) suggest that the pedagogical approaches with the most promise are those that use strategies for conceptual change, and attend to the relevance of understanding evolution (e.g., Pugh, Linnenbrink-Garcia, Koskey, Stewart, & Manzey, 2010; Beardsley et al., 2011; Heddy & Sinatra, 2013). Evolution educators are also still exploring the impact of including examples of evolution in humans. Resistance to learning about human evolution is higher than about evolution in other organisms (e.g., Werth, 2009), but recent summaries of different sources of evidence suggest that using human examples holds promise (Pobiner, 2016).

Moreover, there is wide variation in how teachers choose to address student resistance to learning about evolution and tensions between a religious and scientific worldview. Evolution educators generally suggest that the most promising approaches should include acknowledging, respecting, and being sensitive to students' beliefs. Teachers are encouraged to negotiate this conflict, rather than ignoring it or exacerbating it (e.g., Brem, Ranney, & Schindel, 2003; Sinatra, Southerland, Mcconaughy, & Demastes, 2003; Werth, 2012; Bramschreiber, 2013; Pobiner, 2016). At present, however, few studies have quantitatively explored the impact of using teaching strategies that acknowledge the cultural controversy on students' understanding of natural selection below the collegiate level.

In this chapter, we provide evidence for the impact of using a constructivist, guided inquiry pedagogical approach using human evolution case studies to teach AP Biology high school students about natural selection in an attempt to better understand the most promising approaches to support teachers in helping students learn about evolution in general, and natural selection in particular, and overcome common cognitive biases and misconceptions. Importantly, we also report the effect of these curricular materials when used in tandem—or not—with teaching strategies that explicitly acknowledge the cultural controversy around evolution.

### *1.1 Common Cognitive Biases and Evolution Misconceptions Among Students*

There are three major categories of cognitive biases to learning about natural selection that begin in preschoolers which we will outline here.

**Essentialism** is the belief that individuals and groups have an essential nature that allows them to be placed into categories or kinds with sharp, immutable boundaries (Nehm et al., 2010). In this type of reasoning, membership in a category leads to observable properties which stem from an unobservable, unchangeable core 'essence' that is transmitted from parent to offspring.

**Intentionality** assumes that events are caused by an intentional mental agent and are purposeful, goal-directed, or progressive, including the idea that evolution is progressing toward an ideal (Evans, 2001; Gregory, 2009; Nettle, 2010).

**Teleology** assumes that the characteristics and actions of entities or groups have a goal or are inevitable (Nehm et al., 2010) and that aspects of an object's or organism's form are explained by their ultimate purpose. Teleology includes the beliefs that the traits organisms currently possess perform roles or functions that aid survival (that they 'need' these traits) and that natural phenomena are intentionally designed or created for a purposeful goal (Jensen & Finley, 1995; Kelemen, 2012). Such need-based rationales tend to lead to the conclusion that major changes occur within an individual's lifetime and are heritable (Kelemen, 2012).

In addition to cognitive biases, misconceptions, defined as inaccurate ideas that can predate or emerge from instruction (Andrews et al., 2012), and prior beliefs may impede a correct understanding of evolution and the construction of knowledge in the biology classroom (Sinatra, Brem, & Evans, 2008; Smith, 2010). Unfortunately, there is a 'strikingly high prevalence of misconceptions about evolution' among students of all levels (Gregory, 2009: 163). These misconceptions encompass student understanding of a process of evolution, namely natural selection, common ancestry, deep time, and 'tree-thinking' (see Smith, 2010; Werth, 2012; Pobiner, 2016 for more extensive lists of misconceptions about evolution).

The current study focuses on misconceptions associated with students' understanding of natural selection. A prominent assessment for student understanding of natural selection is the open response instrument called Assessing Contextual Reasoning about Natural Selection, or ACORNS (Nehm et al., 2012) (used in this study; see methods) which assesses the three cognitive biases previously described and six misconceptions associated with the following terms: 'pressure,' 'adapt,' 'need,' 'must,' 'use,' and 'energy' (see Table 5 for more details on these terms).

## *1.2 How Do Different Types of Evolution Instruction Affect Student Understanding of Evolution and the Frequency of Cognitive Biases and Misconceptions?*

Evolution educators are still working to build a consensus about the most promising pedagogical approaches to help students overcome cognitive biases and misconceptions. Beardsley et al. (2012) summarized studies of diverse curricular interventions with students, including software-based instruction to problem-based learning, argumentation-eliciting treatments, and targeting-specific misconceptions. Studies published since that review mostly support these conclusions; for example, Andrews, Kalinowski, and Leonard (2011) examined student learning of natural selection in college from 33 instructors at 28 institutions in 22 states in the USA and showed that most learning gains were modest, but two factors associated with misconceptions were positively related to the gains in understanding ('explaining why misconceptions are incorrect' and 'using active-learning exercises to make a substantial effort toward changing misconceptions'), as was student interest in the biology course.

Fewer studies of misconceptions among high school students have been undertaken. In a study with teachers in Maryland and Illinois, all of the six teachers thought individual versus population-level thinking was the main reason for the low understanding of evolution among their students (Hermann, 2013). One recent study of high school students found that they left an introductory biology course with greater numbers of evolution misconceptions than at the beginning of the course, despite an increase in their confidence in their knowledge of evolution (Yates & Marek, 2014). Results of studies conducted with college students are hopeful for identifying effective pedagogical approaches for supporting high school students' understanding of evolution and helping them overcome cognitive biases and misconceptions, but the gap in the research on these and other approaches with high school students indicates that additional work is needed.

Despite potential cultural controversies and a lower acceptance of evolution in humans than non-humans among at least some college students (Ranney & Thanukos, 2011), a growing body of the literature suggests that using human examples may help students learn core evolutionary concepts (summarized in Pobiner, 2016). College students prefer science courses in which human examples are included in evolution instruction along with non-human examples (Paz-y-Miño & Espinosa, 2009) and some studies have found an increase in understanding and/or acceptance of evolution in college students when including human examples (Wilson, 2005; Werth, 2009; Nettle, 2010; Andrews, Leonard, Colgrove, & Kalinowski, 2011; Borgerding et al., 2015). Additionally, student misconceptions may persist because of non-scientific worldviews (Hermann, 2012), and recent studies suggest that explicitly addressing students' beliefs with respect and sensitivity, with the goal of creating a classroom environment conducive to learning about evolution, is the best strategy for students for whom the subject is controversial (Sinatra et al., 2003; Verhey, 2005; Smith, 2010; Hermann, 2012; Bramschreiber, 2013)—with a possible goal of helping students reconcile their personal beliefs with scientific understanding (Anderson, 2007). Therefore, we think a promising approach to engage students with evolution content includes (1) using human examples and (2) explicitly discussing the relationship between evolution and students' beliefs.

# 2 Methods

## *2.1 The Teaching Evolution through Human Examples (TEtHE) Project*

Teaching Evolution through Human Examples (TEtHE) was a three-year exploratory research and development project funded by the National Science Foundation of the USA. The overall goals of the project were to develop, field-test, and assess the effectiveness of two main components related to the above suggestions about how best to engage students in learning evolution. The first component was four mini-units that use case studies of human evolution to address specific core evolutionary concepts included in the high school AP Biology curriculum. Using a constructivist approach, the units explicitly elicit a range of misconceptions as well as cognitive biases in the broad categories of essentialism, intentionality, and teleology, and then provide opportunities for students to reflect on their prior ideas and gain experience with ways of describing change over time that are scientifically accurate. The second component was two classroom activities that use Cultural and Religious Sensitivity (CRS) Teaching Strategies to create a comfortable classroom environment for learning about human evolution, which also included eliciting possible misconceptions and cognitive biases. This chapter reports on a subset of the results from the national field test for three of the four mini-curriculum units and CRS activities related to addressing core evolution understanding and misconceptions in human and non-human evolution contexts building on two initial publications from the TEtHE project (Pobiner et al., 2018 and Bertka et al., 2019).

TEtHE was primarily a curriculum development project, which led to the decision to identify teachers and students in the 'best-case scenario' for piloting and assessing baseline impact of the mini-units and CRS activities. The sample therefore reflects the intentional selection of well-qualified teachers and their AP Biology students, who are generally more motivated to learn and good at reflecting on their own learning. Overall design for collecting data from the national field test is therefore subject to self-selection bias and the limitations of a non-random and non-systematically selected sample and should be considered exploratory, as our sample did not capture the likely increased variability that would result from a more randomly selected student sample from across the USA. Within these limits of explanatory power, we present the compelling results of our analyses of student understanding of natural selection, cognitive biases, and misconceptions about evolution and encourage additional future research to rigorously test hypotheses that these initial studies elucidate.

## *2.2 Research Questions*

Data from the TEtHE study were analyzed in multiple ways within and between mini-units, using combinations of items with high validity and reliability to address

**Table 1** Research questions for the TEtHE project results reported here

| | |
|---|---|
| Research question 1 | Are patterns of changes in student understanding of key concepts, cognitive biases, and misconceptions about natural selection the same between a non-human and human context? |
| Research question 2 | Are there posttest differences in student understanding of key concepts, cognitive biases, and misconceptions about natural selection between a non-human and human context? |
| Research question 3 | Are changes in student understanding of key concepts, cognitive biases, and misconceptions about natural selection the same when teachers use CRS lessons and when they do not? |

underlying constructs of understanding, cognitive biases, and misconceptions. Our overall findings indicate that student understanding of evolution increases on a statistically significant level from the pretest to the posttest for two of the three mini-units, *Adaptation to Altitude* and *Malaria*, with effect sizes ranging from 0.26 to 1.32 (Pobiner et al., 2018). For this chapter, we have taken a simpler approach through which we seek to describe patterns in student understanding of critical components of evolution, cognitive biases, and misconceptions about evolution in the two contexts (mouse and human) and between students whose teachers used the CRS and those who did not. Research questions for this subset of the TEtHE project are outlined in Table 1.

## 2.3 *Sample*

### 2.3.1 Participating Teachers

Teachers were selected from a pool of teachers recruited by email and word of mouth by project personnel and their colleagues in the evolution education field. Selected teachers either (a) self-identified as interested in the project or (b) were identified by project staff as teaching in schools with student demographics of interest to the project and the funder, i.e., those traditionally underrepresented in Science, Technology, Engineering, and Technology (STEM) careers. Table 2 includes demographic and socioeconomic information for ten schools in eight states (California, Colorado, Connecticut, Maryland, New Jersey, New York, Utah, and Virginia) at which the teachers implemented these mini-units. More details on implementation can be found in Pobiner et al. (2018).

### 2.3.2 Participating Students

Participating students were high school juniors and seniors who were qualified by their schools' criteria to participate in an AP Biology class. Attempts were made in

**Table 2** Summary of school data and implementation characteristics for students of each teacher who taught using the altitude, malaria, or skin color mini-units

| Mini-unit | School type | Low SES (%) | URM (%) | Fidelity: implement | Fidelity: assess | CRS | *n* |
|---|---|---|---|---|---|---|---|
| Altitude | Public | 4 | 12 | High | High | None | 51 |
| Altitude | Public | 8 | 8 | Low | Low | None | 39 |
| Altitude | Private | 4 | 10 | High | High | 1 | 18 |
| Altitude | Public | 13 | 11 | High | Moderate | 1 | 52 |
| Altitude | Public | 20 | 13 | Unknown | Low | 2 | 28 |
| Malaria | Public | 11 | 33 | High | Low | 1 | 24 |
| Malaria | Public | 30 | 52 | High | High | None | 43 |
| Skin color | Public | 22 | 30 | High | High | None | 23 |
| Skin color | Public | 82 | 81 | High | High | 1 | 15 |
| Skin color | Private | 3 | 10 | High | High | 2 | 11 |

*Low SES* indicates the percentage of students at the school who qualify for free or reduced price lunch
*Underrepresented minority* (*URM*) indicates the percentage of students at the school who identify as African American or Hispanic
*Time frame* indicates the teacher-reported month in which the supplement was taught
*Fidelity: Implement* indicates the extent to which teacher reports indicate that the supplement was taught as intended by the developer
*Fidelity: Assess* indicates the extent to which the timing of the assessment administration occurred as directed by project staff
*CRS* indicates whether CRS activity 1 or 2, respectively, was used in the classroom
*n* indicates the number of students in each class from whom data were collected

the selection of participating teachers to identify a student sample that at minimum is demographically representative of the AP Biology classes taught by highly qualified teachers nationwide, both in STEM careers and in AP Biology classes.

## 2.4 Interventions

### 2.4.1 Curriculum Mini-units

The project team developed four curriculum mini-units that focused on using human examples or case studies to teach core evolutionary content. This chapter describes results from the three that focused on natural selection, which are summarized in Table 3. Each unit includes four or five lessons, which were designed to be implemented over five to nine days (depending on whether the full or condensed version is used) and integrated into each teacher's larger instructional sequence for evolution in the AP curriculum. Teachers were asked to implement the lessons 'as intended,' meaning that they were asked to teach all the lessons without modification in the

**Table 3** Titles and descriptions of the three mini-units focused on natural selection in modern humans used in the study

| Title | Description |
|---|---|
| Adaptation to altitude | Students learn how to devise an experiment to test the difference between acclimation and adaptation, investigate how scientific arguments show support for natural selection in Tibetans, design an investigation using a simulation based on the Hardy–Weinberg principle to explore mechanisms of evolution, and devise a test to investigate whether or not other populations of people have adapted to living at high altitudes |
| Evolution of human skin color | Students examine evidence for the relationship between ultraviolet (UV) light and melanin in other animals, investigate the genetic basis for constitutive skin color in humans, learn to test for natural selection in mouse fur color, investigate how interactions between UV and skin color in humans can affect fitness, and explore data on migrations and gene frequency to show convergent evolution of skin color |
| Malaria | Students examine evidence to compare four different explanations for why many malarial parasites are resistant to antimalarial drugs, investigate how scientific arguments using G6PD data show support for natural selection in humans, and apply their understanding to other genes whose allele frequencies have changed in response to malaria |

sequence in which they were provided within recommended time duration, which they did to varying degrees (see Table 3).

### 2.4.2 Cultural and Religious Sensitivity (CRS) Teaching Strategies Resource

The purpose of the CRS resource is to encourage and equip high school teachers to help students manage any tension they may experience between a scientific study of evolution and their religious and cultural beliefs, and create a classroom environment that supports both an increased understanding of the nature of science and a scientific understanding of evolution. It is not meant to specifically resolve any conflict students may see between their personal worldviews and the scientific account of human evolution, but to help create a nonthreatening classroom environment.

The resource includes background information for teachers on: the nature of science as pertinent to managing a conflict between science and cultural or religious beliefs; the range of creationists' views, from those that are anti-evolution in nature to those that are supportive of a scientific understanding of evolution; the variety of possible relationships between science and religion, including examples of how individuals accommodate evolution and religion; and the historical context and background on legal cases dealing with the teaching of evolution. It also includes two activities to engage students in directed classroom discussions for 50–75 min

**Table 4** Titles, timing, classroom setting, and descriptions of the two CRS classroom activities used in the study

| | Activity 1 | Activity 2 |
|---|---|---|
| Title | Directed discussions: 'why study evolution?' | A historical role play: 'how do people think about evolutionary theory?' |
| Timing | Just prior to implementing the mini-unit on evolution | After implementing the mini-unit on evolution (for reinforcement) |
| Classroom setting | Teachers are aware that many of their students have been exposed to only negative and/or mistaken notions of evolutionary theory | Teachers believe that anti-evolutionism is a minority or a nonexistent viewpoint |
| Description | Through three in-class exercises that include small group and class discussions, students reflect on how science as a way of knowing differs from other ways of knowing about the world, classify a collection of statements by individuals and religious groups to illustrate a range of possible relations between science and religious or cultural beliefs, and identify the type of data scientists are collecting in example studies of biologists using evolutionary theory as a tool to solve problems and make testable hypothesis. Before the class meets, students complete an assignment that provides insight into their current knowledge and concerns about evolution | Students are assigned one of eight historical characters and work in groups to envision how their character would reply to questions about Darwin's theory of evolution. Paired character groups work together to draft both a historical and a modern-day response to concerns about evolution highlighted by one of their characters |

(Table 4). The classroom activities use a procedural neutrality approach (Hermann, 2008) in which information about the cultural controversy surrounding evolution and different points of view about this controversy are elicited from students and from resource material. The teacher does not make a value judgment about these views, but help students come to a correct understanding of the nature of science. Teachers could opt into using either of the two (but not both) classroom activities.

## 2.5 Assessments

The Assessing Contextual Reasoning about Natural Selection instrument (ACORNS; Nehm et al., 2010, 2012) is intended to assess increased understanding of evolution concepts, specifically natural selection. It is a short-answer diagnostic test that was

**Table 5** Key concepts, cognitive biases, and misconceptions scored in the ACORNS questions including brief definitions or descriptions

| | | Definition/description or phrases used |
|---|---|---|
| Key concepts | Variation | Presence of variation caused by mutations, genes, or changes in DNA |
| | Heritability | Genes are passed on to the next generation, production of offspring with the same traits, inheritance, heritable |
| | Competition | Competition, struggle |
| | Hyperfecundity | Overproduction of offspring, more individuals born than can survive |
| | Resource limitations | Resources, predation (predator or prey) |
| | Differential survival | Greater or higher survival, others died off, more fit, advantage of a trait, reproduce more, trait/gene selected for or favored, sexual selection |
| | Frequency/distribution | Generational changes in the distribution or frequency of variation, over time, gene or trait became dominant or more common |
| Cognitive biases | Essentialism | Change at a level higher than the individual, assumes no within species variability |
| | Intentionality | Explanation contains mental verb; agent of mental verb is evolving species or nature |
| | Teleology | Organisms change because they 'need' to |
| Misconceptions | Pressure | Pressure (by an external force) or lack thereof causes a mutation or trait to occur |
| | Adapt | Individuals change to adapt to their environment |
| | Need | Need of an organism causing a mutation or trait to occur so it could survive or reproduce and does not include process |
| | Must | Desire or preference caused a change |
| | Use | Traits changed because they were being intensively used or no longer being used |
| | Energy | Energy/resources were reallocated to another trait for better use |

designed with a scoring rubric that standardizes student responses across different contextual variables for evolution (e.g., gain vs. loss of traits, plants vs. animals, within vs. between species differences). The ACORNS scoring instructions and rubrics allow raters to score student responses in seven key features of understanding evolution, three cognitive biases, and six misconceptions (see Table 5, which uses adapted descriptions and examples in Nehm et al., 2010). A score of '1' indicates the presence of a key concept, cognitive bias, or misconception, and a score of '0' indicates the absence of that key concept, cognitive bias, or misconception.

**Table 6** Two ACORNS questions used in the TEtHE study

| | |
|---|---|
| Human evolution question | How would biologists explain how individual people are alive today who can digest lactose originated within a population of people who were all lactose intolerant? |
| Non-human evolution (mouse) question | How would biologists explain how some individuals of a mouse species that have claws originated within a population of a mouse species that lacked claws? |

We used one human-based and one non-human-based question, both of which focused on trait gain (see Table 6). All students answered both questions: first the human context question, which was created by the TEtHE research team, and then the non-human (mouse) context question, which is directly from Nehm et al. (2012). This question was chosen as it also includes gaining a trait and mice is familiar to students. The same versions of the ACORNS instrument were given pre- and post-instruction with the mini-unit. Teachers were asked to distribute the ACORNS as a pretest the day before implementation of the mini-unit and as a posttest the day after implementation.

Each ACORNS instrument was assessed for seven knowledge attributes, three cognitive biases, and six misconceptions (see Table 5), by one of the teams of three raters including two of the authors (Pobiner and Watson) blind to whether or not any assessment was a pretest or a posttest. Inter-rater reliability ranged from 0.71 to 0.99 using a simple comparison of percent agreement across items and raters. The agreement for some items may be skewed by the relatively low percentage of students showing evidence of understanding or misconceptions for those items.

## 2.6 Analyses

All analyses were conducted using a combined dataset that included students who experienced any of the mini-units for whom we had both pretest and posttest ACORNS data ($n = 320$). No student experienced more than one mini-unit. All students took the same ACORNS assessment, as the target concepts and standards addressed by each mini-unit were identical. Combining students across mini-units also helped to mitigate against the results in any condition being based too heavily on any one teacher's abilities or methods.

Analyses were conducted on an item-by-item basis for each research question, resulting in 80 total comparisons, as described below. We recognize that some of the effects reported therefore may be due to chance rather than to the impact of the interventions. We emphasize that their inclusion here is intended to illuminate potential overall patterns and identify compelling areas for future research.

### 2.6.1 Research Questions 1 and 2

To identify potential patterns of changes in understanding of key evolution concepts, cognitive biases, and misconceptions about natural selection across mouse and human contexts, we compared the number of scores of 1 for each item at the pretest to the number of scores of 1 at the posttest, for both the mouse and the human contexts. This was done by conducting a series of Wilcoxon nonparametric significance tests for two paired variables.

To identify potential posttest differences in understanding of key evolution concepts, cognitive biases, and misconceptions about natural selection across mouse and human contexts, we compared the number of scores of 1 at posttest for responses in the human context to the number of scores of 1 at posttest for responses in the mouse context. We again conducted a series of Wilcoxon nonparametric significance tests for two paired variables because the data represented different responses from the same students, not assignment to different conditions. For research questions 1 and 2, because each item was rated as either a 1 or a 0, the overall effect was to compare the percentage of correct responses at the pretest to the percentage of correct responses at the posttest.

### 2.6.2 Research Question 3

To identify potential changes in student understanding of key evolution concepts, cognitive biases, and misconceptions about evolution the same when teachers use CRS lessons and when they do not, we first calculated a pretest–posttest gain score for each item for each student. Resulting scores were either −1 (scored 1 at pretest and 0 at posttest), 0 (no change), or 1 (scored 0 at pretest and 1 at posttest). We then conducted a series of Mann–Whitney nonparametric significance tests for two independent groups, with whether or not a student experienced the CRS as the independent variable.

# 3 Results

## *3.1 Research Question 1*

Patterns of pretest–posttest gain were found to be similar across the mouse and the human contexts, with significant increases in variation, heritability, differential survival, and frequency/distribution in both contexts and significant decreases in teleology and adapt. Both contexts showed a trend toward decreased presence of cognitive bias and misconceptions at the posttest than at the pretest, with some variation in the specific biases and misconceptions. Table 7 presents the results of the significance tests in the mouse and human contexts, respectively.

**Table 7** Pretest–posttest differences by item within the mouse and human contexts; statistically significant differences are in bold

| | | Mouse context | | | Human context | | |
|---|---|---|---|---|---|---|---|
| | | Pretest | Posttest | Z | Pretest | Posttest | Z |
| Key concepts | Variation | 0.71 | 0.87 | **−5.378**** | 0.71 | 0.84 | −4.558** |
| | Heritability | 0.49 | 0.66 | **−4.596**** | 0.45 | 0.64 | **−5.388**** |
| | Competition | 0.03 | 0.03 | 0.000 | 0.02 | 0.01 | −0.632 |
| | Hyperfecundity | 0.00 | 0.00 | 0.000 | 0.00 | 0.00 | 0.000 |
| | Resource limitations | 0.16 | 0.20 | −1.434 | 0.09 | 0.11 | −1.050 |
| | Differential survival | 0.58 | 0.71 | **−3.563**** | 0.34 | 0.55 | **−5.892**** |
| | Frequency/distribution | 0.22 | 0.35 | **−3.866**** | 0.18 | 0.33 | **−5.004**** |
| Cognitive biases | Essentialism | 0.12 | 0.04 | **−3.429**** | 0.08 | 0.04 | −1.667 |
| | Intentionality | 0.04 | 0.03 | −0.626 | 0.03 | 0.02 | −0.535 |
| | Teleology | 0.17 | 0.07 | **−4.160**** | 0.06 | 0.03 | **−2.041*** |
| Misconceptions | Pressure | 0.04 | 0.04 | 0.000 | 0.05 | 0.01 | **−3.153*** |
| | Adapt | 0.12 | 0.07 | **−2.359*** | 0.13 | 0.05 | **−3.501**** |
| | Need | 0.16 | 0.10 | **−2.496*** | 0.04 | 0.03 | −0.943 |
| | Must | 0.09 | 0.06 | −1.667 | 0.03 | 0.02 | −1.069 |
| | Use | 0.01 | 0.01 | −1.000 | 0.06 | 0.03 | **−2.041*** |
| | Energy | 0.00 | 0.00 | 0.000 | 0.00 | 0.00 | 0.000 |

*$p < 0.05$
**$p < 0.001$

**Table 8** Posttest comparisons by item: human context versus mouse context; statistically, significant differences are in bold

| | | Human | Mouse | *Z* |
|---|---|---|---|---|
| Key concepts | Variation | 0.84 | 0.87 | −1.474 |
| | Heritability | 0.64 | 0.66 | −0.577 |
| | Competition | 0.01 | 0.03 | −1.265 |
| | Hyperfecundity | 0.00 | 0.00 | 0.000 |
| | Resource limitations | 0.11 | 0.20 | **−3.414**** |
| | Differential survival | 0.55 | 0.71 | **−5.392**** |
| | Frequency/distribution | 0.33 | 0.35 | −1.068 |
| Cognitive biases | Essentialism | 0.04 | 0.04 | 0.000 |
| | Intentionality | 0.02 | 0.03 | −1.291 |
| | Teleology | 0.03 | 0.07 | **−2.985*** |
| Misconceptions | Pressure | 0.01 | 0.04 | **−3.317**** |
| | Adapt | 0.05 | 0.07 | −1.225 |
| | Need | 0.03 | 0.10 | **−4.116**** |
| | Must | 0.02 | 0.06 | **−2.982*** |
| | Use | 0.03 | 0.01 | −1.897 |
| | Energy | 0.00 | 0.00 | 0.000 |

*$p < 0.05$
**$p < 0.001$

## 3.2 Research Question 2

There were six significant differences between the human and mouse contexts, with students showing evidence of understanding resource limitations and differential survival more frequently in the mouse context than the human context, but greater frequency of teleology cognitive bias and pressure, need, and must misconceptions in the mouse context. Table 8 presents the results of the significance tests comparing the human and mouse contexts for each item.

## 3.3 Research Question 3

When changes in each variable from pretest to posttest were compared between students who experienced the CRS and those who did not, in the mouse context, students who experienced the CRS appear to have a significantly greater gain in variation, heritability, differential survival, and frequency/distribution than students who did not experience the CRS. They also appear to have significantly larger decreases in the teleology cognitive bias and need misconception.

**Table 9** Change, in percentage points, of students answering correctly on each ACORNS component by CRS strategy in the human and mouse contexts; statistically, significant differences are in bold

| | | Mouse context | | | Human context | | |
|---|---|---|---|---|---|---|---|
| | | CRS | No CRS | Z | CRS | No CRS | Z |
| Key concepts | Variation | 0.30 | 0.04 | **−4.641**** | 0.18 | 0.10 | −1.490 |
| | Heritability | 0.28 | 0.08 | **−2.885*** | 0.26 | 0.13 | −1.800 |
| | Competition | −.002 | 0.02 | −1.614 | −0.01 | −0.01 | −0.073 |
| | Hyperfecundity | 0.00 | 0.00 | 0.000 | 0.00 | 0.00 | 0.000 |
| | Resource limitations | 0.06 | 0.02 | −0.713 | 0.07 | −0.01 | −1.703 |
| | Differential survival | 0.26 | 0.02 | **−3.490**** | 0.25 | 0.16 | −1.434 |
| | Frequency/distribution | 0.23 | 0.06 | **−2.577**** | 0.15 | 0.15 | −0.079 |
| Cognitive biases | Essentialism | −0.10 | −0.05 | −1.263 | −0.02 | −0.04 | −0.470 |
| | Intentionality | 0.01 | −0.02 | −0.964 | −0.01 | −0.01 | −0.058 |
| | Teleology | −0.21 | −0.01 | **−4.651**** | −0.04 | 0.02 | −0.650 |
| Misconceptions | Pressure | 0.02 | −0.02 | −1.288 | −0.05 | −0.03 | −0.635 |
| | Adapt | −0.09 | −0.02 | −1.841 | −0.10 | −0.06 | −0.907 |
| | Need | −0.15 | 0.02 | **−3.776**** | −0.02 | −0.01 | −0.589 |
| | Must | −0.05 | −0.02 | −0.936 | −0.02 | −0.01 | −0.661 |
| | Use | −0.01 | −0.01 | −0.131 | −0.04 | −0.02 | −0.663 |
| | Energy | 0.00 | 0.00 | 0.000 | 0.00 | 0.00 | 0.000 |

*$p < 0.05$
**$p < 0.001$

In contrast, there were no significant differences in the human context (see Table 9). Although the trend appears to be toward greater increase in frequency of responses that show evidence in understanding key elements of natural selection and greater decrease in misconceptions when the CRS was used, the results are not significant. Table 9 shows the changes in each variable in students who did and did not experience the CRS.

## 4 Discussion

This is the first study of which we are aware that assessed high school students' understanding of natural selection before and after using curriculum materials that use human examples to teach evolution. It is also the first study in the USA to assess quantitatively high school students' understanding of natural selection before and after using teaching strategies that acknowledge the cultural controversy around teaching and learning evolution that exists in many contexts. The overall increases in understanding of natural selection suggest that combining human examples as the context for evolution instruction with classroom activities that acknowledge the cultural controversy and help manage students' tension around the topic of evolution hold promise as an effective strategy for high school evolution education.

In this study, we saw significant gains in evolution understanding in high school students from pretest to posttest in four aspects of understanding evolution: variation, heritability, differential survival, and frequency/distribution, in both human and mouse contexts. These results are important because variation, heritability, and differential survival are considered by many to be the three main essential components for natural selection. These results suggest that the TEtHE materials may be contributing to changes in understanding of natural selection where it counts the most conceptually. Interestingly, at both pretest and posttest, students were more likely to use resource limitations and differential survival in the mouse context than the human context. Perhaps, these ideas are more difficult realities and/or processes for students to associate with humans.

We also found significant reductions in cognitive biases and misconceptions across both mouse and human contexts, indicating the utility of the TEtHE materials for this purpose as well. Our findings (Tables 6 and 7) agree with previous studies indicating that the idea that individual organisms change in response to 'need' is the most common misconception in secondary and postsecondary students (Gregory, 2009). Interestingly, we saw a higher proportion of students with a teleology cognitive bias and more misconceptions generally in the mouse context than the human context. While this may be the opposite of an intuitive prediction which assumes students will have greater cognitive biases and misconceptions when it comes to humans, it could be interpreted as students still not grasping that evolutionary processes apply to humans at all and are therefore less likely to even have cognitive biases and misconceptions in that context.

We were surprised to find that the CRS activities seemed to pave the way for greater increases in understanding and decreases in cognitive biases and misconceptions in the mouse context, but not the human context. Perhaps, some students who experience the cultural controversy personally and participated in the CRS activities increased their openness to considering natural selection in a non-human context but still have some barriers to learning correct evolution concepts when it comes to humans. It is possible for students to create cognitive walls between things they believe and things they do not believe in order to understand evolution but not 'believe' it or accept it (Coburn, 1996; McKeachie, Lin, & Strayer, 2002; Ingram & Nelson, 2006; Hermann, 2012). Rather than a true lack of conceptual understanding, many students' misconceptions may be the result of this compartmentalizing or dismissal of scientific knowledge, especially if they feel that it contradicts their faith (Coburn, 1996; Hermann, 2012). Overall, the case for the CRS activities attributing to a decrease in some misconceptions and cognitive biases is a compelling finding that is worthy of additional research.

The data reported here support the general findings of educational research on college students in terms of effective pedagogical approaches. Constructivist-learning approaches that elicit student ideas and then give them multiple opportunities and experiences to engage in science practices to build explanations consistent with scientific understandings show promise to achieve the goals of evolution educators. The data also suggest that using examples of evolution in humans, which may be more relevant and interesting to students than examples of evolution in more distantly related organisms (and often ones they have never encountered), also shows important promise (Pobiner et al., 2018). Finally, the data suggest that at least in some contexts, eliciting students' cultural concerns through an explicit discussion of non-scientific views can pave the way to greater correct understanding of evolution (Bertka et al., 2019).

## References

AAAS. (2011). *Vision and change in undergraduate biology education: A call to action*. Washington, DC: AAAS.

Anderson, R. D. (2007). Teaching the theory of evolution in social, intellectual, and pedagogical context. *Science Education, 91,* 664–677.

Andrews, T. M., Kalinowski, S. T., & Leonard, M. J. (2011a). "Are humans evolving?" A classroom discussion to change student misconceptions regarding natural selection. *Evolution: Education and Outreach, 4,* 456–466.

Andrews, T. M., Leonard, M. J., Colgrove, C. A., & Kalinowski, S. T. (2011b). Active learning not associated with student learning in a random sample of college biology courses. *CBE-Life Sciences Education, 10,* 394–405.

Andrews, T. M., Price, R. M., Mead, L. S., McElhinny, T. L., Thanukos, A., Perez, K. E., et al. (2012). Biology undergraduates' misconceptions about genetic drift. *CBE-Life Sciences Education, 11,* 248–259.

Australian Curriculum, Assessment and Reporting Authority [ACARA]. (2014). *Foundation to year 10 curriculum: Language for interaction (ACELA1428).* Available at: http://www.australiancurriculum.edu.au/english/curriculum/f-10?layout=1#cdcode=ACELA1428&level=F. Retrieved December 15, 2017.

Beardsley, P. M., Stuhlsatz, M. A. M., Kruse, R. A., Eckstrand, I. A., Gordon, S. D., & Odenwald, W. F. (2011). Evolution and medicine: An inquiry-based high school curriculum supplement. *Evolution: Education and Outreach, 4,* 603–612.

Beardsley, P. M., Bloom, M. V., & Wise, S. B. (2012). Challenges and opportunities for teaching and designing effective K-12 evolution curricula. In K. D. Rosengren, S. K. Brem, E. M. Evans, & G. M. Sinatra (Eds.), *Evolution challenges: Integrating research and practice in teaching and learning about evolution* (pp. 287–310). New York: Oxford University Press.

Berkman, M. B., Pacheco, J. S., & Plutzer, E. (2008). Evolution and creationism in America's classrooms: A national portrait. *PLoS Biology, 6*(5), e124. https://doi.org/10.1371/journal.pbio.0060124.

Bertka, C., Pobiner, B., Beardsley, P., & Watson, W. (2019). Acknowledging students concerns above evolution: a proactive teaching strategy. *Evolution: Education and Outreach, 12,* 3. https://doi.org/10.1186/s12052-019-0095-0

Borgerding, L. A., Klein, V. A., Ghosh, R., & Eibel, A. (2015). Student teachers' approaches to teaching biological evolution. *Journal of Science Teacher Education, 26,* 371–392.

Bramschreiber, T. L. (2013). Teaching evolution: Strategies for conservative school communities. *Race Equality Teaching, 32,* 10–14.

Brem, S. K., Ranney, M., & Schindel, J. (2003). Perceived consequences of evolution: College students perceive negative personal and social impact in evolutionary theory. *Science and Education, 87,* 181–206.

Coburn, W. W. (1996). Worldview theory and conceptual change in science education. *Science Education, 80,* 579–610.

Department of Education. (2014). *The national curriculum in England: Key stages 3 and 4 framework document.* Available at: https://www.gov.uk/government/publications/national-curriculum-in-england-secondary-curriculum. Retrieved December 15, 2017.

Evans, E. M. (2001). Cognitive and contextual factors in the emergence of diverse belief systems: Creation versus evolution. *Cognitive Psychology, 42,* 217–266.

Gregory, T. R. (2009). Understanding natural selection: Essential concepts and common misconceptions. *Evolution: Education and Outreach, 2,* 156–175.

Heddy, B. C., & Sinatra, G. M. (2013). Transforming misconceptions: Using transformative experience to promote positive affect and conceptual change in students learning about biological evolution. *Science Education, 97,* 723–744.

Hermann, R. S. (2008). Evolution as a controversial issue: A review of instructional approaches. *Science and Education, 17,* 1011–1032.

Hermann, R. S. (2012). Cognitive apartheid: On the manner in which high school students understand evolution without believing in evolution. *Evolution: Education and Outreach, 5,* 619–628.

Hermann, R. S. (2013). High school biology teachers' views on teaching evolution: Implications for science teacher educators. *Journal of Science Teacher Education, 24,* 597–616.

Ingram, E. L., & Nelson, C. E. (2006). Relationship between achievement and students' acceptance of evolution or creation in an upper-level evolution course. *Journal of Research in Science Teaching, 43,* 7–24.

Jensen, M. S., & Finley, F. N. (1995). Teaching evolution using historical arguments in a conceptual change strategy. *Science Education, 79,* 147–166.

Kelemen, D. (2012). Teleological minds: How natural intuitions about agency and purpose influence learning about evolution. In K. D. Rosengren, S. K. Brem, E. M. Evans, G. M. Sinatra (Eds.), *Evolution challenges: Integrating research and practice in teaching and learning about evolution* (pp. 66–92). New York: Oxford University Press.

McKeachie, W. J., Lin, Y. G., & Strayer, J. (2002). Creationist vs. evolutionary beliefs: effects on learning biology. *American Biology Teacher, 64,* 189–192.

Nehm, R. H., Ha, M., Rector, M., Obfer, J., Perrin, L., Ridgway, J., & Mollohan, K. (2010). *Scoring guide for the open response instrument (ORI) and evolutionary gain and loss test (EGALT).* Technical Report of National Science Foundation REESE Project 0909999.

Nehm, R. H., Beggrow, E. P., Opfer, J. E., & Ha, M. (2012). Reasoning about natural selection: Diagnosing contextual competency using the ACORNS instrument. *American Biology Teacher, 74,* 92–98.

Nettle, D. (2010). Understanding of evolution may be improved by thinking about people. *Evolutionary Psychology, 8,* 205–228.

NGSS Lead States. (2013). *Next generation science standards: For states, by states.* Washington, DC: The National Academies Press.

Paz-y-Mino, C. G., & Espinosa, A. (2009). Assessment of biology majors 'versus nonmajors' views on evolution, creationism, and intelligent design. *Evolution: Education and Outreach, 2,* 75–83.

Pobiner, B., Beardsley, P. M., Berkta, C. M., & Watson, W. A. (2018). Using human case studies to teach evolution in high school A.P. biology classrooms. *Evolution: Education and Outreach, 11,* 3. https://doi.org/10.1186/s12052-018-0077-7.

Pobiner, B. L. (2016). Accepting, understanding, teaching, and learning (human) evolution: Obstacles and opportunities. *American Journal of Physical Anthropology, 159,* 232–274.

Pugh, K. J., Linnenbrink-Garcia, E. A., Koskey, K. L. K., Stewart, V. C., & Manzey, C. (2010). Teaching for transformative experiences and conceptual change: A case study and evaluation of a high school biology teacher's experience. *Cognition and Instruction, 28,* 273–316.

Ranney, M. A., & Thanukos, A. (2011). Accepting evolution or creation in people, critters, plants, and classrooms: the maelstrom of American cognition about biological change. In R. Taylor, & M. Ferrari (Eds.) *Epistemology and science education: Understanding the evolution versus intelligent design controversy* (pp. 143–172). Oxford: Routledge.

Sinatra, G. M., Southerland, S. A., Mcconaughy, F., & Demastes, J. W. (2003). Intentions and beliefs in students' understanding and acceptance of biological evolution. *Journal of Research in Science Teaching, 40,* 510–528.

Sinatra, G. M., Brem, S. K., & Evans, E. M. (2008). Changing minds? Implications of conceptual change for teaching and learning about biological evolution. *Evolution: Education and Outreach, 1,* 189–195.

Smith, M. U. (2010). Current status of research in teaching and learning evolution: II. Pedagogical issues. *Science and Education, 19,* 539–571.

The College Board. (2011). *AP biology curriculum framework 2012–2013.* New York, NY: The College Board.

Verhey, S. D. (2005). The effect of engaging prior learning on student attitudes towards creationism and evolution. *BioScience, 55,* 996–1003.

Werth, A. J. (2009). Clearing the highest hurdle: Human-based case studies broaden students' knowledge of core evolutionary concepts. *The Journal of Effective Teaching, 9,* 38–53.

Werth, A. J. (2012). Avoiding the pitfall of progress and associated perils of evolutionary education. *Evolution: Education and Outreach, 5,* 249–265.

Wilson, D. S. (2005). Evolution for everyone: How to increase acceptance of, interest in, and knowledge about evolution. *PLoS Biology, 3*(12), e364.
Yates, T. B., & Marek, E. A. (2014). Teachers teaching misconceptions: A study of factors contributing to high school biology students' acquisition of biological evolution-related misconceptions. *Evolution: Education and Outreach, 7*, 7. https://doi.org/10.1186/s12052-014-0007-2.

**Briana Pobiner** is Paleoanthropologist and Museum Educator in the Human Origins Program at the Smithsonian National Museum of Natural History (NMNH), where she was on the core development team for the Hall of Human Origins. Her archeology research in Africa and Asia focuses on the evolution of early human diet. She also leads the Human Origins Program's education and outreach efforts and is engaged in research on evolution education in informal and formal settings, and she is Adjunct Research Professor at George Washington University. She was PI on the *Teaching Evolution through Human Examples* NSF project.

**Bill Watson** is Director of Curriculum and Assessment for the Diocese of Camden Catholic Schools. He is responsible for the development and evaluation of curriculum and assessment for 34 elementary and high schools. In that capacity, he led a team of 40 administrators and teachers to redevelop the K-12 science curriculum to fully align with the Next Generation Science Standards. He was Co-PI on the Teaching Evolution through Human Examples project. He holds a doctorate from the George Washington University, a Master of Arts in Teaching from the University of Portland, and an undergraduate minor in evolutionary biology.

**Paul Beardsley** is Associate Professor of Biological Sciences at Cal Poly Pomona and Senior Educational Media Fellow for HHMI BioInteractive. He has developed evolution-related curricula for K College, including projects funded by NSF, NIH, USDOE, NASA, private foundations, and HHMI. Highlights include being Lead Curriculum Developer for the *Learning Unity and Diversity in Alabama* NSF project and *Teaching Evolution through Human Examples* NSF project, PI and Lead Author of NIH's *Evolution and Medicine*, and Author and Lead Science Educator for *BSCS Biology: A Human Approach* (4th edition). As a biologist, he works on systematics and ecological genetic questions in monkeyflowers.

**Constance Bertka** Science and Society Resources, holds a Ph.D. in geology, Arizona State University, and a master of theological studies, Wesley Theological Seminary. Connie is a Co-Chair of the Broader Social Impacts Committee of the Smithsonian's Human Origins Program. She is Past Director of the AAAS Program of Dialogue on Science, Ethics, and Religion, and the Deep Carbon Observatory at the Carnegie Institution of Washington. She is Editor of *Exploring the Origin, Extent and Future of Life: Philosophical, Ethical and Theological Perspectives* (Cambridge University Press, 2009) and Lead Author of the Smithsonian's *Cultural and Religious Sensitivity (CRS) Teaching Strategies Resource*.

# Models and Modeling in Evolution

**Kathy L. Malone, Anita M. Schuchardt and Zakee Sabree**

## 1 Introduction

The effective teaching of evolution in secondary schools has been challenging and not truly successful in terms of student content gains (Beardsley, Bloom, & Wise, 2012). This is unfortunate given that evolution is a fundamental concept (Dobzhansky, 1973; Tansey et al., 2013). The consensus is that evolution is the key to understanding core concepts in biochemistry and molecular biology while natural selection is expressed throughout biology in areas such as ecosystems, population interactions, variation, physiology, and genetics (Tansey et al., 2013). However, true success in the secondary classroom remains elusive because of students' difficulties with conceptual change due to deeply ingrained alternative conceptions. These alternative conceptions are widely held across ethnic, socioeconomic, and cultural settings (Gregory, 2009; Nehm & Schonfeld, 2007), and across multiple grade bands. Some of the most common alternative conceptions include as follows: Speciation is directly caused by a need for change, acquired traits can be inherited (i.e., Lamarckian), and that use and disuse cause population traits to disappear from populations (Deadman & Kelly, 1978; Bishop & Anderson, 1990; Yates & Marek, 2013).

In their recent meta-analysis, Furtak, Seidel, Iverson, and Briggs (2012) reported that the use of inquiry methods in kindergarten to twelfth-grade (K12) schools produced medium effect sizes over that of traditional methods. However, evolution is a

K. L. Malone (✉)
Nazarbayev University, Astana, Kazakhstan
e-mail: klmalone60@gmail.com

A. M. Schuchardt
University of Minnesota—Twin Cities, Minneapolis, MN, USA
e-mail: aschucha@umn.edu

Z. Sabree
The Ohio State University, Columbus, OH, USA
e-mail: sabree.8@osu.edu



U. Harms and M. J. Reiss (eds.), *Evolution Education Re-considered*,
https://doi.org/10.1007/978-3-030-14698-6_12

**Table 1** Empirical studies focused on evolution

| Study | Study age | Control group | Data sources | Pre-testing | Post-testing | Pedagogy | Focus of study |
|---|---|---|---|---|---|---|---|
| Donnelly, Kazempour, and Amirshokoohi (2009) | HS | n/a | Survey & Int. | | ✓ | UNK | Evolution acceptance and perceptions of students |
| Donnelly, Namdar, Vitale, Lai, and Linn (2016) | 7th grade | n/a | Test, Obs. & Int. | ✓ | ✓ | CS | Elaborating prompts to support Darwin versus competing prompts to distinguish between Darwin and Lamarckian |
| Malone, Schuchardt, and Schunn (2018) | HS | ✓ | Test | ✓ | ✓ | M & EMT | Effect of modeling and excel modeling tools on students knowledge of natural selection |
| Wilensky and Reisman (2006) | HS-1 student | n/a | Video and field notes | ✓ | ✓ | M & CS | Effect of NetLogo programming in on predation model |
| Xiang and Passmore (2015) | 8th grade | n/a | Video | ✓ | ✓ | M & CS | Effect of CS & M on programming and use of M during programming |
| Abraham et al. (2009) | College Bio | n/a | Test | ✓ | ✓ | CS | Effect of CS & M on alternative conceptions and process of natural selection |
| Daurer, Momsen, Speth, Markohon-Moore, and Long (2013) | College bio majors | n/a | Concept maps | ✓ | ✓ | MB | Effect of MB on model correctness and linkages |

(continued)

**Table 1** (continued)

| Study | Study age | Control group | Data sources | Pre-testing | Post-testing | Pedagogy | Focus of study |
|---|---|---|---|---|---|---|---|
| Nehm and Reilly (2007) | College bio majors | n/a | Test | ✓ | ✓ | AL | Effect of AL on NOS discussions, problem-solving, and small group discussions on alternative conceptions |
| Nehm and Schonfeld (2007) | Teacher (in-service) | n/a | Survey | ✓ | ✓ | ACE, CL, CM, F, L, M and R | Teacher acceptance and knowledge of evolution |
| Robbins and Roy (2007) | College nonmajors college non-STEM majors | n/a | Test | ✓ | ✓ | ACE, P | Effect of lecture and peer instruction and data interpretation on open-ended and objective questions |
| Sandoval and Reiser (2004) | HS | n/a | Videos | | | AL, L, CS | Effect of scaffolds on natural selection sci. explanations |
| Geraedts and Boersma (2006) | HS | n/a | Test, Int | ✓ | ✓ | CL, S | Effect of reinvention of Darwinian theory using worksheets and hands-on simulation |
| Cavallo and McCall (2008) | HS | n/a | Test | ✓ | ✓ | AL | Effect of instruction on NOS, evolution (understanding and evolution) |
| Spindler and Doherty (2009) | HS | n/a | Test, survey | ✓ | ✓ | L, S | Effect of predator–prey simulation on learning |

*ACE*—alt conceptions exposure, *CL*—collaborative learning, *CS*—computer simulation, *L*—lecture, *M*—modeling, *P*—peer instruction, *UNK*—unknown, *AL*—active learning/inquiry, *CM*—concept mapping, *EMT*—excel modeling tools, *MB*—model based, *R*—reading, *S*—simulation, hands-on process that occurs over long periods of time and the objects of interest (the genes) are hidden from view. Thus, it is difficult to use inquiry methods to study evolution as a real-world phenomenon. Students rarely design experiments, analyze data, and discuss their conclusions with their peers in the context of studying evolution through the process of natural selection. Recent empirical evolution studies focused on classroom interventions have shown promise but the conceptual gains were either very modest and/or the study lacked a comparison group (refer to Table 1). Literature reviews of K12 evolution studies show mixed results on student learning (conceptual as well as acceptance) and highlight the need for rigorous experimental design studies in this area (Beardsley et al., 2012; Glaze & Goldston, 2015). Thus, additional quasi-experimental studies at the high school level are clearly needed.

## 2 Models and Modeling in the Context of Science Education

Internationally, the use of authentic science practices in science classrooms is on the rise (Khine & Saleh, 2011; KMK, 2005; NGSS Lead States, 2013). One approach to teaching authentic science practices is the development and use of scientific models (Lehrer & Schauble, 2015; Quinn, Schweingruber, & Keller, 2012; Svoboda & Passmore, 2013; Windschitl, Thompson, & Braaten, 2008). However, there are many different ways to conceptualize models and modeling in the secondary classroom. In the USA, the Framework for K12 Science Education defines conceptual models as explicit representations (i.e., graphs, mathematical equations, pictures, and physical models) that students use to make science phenomena more understandable and predictable (Quinn et al., 2012). Quinn et al. (2012) identify modeling as not only the development of models but also the refinement and use of models. Given the increased international emphasis on the use of models and modeling, it is imperative that fully tested materials should be made available to guide teachers during implementation in the science classroom.

This paper describes a project designed to fill the need for not only quasi-experimental studies focused on student learning in evolution but also the development of modeling-based curricula materials in biology. The curricula materials focus on the use of modeling instruction, a model-based pedagogy grounded in guided inquiry.

### *2.1 Research and Curriculum Efforts in Models and Modeling*

Models are idealized representations of the world used to communicate ideas about science as well as make predictions about biological systems (Buckley et al., 2004; Giere, 2004; Svoboda & Passmore, 2013). Consisting of multiple representations

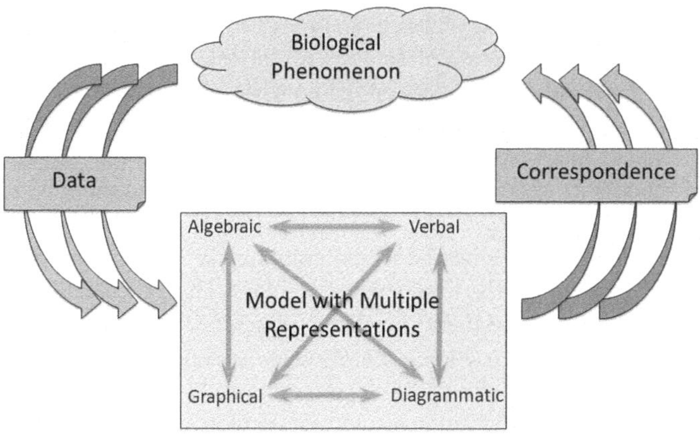


**Fig. 1** Modeling cycle

that work together to portray a comprehensive understanding, models can be empirically developed based upon data analysis or theoretical underpinnings and are iterative. During authentic problem-solving, experts are known to switch fluidly between model representations (Harrison & Treagust, 2000). Multiple model representations evolve as experts move through the problem space. The model and its representations are tested by determining if they are predictive of the original data as well as data from experiments in different contexts. As the model is tested, revisions are made to allow it to be more predictive. This cycle is known as a modeling cycle. Figure 1 illustrates the modeling cycle where not when a model is described as multiple representations.

Educators often communicate to their students only about physical representations (e.g., a 2D or 3D model of a cell) (Krell & Krüger, 2016; Ware, Malone, Irving, & Mollohan, 2017) and present these models as a fixed final product. This restricted perspective can hinder students' abilities to master science. The implementation of a modeling cycle in the classroom should allow students to refine and develop model representations as they move through a unit which includes data collection, data analysis, controlling variables, and deductive reasoning (Dukerich, 2015; Malone, 2014; Posthuma-Adams, 2014; Svoboda & Passmore, 2013). This iterative refinement allows students to not only develop a deeper understanding of the relevant science concepts but also to confront their alternative conceptions (Halloun, 2007).

## 2.2 *Modeling Instruction in Science, a Guided Inquiry Pedagogy*

Modeling instruction (MI) was first developed for physics instruction in the USA and uses models and their multiple representations as the organizing theme (Jackson, Dukerich, & Hestenes, 2008). MI conceptualizes the development and refinement of models within repeated use of the modeling cycle (see Fig. 1). This pedagogy engages students in the practice of science by having them design experiments, collect and analyze data, and use data modeling to develop multiple representations of the science phenomena being studied. Students report their representations to their peers and arrive at a class consensus model through argumentation. Students deploy these initial models in new contexts where they further refine them as they move along the modeling cycle.

As new models are developed, students consistently look for connections and linkages between other models developed in the class. This practice allows students across disciplines to organize their knowledge of science in terms of basic science models while increasing their content knowledge as well as their problem-solving and metacognitive skills (Malone, 2008). In addition, Malone and Schuchardt (in review) demonstrated that MI students increased their scientific reasoning skills when exposed to MI in multiple contexts (e.g., physics, chemistry). The use of MI has shown concept gains in physics (Liang, Fulmer, Majerich, Clevenstine, & Howanski, 2012; Malone, 2008) and chemistry (Malone, 2014) as compared to traditionally taught students.

### *2.3 Use of Models and Modeling in Biology and Evolution—Past Research*

The use of models and modeling has been linked to success in biology in case studies lacking a comparison group. In the Models for Understanding in Science Education (MUSE) project, secondary students evaluated existing models of evolution and while supporting empirical data were not reported, the authors suggested the students had a richer understanding of the model of natural selection (Passmore & Stewart, 2002). These materials were combined with computer-based modeling simulations in an eighth-grade class; however, students still produced incomplete models of natural selection (Xiang & Passmore, 2015).

Recently, the usage of models and modeling in biology, specifically genetics and evolution, has been assessed via quasi-experimental studies (Malone et al., 2018; Schuchardt & Schunn, 2016). The effects of a modeling-based curriculum focused on the development of population growth and evolution models (Malone et al., 2018) and models of inheritance (Schuchardt & Schunn, 2016) both with multiple representations demonstrated significant concept gains over that of traditionally taught control students. In addition, the treatment groups demonstrated greater ability to use multiple representations. When students' problem-solving strategies during semi-structured interviews were analyzed, students who had participated in the model-based inheritance unit showed a tendency to make more connections between the biological phenomenon and the mathematical representation as well as easily switch problem-solving strategies from conceptually driven to mathematically driven in ways that were likely to facilitate success (Schuchardt, 2016).

While these previous units in model-based instruction in biology incorporated computer modeling and mathematical modeling, a quasi-experimental study on the effect of hands-on simulations as the basis for model building has not yet been done. This chapter will fill a gap in the research by determining the effect of MI in biology and hands-on simulations on student conceptual gains as well as their use of representations in a quasi-experimental study.

## 3 Modeling Instruction in Evolution and Natural Selection

The MI unit in evolution and natural selection was developed over the course of a year and a half. The developers consisted of the authors and two master biology teachers. The master biology teachers piloted the unit during the school year with rapid revisions between implementations.

### *3.1 Modeling Instruction in Biology Evolution Unit Overview*

An MI unit builds upon students' initial predictions concerning the unit's paradigm lab. A paradigm lab is the students' initial experience with the biological phenomena and is carefully designed so that students engage with key aspects of the phenomenon as well as confront common alternative conceptions. Prior to the paradigm lab, the students are not informed that the new unit is about evolution. They have been asked to do *no* prior readings nor engage in discussions to bring out their prior knowledge. This was done because of the resistance on students' part to looking at evolution from a scientific viewpoint (Glaze & Goldston, 2015; Reiss, 2010). The unit was conceptualized to allow students to collect data, analyze the data, and reach conclusions that are supported by evidence. The power of this approach is that students are not confronted with a reject or accept decision until after they have built a model from evidence. Only after the model is developed is it named and if they are to reject it, they also have to reject the consensus building and evidence evaluation that they and their classmates have been engaged in. This moves students away from a reflexive response and forces them to engage with the evidence.

### *3.2 Paradigm Lab Description*

This activity allows students to simulate the process of natural selection over the course of many generations of lizards. While this simulation is complex, it is also flexible. Along with addressing the evolutionary core concepts of the requirement for pre-existing variation, and the role of selective pressures in the environment, the activity can address numerous other biological concepts including sexual selection, genetic drift, and predator–prey relationships, as well as preview important concepts in inheritance.

### 3.2.1 Simulation Overview

Students simulate the life and reproduction cycles of a lizard population with two traits of interest: mouthpart and skin color. They undertake multiple rounds of feeding and reproducing (described below), under conditions of no selective pressure and selective pressure (i.e., no drought vs drought). The data collected over these multiple rounds (generations) are analyzed to permit students to develop a model of what happens to a population's traits with and without selective pressure.

*Feeding simulation*

To simulate feeding, students transfer water from a large shared container to their 'stomach' (i.e., a small beaker). At the end of each round of feeding, students measure and record the amount of water they have collected in their 'stomach'.

*Mating simulation*

Following feeding, students 'mate' randomly with a classmate to produce the next generation of lizards. Students' traits (color and mouthpart) are determined by the combination of cards that they are given. Each student has two cards for each trait. Figure 2 shows the cards and the combinations that are possible. During mating, students pair up and randomly and blindly pick one card from each other's hands, recording the combination of selected cards. They repeat the process for the mouth cards. These will be the mouth and color 'genes' for the first offspring. The students then pick up the cards they originally had and repeat the process to produce a second offspring. In the next generation, each student will become one of these offspring by picking up new cards to represent the new set of genes. A new cycle of feeding and mating occurs.

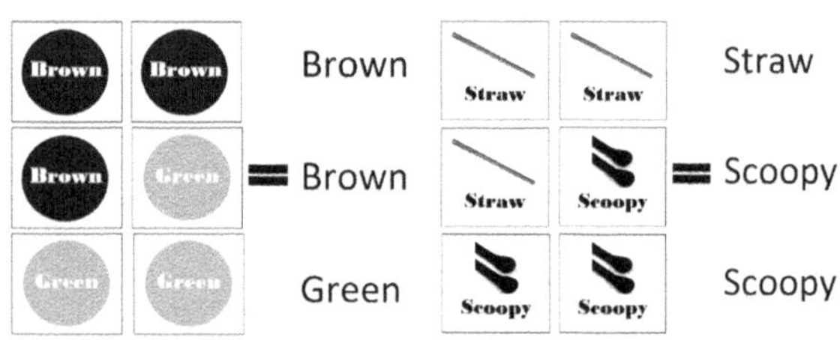


**Fig. 2** Cards denoting lizard coloration and mouthpart

### 3.2.2 Addressing Key Concepts Through Paradigm Lab Design

This simulation has been designed to target four key concepts about evolution by the process of natural selection while allowing students to collect evidence to build a model of the process. The four key concepts that are addressed are:

(1) The need for pre-existing variation in the population,
(2) The ability of an environmental factor to exert a differential selective pressure on a trait,
(3) The role of reproduction and inheritance,
(4) The shift in allelic and trait frequencies that are the result of the process of natural selection.

*Need for pre-existing variation.* Students begin the simulation with a mouthpart (straw) that is adequate for obtaining water from deep basins of water. They undergo several rounds of feeding and mating and note that no change occurs in the population. A selective pressure is introduced by simulating drought conditions through: (a) drastically reducing the amount of water in each basin; and (b) having students who did not acquire a preset amount of water in each feeding cycle sit out subsequent cycles of feeding and mating. To simulate reproductive advantage, the two top feeders in each cycle produce extra offspring whose roles are assumed in subsequent generations by some of those students who did not survive. After several rounds in these new conditions, students note that while the population is decreasing, there are still only 'straw' mouthparts in the population.

*Ability of an environmental factor to exert differential selective pressure on specific traits.* The next part of the simulation involves introduction of a new mouthpart by immigration. This new 'scoopy straw' mouthpart is better able to acquire food under drought conditions—when the water in the basin is shallow. However, it is no more adept when the water in the basins is deep. Teachers reset the environment to one of ample resources, with deep water in the basins, and all students surviving after feeding. Under these conditions, the scoopy mouthpart may increase slightly in frequency in the population or it may die out. It has no advantage over the straw mouthpart. After several rounds of mating and feeding, drought conditions are reintroduced. Under these conditions, the scoopy straw mouthparts increase in frequency in the population. It is worth noting that in all rounds, color traits vary and are not affected by the depth and availability of water. Therefore, the color trait serves as a comparative illustration that the environment must act differentially on the trait for a change in frequency to occur in a particular direction. Figure 3 diagrams the simulation cycle.

*The role of reproduction and inheritance.* Because top feeders produce more offspring, an association is formed between acquiring extra water and reproductive success. The transmission of cards from 'parents' to 'offspring' in a structured manner that follows Mendel's Laws of Inheritance reinforces the role of inheritance, even if students have not yet studied this topic formally. To further reinforce the need for reproductive success and not just survival, teachers randomly deform some of the straw mouthparts, slitting and flattening them, during one of the drought conditions.

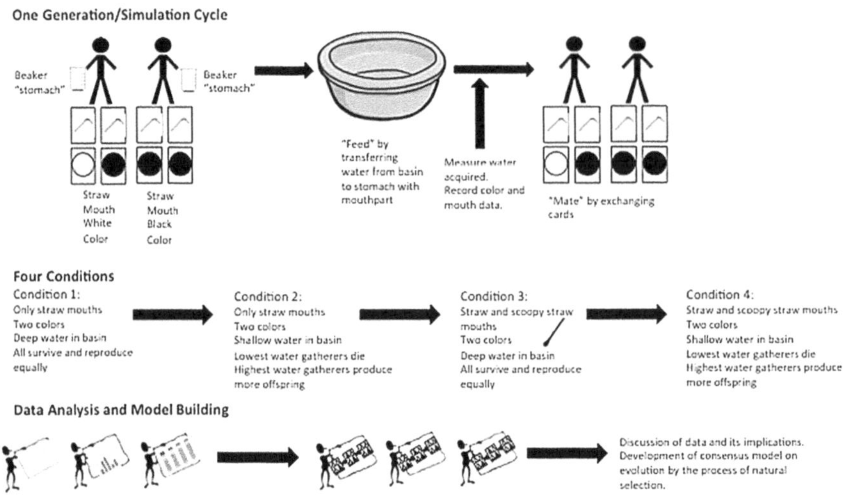


**Fig. 3** Simulation cycle

These deformed straws may transfer water better under the drought conditions. The deformities are presented as caused by the environment and are not able to be passed on by offspring. In the class discussions following the simulation, the teacher can return to this intervention to allow students to discuss the role of reproduction and inheritance.

*Shift in allelic and trait frequencies*. At the end of each cycle of feeding, class data are collected on the number of cards of each color and mouthpart, and the number of students who are expressing each mouthpart and color. This class dataset will form the basis for discussions that reveal the shift in allelic and trait frequencies in the four conditions discussed above. After students graph their data for each of the four conditions and discuss their results, they reach the consensus that the only condition in which a trait and its allele increased in frequency in the population over time was the one where there was pre-existing variation in the population, the trait was subject to differential survival due to an environmental condition, and the trait was inheritable. In this way, students have developed a model of the conditions necessary for evolution (the shift in allelic/trait frequencies in a population) as a result of natural selection.

It is only after the students reach their final consensus that this activity is labeled in terms of natural selection and evolution.

### 3.2.3 Developing Representational Fluency

Students develop representational fluency during the data analysis stage. Students are not told how to graph the data, but are asked to represent their data in a way that will communicate the results. Through discussion, they come to the realization that bar graphs are the best representation. From their graphs, students are asked to develop pictorial representations that summarize the changes that have taken place over time. Therefore, by the end of the data discussion, students will have a pictorial, verbal, and graphical representation that are interconnected.

*Simulation modifications*

As the simulation moves forward, students are generating questions they have about their population of lizards, which can include the effect of acquired traits, predators, etc. on the traits in a population. At this point, students design modifications to the simulation to answer their questions.

### *3.3 Deployment of Initial Model*

After the development of the initial model, students must next deploy their model to determine its ability to be predictive in multiple contexts. In addition, the initial activity does not develop a full model of evolution as no one activity could do so. Thus, the rest of the unit includes simulations, worksheets, and projects designed to further develop the model. For example, during a simulation which allows students to select mates based upon size of lizard dewlaps, students discover that their model of natural selection is not predictive in the case of sexual selection. This leads to the development of a model of evolution through the process of sexual selection. Other activities allow students to include the core idea that new traits are continuously produced via the reassortment of gene combinations during fertilization as well as mutation. Through these activities, students also tackle a common alternative concept that mutation occurs in response to the environment. Through the iterative process of MI, students implicitly learn that models are fluid and continually changing.

## 4 Research Design

The goals of this study were (1) to develop a MI in biology unit on evolution and natural selection that uses hands-on modeling simulations; and (2) to examine the effects of the unit on students' conceptual knowledge, and representations of the phenomenon. A quasi-experimental study was designed with treatment and comparison groups using pre- and post-conceptual assessments within secondary science classrooms. The following research questions guided the study:

1. Do modeling instruction students develop greater conceptual understanding of evolution and natural selection than comparison students?
2. Do students incorporate multiple representations into their depiction of evolution?

## 4.1 Participants and Settings

The participants, 15- and 16-year-olds, are from the Midwestern USA. An attempt was made to conservatively match the treatment and control teachers in terms of years of experience, type of district, and educational level. The MI cohort did have three participating teachers from inner city schools while all of the control cohort teachers were from suburban and rural districts. Students in the control group performed higher on external standardized tests such as the SAT, were less socioeconomically disadvantaged, and should have outperformed the MI students.

Fourteen teachers participated in the study: seven comparison teachers with 425 students from three different school districts, and seven treatment teachers with 535 students from six different school districts.

## 4.2 Research Instruments

*Conceptual Assessment*

To assess conceptual understanding, the study utilized the Secondary-Biology Concept Inventory (S-BCI), which has been previously assessed for validity and reliability using a sample of 1016 students which were similar in age, SES, and to the sample in this study (Malone et al., 2017; Stammen et al., 2016). This multiple-choice instrument consists of 25 questions that cover a full year of instruction and the answer stems target the most salient alternative conceptions. Only the five questions that specifically focused on evolution were used in this study. The S-BCI was given as a pre-test to students within the first two weeks of the school year. The post-test was administered within the last month of the school year. This scheduling favored the comparison group since the MI cohort taught evolution at the beginning of the year while the comparison group taught evolution near the end of the school year. Because of the greater length of time between instruction and the post-assessment for the MI group, it was thought that the MI could have lower post-test scores than the comparison group?

*Case Study Qualitative Analysis*

A subset of three teachers (two control and one treatment teacher) administered the following prompt to students at the end of the unit: 'Draw what you currently know about biology and its main ideas. Focus on the main ideas and show how and if they are connected to one another. Include as much detail as possible. You can use any representation you wish including: words, graphs, pictures, math equations, etc.' The prompt was open-ended to determine how often ideas about evolution surfaced and what kinds of representations students would use. All student responses were coded for Teacher A. Teachers B and C had over 100 responses so thirty responses were selected at random (illegible responses were discarded). Responses were coded for whether students mentioned evolution or a topic associated with evolution (e.g., natural selection, genetic drift, and change in alleles or traits over time). Any response that contained a representation of evolution was coded for the use of verbal, graphical, or pictorial representations in reference to evolution. The graphical and pictorial responses were analyzed for the depiction of key concepts of evolution: either one of the mechanisms driving evolution (genetic drift, sexual selection, natural selection) or for the outcome of these mechanisms, a change in trait or allele frequency over time.

## 4.3 *Data Analysis and Results*

### 4.3.1 Conceptual Assessment

Conceptual Assessment—Mean Scores

Statistical significance of assessment scores was determined using ANOVA on pre- and post-test scores by condition. All raw scores were converted to RASCH measures for analysis. The pre-test scores were not significantly different between conditions ($F(1, 862) = 0.67, p = 0.41$). Thus, even though the MI cohort has a slightly higher pre-test average, they can be considered identical statistically.

An ANOVA showed significant differences between groups at the time of post-testing, favoring the MI cohort ($F(1{,}1005) = 6.21$; $p < 0.01$). Mean scores and standard error bars can be seen in Fig. 4. The comparison group demonstrated a percent gain pre- to post-assessment of 8% and Cohen's *d* effect size of 0.11 (i.e, no effect), while the MI cohort had a 14% gain and a Cohen's *d* effect size of 0.24 (i.e., a small effect).

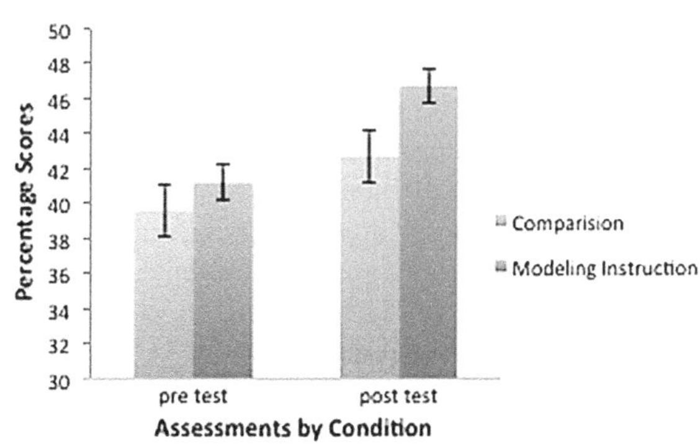


**Fig. 4** Graph of score by condition

Conceptual Assessment—Item Analysis

In order to develop a more in-depth understanding of the conceptual differences and similarities between the two groups, an item analysis of the assessment was conducted.

Both cohorts have a solid understanding that all life forms arose from pre-existing organisms with each cohort scoring in the mid-60% on the pre-assessment question. The Lamarckian idea that acquired traits can be passed onto the next generation showed no movement in the comparison cohort with about 45% of the students maintaining this belief on the post-assessment. However, the MI cohort demonstrated a 10% decline in this belief on the post-assessment.

In two areas, students in both cohorts maintained similar pre- to post-understandings. Those areas included the idea that 'fittest' meant that the organisms were the strongest (~30%) as well as the idea that the chimpanzees share a common ancestor with humans and other organisms such as bees do not (~80%). The lack of pre-post change in these areas suggests that the MI curriculum could be modified so that the MI cycle tackles these ideas more directly through the inclusion of deployments targeted to confront these ideas.

### 4.3.2 Biological Connections

Table 2 summarizes the results. Students from at least one of the comparison teachers made frequent reference to the concept of evolution, as did students from the treatment teacher, suggesting that method of instruction does not necessarily determine how pertinent the topic of evolution is to students. As seen in Table 2, MI students tend to produce multiple representations when prompted to use them. Eighteen modeling students represented their ideas about evolution with a picture (67%), and fourteen students used a graph (52%). Comparison students tended not to use multiple representations to express their ideas about evolution as only seven students across both teachers (21%) used a pictorial representation and one student (3%) used a graphical representation. None of the comparison students used more than one representation, whereas twelve MI students did.

All but one of the 27 modeling students used the graphical or pictorial representation to express specific ideas about evolution; either the mechanisms or the direct outcome of those mechanisms, or a change in frequency of traits or alleles in the population. Five of the eight students in the comparison group with graphical or pictorial representations used them to depict ideas about the mechanism of evolution or the change in frequency of traits or alleles. The other comparison pictures were general depictions of an idea associated with evolution (e.g., a monkey connected by an arrow to a man; a fish connected by an arrow to a man; bird beaks of different sizes) but not clearly related to the mechanisms or change in frequency of traits or alleles.

Most of the nonverbal representations in the MI cohort, but not many in the comparison group, depicted changes in trait or allele frequency over time. Across

**Table 2** Evolution concepts described by different types of student representations, by teacher and condition

| | Control condition | | Treatment condition |
|---|---|---|---|
| Representation and concept | Teacher A ($N = 23$) | Teacher B ($N = 30$) | Teacher C ($N = 29$) |
| Evolution represented | 9 | 24 | 27 |
| *Pictorial representation* | 3 | 4 | 18 |
| Natural selection | 0 | 2 | 4 |
| Sexual selection | 0 | 0 | 1 |
| Genetic drift | 0 | 0 | 3 |
| Change in trait/allele frequency | 2 | 0 | 15 |
| *Graphical representation* | 0 | 1 | 14 |
| Natural selection | 0 | 1 | 4 |
| Sexual selection | 0 | 0 | 5 |
| Genetic drift | 0 | 0 | 0 |
| Change in trait/allele frequency | 0 | 1 | 13 |
| *Verbal representation* | 9 | 19 | 21 |
| Natural selection | 2 | 13 | 12 |
| Sexual selection | 0 | 0 | 10 |
| Genetic drift | 3 | 0 | 8 |
| Change in trait/allele frequency | 1 | 1 | 4 |

both conditions, only six students (four MI, two comparison) verbally expressed the idea that evolution is associated with a change in trait or allele frequency over time. Out of those six instances, only one explicitly referenced a change in the allele or trait frequency without being supported by reference to a picture or graph that showed allelic or trait change in frequency within a population. Many students referenced 'change over time' with respect to evolution without being clear as to what was changing and how change was occurring. These findings suggest that pictorial and graphical representations may focus students' attention on a specific and key feature of the evolutionary process: the change in allele or trait frequency over time within a population. Thus, building these representations within the curriculum may help to build conceptual understanding by making it clear to students what is changing (traits/alleles) and how it is changing (proportionately and gradually over time).

# 5 Discussion

## 5.1 *Students Develop Greater Conceptual Understanding of Natural Selection*

The evolution by natural selection unit grounded in MI pedagogy using a hands-on simulation for data collection supported higher accuracy on a conceptual assessment for implementing students over that of more traditionally taught students. Overall, the effect size pre-score to post-score on the conceptual assessment for MI students was over 50% larger than the comparison cohort. This gain is impressive due to the difference in timing between instruction and post-testing between the two cohorts (6 months for MI cohort and 1 month for comparison cohort).

These findings demonstrate that evolution by natural selection can be effectively taught using MI pedagogy in which the students discover the model of natural selection by analysis of simulation data they personally collected.

## 5.2 *Students Have Fewer Alternative Conceptions Using Modeling Instruction*

The traditionally taught comparison students continued to maintain the belief that acquired traits could be passed down to future generations. MI allows for students to revise their alternative conceptions by confronting their beliefs when analyzing data collected for multiple generations of organisms. This revision of conceptions through modeling allows students to move toward a more expert-like ability in those areas (Malone, 2008; Schuchardt & Schunn, 2016).

## 5.3 *Students Increase Usage of Multiple Representations*

A greater percentage of modeling students not only drew pictorial and graphical representations of evolution but they also used them to make sense of and depict a key feature of the evolutionary process: the change in allele/trait frequency over time within a population. The use of multiple representations allows students to move toward a more expert-like understanding of this area of biology while possibly allowing students to test their conceptions in such a way as to allow for conceptual change.

### 5.4 *Limitations*

The main limitation of this study is that the participating students were all from the Midwest of the USA. In addition, there was a marked difference between the timing of the evolution unit for the two cohorts, thus giving an advantage to the comparison cohort. Moreover, the biology connections study focused on a small subset of the participants so the results of this part of the study should be repeated with a larger number of students.

## 6 Concluding Remarks and Implications

This study determined that teacher use of MI pedagogy does seem to increase student knowledge and shift alternative conceptions held in the area of evolution and natural selection while increasing student use of multiple representations over that of more traditionally taught students. The development of models and their multiple representations may be the key to the decline in alternative conceptions held by students. The quantitative assessment pointed to the need to develop additional activities to scaffold student understanding about common ancestry as well as the biological meaning of survival of the fittest. In the future, the effect of MI should be tested in additional areas of biology as well as other sciences.

**Acknowledgements** Funded by a grant under the federally funded Math Science Partnership State Grants Program, administered by the Ohio Department of Education. Any opinions, findings, and conclusions or recommendations expressed in this material are those of the authors and do not necessarily reflect the views of the funding organizations.

## References

Abraham, J. K., Meir, E., Perry, J., Herron, J. C., Maruca, S., & Stal, D. (2009). Addressing undergraduate student misconceptions about natural selection with an interactive simulated laboratory. *Evolution: Education and Outreach, 2*(3), 393–404.

Beardsley, P. M., Bloom, M. V., & Wise, S. B. (2012). Challenges and opportunities for teaching and designing effective K-12 evolution curricula. In *Evolution challenges: Integrating research and practice in teaching and learning about evolution* (pp. 287–310). New York: Oxford University Press.

Bishop, B. A., & Anderson, C. W. (1990). Student conceptions of natural selection and its role in evolution. *Journal of Research in Science Teaching, 27*(5), 415–427.

Buckley, B. C., Gobert, J. D., Kindfield, A. C. H., Horwitz, P., Tinker, R. F., & Gerlits, B. (2004). Model-based teaching and learning with Biologica: What do they learn? How do they learn? How do we know? *Journal of Science Education and Technology, 13,* 23–41.

Cavallo, A. M., & McCall, D. (2008). Seeing may not mean believing: Examining students' understandings & beliefs in evolution. *The American Biology Teacher, 70*(9), 522–530.

Dauer, J. T., Momsen, J. L., Speth, E. B., Makohon-Moore, S. C., & Long, T. M. (2013). Analyzing change in students' gene-to-evolution models in college-level introductory biology. *Journal of Research in Science Teaching, 50*(6), 639–659.

Deadman, J. A., & Kelly, P. J. (1978). What do secondary school boys understand about evolution and heredity before they are taught the topics? *Journal of Biological Education, 12*(1), 7–15.

Dobzhansky, T. (1973). *Genetic diversity and human equality* (p. 129). New York: Basic Books.

Donnelly, L. A., Kazempour, M., & Amirshokoohi, A. (2009). High school students' perceptions of evolution instruction: Acceptance and evolution learning experiences. *Research in Science Education, 39*(5), 643–660.

Donnelly, D. F., Namdar, B., Vitale, J. M., Lai, K., & Linn, M. C. (2016). Enhancing student explanations of evolution: Comparing elaborating and competing theory prompts. *Journal of Research in Science Teaching, 53*(9), 1341–1363.

Dukerich, L. (2015). Applying modeling instruction to high school chemistry to improve students' conceptual understanding. *Journal of Chemical Education, 92*(8), 1315–1319.

Furtak, E. M., Seidel, T., Iverson, H., & Briggs, D. C. (2012). Experimental and quasi-experimental studies of inquiry-based science teaching: A meta-analysis. *Review of Educational Research, 82,* 300–329.

Geraedts, C. L., & Boersma, K. (2006). Reinventing natural selection. *International Journal of Science Education, 28*(8), 843–870.

Giere, R. N. (2004). How models are used to represent reality. *Philosophy of Science, 71,* 742–752.

Glaze, A. L., & Goldston, M. J. (2015). US science teaching and learning of evolution: A critical review of the literature 2000–2014. *Science Education, 99*(3), 500–518.

Gregory, T. R. (2009). Understanding natural selection: Essential concepts and common misconceptions. *Evolution: Education and Outreach, 2*(2), 156–175.

Halloun, I. A. (2007). Mediated modeling in science education. *Science & Education, 16,* 653–697.

Harrison, A. G., & Treagust, D. F. (2000). Learning about atoms, molecules, and chemical bonds: A case study of multiple-model use in grade 11 chemistry. *Science Education, 84,* 352–381.

Jackson, J., Dukerich, L., & Hestenes, D. (2008). Modeling instruction: An effective model for science education. *Science Educator, 17*(1), 10–17.

Khine, M. S., & Saleh, I. M. (Eds.). (2011). *Models and modeling: Cognitive tools for scientific enquiry* (Vol. 6). Springer Science & Business Media.

KMK [Sekretariat der Ständigen Konferenz der Kultusminister der Länder in der BRD] (Ed.). (2005). *Bildungsstandards im Fach Biologie für den Mittleren Schulabschluss* [*Biology education standards for the Mittlere Schulabschluss*]. München & Neuwied: Wolters Kluwer.

Krell, M., & Krüger, D. (2016). Testing models: A key aspect to promote teaching activities related to models and modelling in biology lessons? *Journal of Biological Education, 50*(2), 160–173.

Lehrer, R., & Schauble, L. (2015). The development of scientific thinking. In *Handbook of child psychology and developmental science* (Vol. 2, pp. 671–714). Hoboken: Wiley.

Liang, L. L., Fulmer, G. W., Majerich, D. M., Clevenstine, R., & Howanski, R. (2012). The effects of a model-based physics curriculum program with a physics first approach: A causal-comparative study. *Journal of Science Education and Technology, 21,* 114–124.

Malone, K. L. (2008). Correlations among knowledge structures, force concept inventory, and problem-solving behaviors. *Physical Review Special Topics-Physics Education Research, 4*(2), 20107.

Malone, K. L. (2014). Modeling instruction: Empowering students in the 21st century science classroom. In J. Leach, N. J. Ahmad, and S. Tahir (Eds.), *Learning science and mathematics. In the classroom: Case studies of successful practices* (pp. 61–70). Penang, Malaysia: SEAMEO RECSAM.

Malone, K. L., Schunn, C. D., & Schuchardt, A. M. (2018). Improving conceptual understanding and representation skills through Excel-Based modeling. *Journal of Science Education and Technology, 27*(1), 30–44.

Malone, K. L., Stammen, A., Ding, L., Schuchardt, A., Boone, W., & Sabree, Z. (2017). *Development of a concept inventory to measure high school biology students' concept knowledge.* Paper presentation at NARST 2017: The 2017 International Conference of the National Association for Research in Science Teaching, San Antonio, TX.

Nehm, R. H., & Reilly, L. (2007). Biology majors' knowledge and misconceptions of natural selection. *BioScience, 57*(3), 263–272.

Nehm, R. H., & Schonfeld, I. S. (2007). Does increasing biology teacher knowledge of evolution and the nature of science lead to greater preference for the teaching of evolution in schools? *Journal of Science Teacher Education, 18*(5), 699–723.

NGSS Lead States. (2013). *Next generation science standards: For states, by states*. Washington, DC: The National Academies Press.

Passmore, C., & Stewart, J. (2002). A modeling approach to teaching evolutionary biology in high schools. *Journal of Research in Science teaching, 39*(3), 185–204.

Posthuma-Adams, E. (2014). How the chemistry modeling curriculum engages students in seven science practices outlined by the college board. *Journal of Chemical Education, 91*(9), 1284–1290.

Quinn, H., Schweingruber, H., & Keller, T. (Eds.). (2012). *A framework for K-12 science education: Practices, crosscutting concepts, and core ideas*. National Academies Press.

Reiss, M. J. (2010). Science and religion: Implications for science educators. *Cultural Studies of Science Education, 5*(1), 91–101.

Robbins, J. R., & Roy, P. (2007). The natural selection: Identifying & correcting non-science student preconceptions through an inquiry-based, critical approach to evolution. *The American Biology Teacher, 69*(8), 460–466.

Sandoval, W. A., & Reiser, B. J. (2004). Explanation-driven inquiry: Integrating conceptual and epistemic scaffolds for scientific inquiry. *Science Education, 88*(3), 345–372.

Schuchardt, A. (2016). *Learning biology through connecting mathematics to scientific mechanisms: Student outcomes and teacher supports* (Order No. 10298845). Available from ProQuest Dissertations & Theses A&I; ProQuest Dissertations & Theses Global. (1847567134). Retrieved from http://proxy.lib.ohio-state.edu/login?url=http://search.proquest.com.proxy.lib.ohio-state.edu/docview/1847567134?accountid=9783.

Schuchardt, A. M., & Schunn, C. D. (2016). Modeling scientific processes with mathematics equations enhances student qualitative conceptual understanding and quantitative problem solving. *Science Education, 100*(2), 290–320.

Spindler, L. H., & Doherty, J. H. (2009). Assessment of the teaching of evolution by natural selection through a hands-on simulation. *Teaching Issues and Experiments in Ecology, 6,* 1–20.

Stammen, A., Lan, D., Schuchardt, A., Malone, K., Ding, L., Sabree, Z. & Boone, W. (2016). Development of the secondary-biology concept inventory (S-BCI): A study of content and construct validation. In ICMST Conference Committee (Ed.), *Education research highlights in mathematics, science and technology 2016*. Egiten Publishing: Turkey.

Svoboda, J., & Passmore, C. (2013). The strategies of modeling in biology education. *Science & Education, 22*(1), 119–142.

Tansey, J. T., Baird, T., Cox, M. M., Fox, K. M., Knight, J., Sears, D., et al. (2013). Foundational concepts and underlying theories for majors in "biochemistry and molecular biology". *Biochemistry and Molecular Biology Education, 41*(5), 289–296.

Wagh, A., & Wilensky, U. (2014). Seeing patterns of change: Supporting student noticing in building models of natural selection. In *Proceedings of Constructionism 2014*, Vienna, Aug 19–23.

Ware, T., Malone, K. L., Irving, K., & Mollohan, K. (2017, January). Models and modeling: An evaluation of teacher knowledge. In *Proceedings from HICE 2017: The 15th Annual Hawaii International Conference on Education*, Honolulu, HI.

Wilensky, U., & Reisman, K. (2006). Thinking like a wolf, a sheep, or a firefly: Learning biology through constructing and testing computational theories—An embodied modeling approach. *Cognition and instruction, 24*(2), 171–209.

Windschitl, M., Thompson, J., & Braaten, M. (2008). Beyond the scientific method: Model-based inquiry as a new paradigm of preference for school science investigations. *Science Education, 92*(5), 941–967.

Xiang, L., & Passmore, C. (2015). A framework for model-based inquiry through agent-based programming. *Journal of Science Education and Technology, 24*(2–3), 311–329.
Yates, T. B., & Marek, E. A. (2013). Is Oklahoma really OK? A regional study of the prevalence of biological evolution-related misconceptions held by introductory biology teachers. *Evolution: Education and Outreach, 6*, 6. https://doi.org/10.1186/1936-6434-6-6.

**Kathy L. Malone** is an associate professor of education in the Graduate School of Education at Nazarbayev University in Kazakhstan. She served as an assistant professor in the Department of Teaching and Learning at The Ohio State University. Her research currently focuses on the affordances of modeling instruction on student learning within the context of secondary schools. Kathy received her Ph.D. from Carnegie Mellon University's Center for Innovation in Learning and served as a postdoctoral scholar at the University of Pittsburgh's Learning Research and Development Center.

**Anita Schuchardt** is an assistant professor of biology education in the Department of Biology Teaching and Learning at the University of Minnesota. She has published articles on the effect of mathematical modeling in high school biology on student problem-solving and conceptual understanding. Her research interests include developing college-level biology curricula that include model-based instruction, particularly developing and using mathematical and computational models, and researching the effect on conceptual understanding. Anita has received a Ph.D. from Columbia University in genetics and development and from the University of Pittsburgh in learning sciences and policy.

**Zakee Sabree** is an assistant professor in the Department of Evolution, Ecology and Organismal biology at The Ohio State University. He is generally interested in, and has published on, the functional and trophic relationships that forge intimate host–microbe interactions and shape bacterial communities, and the evolutionary outcomes of these symbioses. Zakee received his Ph.D. from the University of Wisconsin-Madison in the Department of Microbiology.

# Cultural Diversity and Evolution: Looking for a Dialogical Teaching Perspective

**A. A. Gómez Galindo, Alejandra García Franco, Leonardo Gonzáles Galli and José de la Cruz Torres Frías**

## 1 Introduction

Learning evolution is challenging for students for various reasons, including the counterintuitive character of scientific models, the existence and resistance of alternative conceptions, the not directly discernible character of evolutionary phenomena (distal causes according to Mayr, 1988), the inadequacy of teaching materials and the conflict with religious worldviews. Over four decades of research on didactics have shown that many students and some teachers do not reach a meaningful level of learning of the basic content of evolutionary biology (e.g. Bergstrom & Dugatkin, 2012; Bishop & Anderson, 1990; Jiménez Aleixandre, 1992; Jungwirth, 1975; Keown, 1988; Lucas, 1971; Settlage, 1994; WGTE, 1998).

In the many different approaches that research about evolution education has undertaken, studies in culturally diverse contexts have been left out. In countries like Mexico, home to more than sixty different indigenous groups, which speak more than 365 dialects of 68 languages, there is a need for a deep reflection on what is relevant in the classrooms of indigenous people, what problems need to be understood and to what ends.

A. A. Gómez Galindo (✉)
Unidad Monterrey, Cinvestav, Monterrey, Mexico
e-mail: agomez@cinvestav.mx

A. García Franco
Universidad Autónoma Metropolitana—Cuajimalpa, Mexico City, Mexico
e-mail: agarcia@correo.cua.uam.mx

L. Gonzáles Galli
Consejo Nacional de Investigaciones Científicas y Técnicas, Buenos Aires, Argentina
e-mail: lmgonzales@gmail.com

J. C. Torres Frías
UdeG, Guadalajara, Mexico
e-mail: cruzfrias@gmail.com



U. Harms and M. J. Reiss (eds.), *Evolution Education Re-considered*,
https://doi.org/10.1007/978-3-030-14698-6_13

In this chapter, we seek to contribute to a field that has been relatively forgotten in research on evolution education: the consideration of cultural diversity. We aim to develop an intercultural dialogic approach to evolution education that includes context and culture as central aspects in the analysis of learning difficulties and in the generation of teaching proposals. The research question addressed in this chapter is: What is the knowledge of Tsotsil students about artificial selection of maize and what is its potential to promote an intercultural dialogic perspective for teaching evolution?

# 2 Teaching Evolution from an Intercultural Dialogic Perspective

## *2.1 Scientific and Traditional Knowledge*

The idea that there should be communication between science and local or traditional knowledge has been around for some time now. The International Council for Science (ICSU) declared that in order to face the current challenges posed to humanity, it is necessary to acknowledge that *we need many partners* and that, whilst it is important to recognize that scientific knowledge has led to innovations that have benefited humankind, it is equally important to recognize that traditional knowledge systems have made valuable contributions to science and technology. Hence, there is a call to preserve, protect, research and promote this cultural heritage and empirical knowledge (ICSU, 2002).

The relationship between scientific knowledge and traditional knowledge is complex. The way in which they interact can be understood from a pluralist epistemology, according to which each kind of knowledge is valid in its own realm and has its own epistemology. We need not only to recognize the diversity of knowledge but also to find spaces and forms in which these can establish a dialogue with each other. The proposal for "knowledge dialogue" that we are advocating recognizes the epistemological status of traditional knowledge as a legitimate way of knowing the world (Olivé, 2007). It also holds that traditional knowledge is efficient for solving problems either on its own or alongside scientific and technological knowledge, as has been shown in problems related to conservation and adaptation to climate change (Parrota & Trosper, 2012).

We understand traditional knowledge as the knowledge that has been developed by people with ample histories of interaction with the natural environment and that originated independently from science in a particular cultural environment (Pérez Ruiz & Argueta Vilamar, 2011). In the present work, we use the term "indigenous knowledge" in the same sense as "traditional knowledge".

## 2.2 *Intercultural Dialogic Education*

The proposal to establish a dialogue between both types of knowledge comes from the recognition that scientific knowledge is constructed and validated socially and that learning science involves learning to participate in a community that has its own rules and communication methods. Aikenhead and Michell (2011, p. 28) express that "a scientific culture depends on expectations shared by its practitioners and these expectations are cultural performances that express certain values and demonstrate expertise". If we recognize science as a culture, we can establish a dialogue with other kinds of knowledge that were generated in a different culture which are related to the same phenomena. An intercultural science education can expand students' and teachers' discourse universe by incorporating different ways of understanding nature (García Franco & Lazos Ramírez, 2016).

Culturally relevant educational practices look to facilitate students' dialogue between their own ways of knowing and scientific ways of knowing (Aikenhead, 2001). Such practices give space to focus more on real-world issues that consider students' lives (Chinn, 2007), making science education a space where there are opportunities for personally significant experiences.

There have been several efforts in different parts of the world to include traditional knowledge in science education (e.g. Aikenhead, 2006). But when it comes to evolution education, attention to context and culture have been absent and we are not aware of any effort that incorporates traditional knowledge in dialogue with science knowledge.

Recognizing and valuing what students and their communities know and bring to school to establish a dialogue with scientific knowledge can empower indigenous students. Aikenhead and Michell (2011) have put forward a series of reasons for integrating indigenous knowledge into the school science curriculum. Amongst the most relevant are equity and social justice, the strength of a nation's economy, improvement of Eurocentric science, indigenous sovereignty and cultural survival.

Our intention when we undertook this project was to establish a "knowledge dialogue" to enhance students' learning of evolution theory and recognize the value of the knowledge generated by indigenous communities by bringing it into the science classroom.

# 3 Evolution Education and Natural Selection

## 3.1 *Importance of Teaching Evolution Theory and Natural Selection*

The theory of evolution is a central framework in biology instruction. This relevance comes from its central place in all biological sciences because every system and biological process is the product of its evolutionary history (Kampourakis & Zogza,

2008; Kampourakis, 2014) and because it has a profound influence in many fields of knowledge. There is a wide consensus that compulsory schooling should aim to convey the basic principles of evolution (Smith, 2010), so every citizen recognizes this theory and its relevance. Among these principles, one that is very relevant for students is the theory of natural selection.

There is ample research that shows that learning evolution poses multiple challenges (Sinatra, Brems & Evans, 2008; Smith, 2010). In particular, different studies show that students hold conceptions that are at odds with evolutionary theory and these conceptions are highly resistant to change through instruction (Kampourakis, 2014). Resistance to changing these conceptions can be attributed to the cognitive bias that is deeply entrenched in the cognitive structure of subjects (Rosengren, Brem, & Sinatra, 2012). Among such bias, we can identify teleology—the tendency to explain phenomena recurring to the notion of goal-directedness—and essentialism—the tendency to assume that everything has an immutable essence that defines it and determines its properties (Coley & Tanner, 2012; Inagaki & Hatano, 2006). One of the main conclusions of current research about teaching and learning evolutionary biology is that evolution theory supposes a way of understanding the biological world, which is strongly counterintuitive. This is the reason why learning it is a difficult process that requires, among other things, an elicitation and conscientious revision of our intuitive suppositions about the biological domain.

In different countries students who have finished secondary education (and even biology teachers and students with bachelors' degrees in biology) have an insufficient understanding of such contents (Evans, 2008). Such research has motivated many proposals to teach evolution (see Sickel & Friedrichsen, 2013). However, contextual aspects and specific examples that have relation to students' lives have remained almost unexplored and the cultural diversity approach is practically absent from the research on evolution education.

## 3.2 The Analogy Between Natural and Artificial Selection

The use of metaphors and analogies is amongst the innovative strategies proposed to improve learning of the theory of evolution. Guided by historical analysis, some authors defend the benefits of using the analogy between natural and artificial selection (that was used by Darwin himself) (Gregory, 2009). However, research on the design and evaluation of teaching–learning sequences that use this analogy is scarce.

The use of analogies in teaching has been widely discussed (Harrison & Treagust, 2005). In terms of the analogy between natural and artificial selection, it is necessary to elucidate if these are two different processes or if one is a particular case of the other (see, e.g., Burnett, 2009). It is clear, in any case, that both processes have a lot in common (variability, for example). Table 1 shows our proposal to identify the main similarities and differences between these two processes.

**Table 1** Comparison between the processes of natural and artificial selection

| Aspects of the process | Natural selection | Artificial selection |
|---|---|---|
| 1. Origin of inheritable inter-individual variants<br>It includes:<br>A. Existence of phenotypical differences<br>B. Differences are partially inheritable because they are due to genetic differences | Mutations and genetic recombination | Mutations and genetic recombination[a] |
| 2. Selective factor | Some environmental factor (pathogens, predators, temperature, etc.) | A single environmental factor: the choice of breeding individuals by the human breeder |
| 3. Consequences of differential reproductive success | Change in the proportion of inheritable variants within the population from one generation to the next | Change in the proportion of inheritable variants within the population from one generation to the next |
| 4. Nature of the differential reproductive success | Probabilistic: not all of the holders of the most advantageous variant breed more than the holders of alternative variants | Deterministic: all of the holders of the advantageous variant breed more than the holders of alternative variants |
| 5. Nature of differential reproductive success | Advantage of the selected trait variant in relation to some selective factor<br>This advantage results from the causal interaction at the individual level between the selected trait variant and the selective factor. For example, thick fur is a better thermal insulator than sparse fur | Advantage of the selected trait variant in relation to some selective factor<br>This advantage results from the coincidence between the selected trait variant and the goal of the human breeder. For example, thicker fur is preferred because it is more useful for the wool industry |
| 6. Intentionality | Absent<br>The entire process is a result of efficient physical causes acting at the level of individuals in interaction with environmental selective factors | Present<br>The entire process is a result of the intentional and conscious choice of the human breeder in relation to his or her objectives |

[a]Here, we exclude the use of technologies, from the mid-twentieth century onwards, that help create new variants through methods such as mutagenesis by radiation, directed mutagenesis or transgenesis. This exclusion is because the diversity of maize used in this work is composed of old variants obtained by the traditional artificial selection of variants resulting from spontaneous mutations. In a broader model of artificial selection, any source of new heritable variants can be included

Beyond these difficulties that point to the need for a conceptual clarification before proposing didactic interventions, we consider that it is indeed an analogy with a high didactical potential for several reasons:

- Artificial selection points towards phenomena that are known and that can be interesting for many students (dog breeds and crop diversity, for example), and this motivation and interest is crucial for understanding (Kayumova & Tippins, 2016).
- It favours the explicit discussion of intentionality and the directed character of the process of change. Intentionality is one of the differences between natural and artificial selection (Table 1) and is related to many of students' alternative conceptions that usually conceive of biological evolution as a directed process, aimed at predetermined ends (Kelemen, 2012).

## 4 Introducing Evolution Education Using an Intercultural Dialogical Teaching Perspective

### *4.1 The Artificial Selection of Maize*

One of the central tenets of intercultural education is that, in order to establish a dialogue, it is necessary to choose phenomena or problems that are relevant for both cultures, where both cultures have an interest in the discussion. In order to bring to the fore traditional knowledge and establish a dialogue with scientific knowledge, we are developing the analogy of artificial selection (domestication) of maize in order to use it as a bridging analogy (Clement, 1993) to learn about natural selection. We are using domestication of maize because maize is a central element of culture in Mexico. Different myths of origin relate to maize, and food has maize as its central component (Carrillo Trueba, 2010).

The evolutionary history of maize has been a riddle for biologists and botanists. The consensus is that the current variety of maize (*Zea mays*) was obtained from the wild ancestor known as teosinte (*Zea* spp.) by a process of artificial selection that started around 9000 years ago in Mexico (Matsuoka et al., 2002).

The artificial selection of maize seeds is a millenary tradition thanks to which many local varieties of the grain have been generated and adapted to very different climatic conditions (CCA, 2004). Currently, more than 3000 races are documented in Latin America. Maize lends itself as a great educational tool because it is one of the best examples of crop domestication and a visual example of genetic inheritance (Fulton, Buckler, & Kissel, 2011).

## 4.2 The Milpa

To introduce ideas of artificial selection for indigenous students, we have to recognize the context in which this process takes place. In indigenous communities, maize is produced in *milpa*, an agro-ecological system considered as a fundamental factor in the constitution of Mesoamerican societies (Carrillo Trueba, 2010). *Milpa* is a policrop based on maize (corn), squash and beans. Besides these three main crops, there are a number of edible vegetables (*quelites*), flowers and trees that form a system that has been recognized as strategic to guarantee food sovereignty, promote biodiversity conservation and allow adaptation to climate change (Álvarez-Buylla, Carreón, & San Vicente, 2011).

*Milpa* is also the centre of quotidian life in indigenous communities, and all festivities are related to *milpa* (Álvarez-Buylla et al., 2011). Currently, over 59 varieties of maize have been identified in Mexico. Knowledge associated with such diversity should be valued and recognized because it could be very relevant to face events such as climate change. It has been recognized that diversity is declining due in part to the knowledge lost among peasants (Dyer, López-Feldman, Yúnez-Naude, & Taylor, 2014).

Indigenous people have a set of knowledge about the ecology, species and variability of maize that can be used to establish a dialogue with models of natural selection via an analogy based on artificial selection.

# 5 Implementation of Evolution Education Activities in Indigenous Groups

The work we are presenting here comes from our different experiences. One of us had previously worked with indigenous teachers and students in the south-eastern part of Mexico (Mountain of Guerrero and Chiapas Highlands). The other three authors have worked in model-based science teaching, and one of them has specialized in evolution. The interest in intercultural education led us to look for ways in which we could relate both fields.

We started by proposing a sequence of activities that are consistent with the conceptual clarification that was the result of comparing natural and artificial selection (Table 1). We also considered how models of natural selection could be related to what students know about maize and the selection process (García Franco, 2015). We designed a series of activities considering students' previous knowledge as well as the theoretical underpinnings for model-based teaching evolution.

The evolution education activities for indigenous groups were carried out in two stages: (1) an exploratory study of secondary students' knowledge in two different

**Table 2** Data collected in two schools in the exploratory study (first stage)

| School 1 | School 2 |
|---|---|
| *Knowledge about milpa* | |
| Flipchart (FE1) with drawings and text | Drawings and text |
| *Explanations about diversity of maize* | |
| Audiotaped oral answers | Individual written answers |

groups of Tsotsil[1] students in the Chiapas Highlands and (2) implementation of activities using an intercultural dialogic teaching approach.

## *5.1 Exploratory Study, 1st Stage*

In June 2015, three of us carried out two activities of two hours each in two secondary[2] schools of a rural zone of the Mayan Highlands of Chiapas, Mexico. The exploratory study included ideas about *milpa* and the sowing process, as well as the origin of the diversity of the maize, and students' knowledge about artificial selection. The students were in the second degree of the secondary level (14 years old) and in the previous year had studied the subject of Biology, with Natural Selection included in the curricular topics. The mother tongue of these students is Tsotsil or Tzeltal, and their second language is Spanish. School 1 (E1) had nine students (6 female, 3 male), and school 2 (E2) had 29 students (16 female, 13 male). In this stage, we compiled artefacts of students' work and took photographs of flipcharts produced (Table 2). In the following section, we present examples of explanations written by students in their flipcharts (FE1), drawings and texts (DE2).

### 5.1.1 *Results: Knowledge About* Milpa

Students establish extensive relationships between *milpa* and various aspects of their life and activities in the community. *Milpa* in general and maize in particular are identified as a central component of the way of life and sustenance: "Maize is part of my life because it gives us food" (Laura-FE1); "The *milpa* is a source of food, in essence what we almost all consume during a meal or breakfast" (Oliver-DE2). We found evidence of an affective relationship with *milpa* and maize: "My maize is so beautiful and there (in the drawing) it's raining" (Lucia-DE2); "I like to take care of

[1]Tsotsil is the language spoken by one of the indigenous groups in Chiapas Highlands. It is spoken by over 400,000 people and is the largest ethnic group, only after Tzeltales.

[2]The type of schools we worked in are *telesecundarias*. This is a particular system in Mexico, where there is only one teacher per group (in regular secondary schools there is a teacher for every subject). The creation of this type of school was seen as a solution to reach distant populations through technological support.

**Fig. 1** Detail of flipchart (left). We can appreciate the decoration of the schoolwork, similar to those of the textile production (see students in the right photograph). *Photograph* G6mez Galindo A.A.

*milpa*, and take good care of myself and I like to grow beans and squash because I really like *milpa*" (Martha-DE2). We identified that this emotional bonding occurs in both girls and boys. Likewise, maize is recognized as a legacy of the original peoples: "[Maize] was known in America before the ancients, it is life" (Fernanda-FE1).

We identified extensive agro-ecological knowledge, for example, cultivation and use of tools, use of policrop (association of maize, beans and squash), quantity of water required and time of planting. Fertilizer use is considered harmful: "Maize treated [with fertilizers] does not grow enough, loses many nutrients, natural corn does not lose anything, it gives us more vitamins" (Oliver-DE2). In addition, they know the varieties of maize in colour and forms—"yellow, black, red, white" (Macrina-FE1)—and recognize the morphology of the grain: "This seed is from a corn as a little crushed, round" (Luis-DE2).

The students, when producing their school work, use a large part of their time in producing drawings, graphic organization and decoration; this could be related to the textile production of these groups, undertaken especially by women but valued by the whole community. These textiles, richly decorated with flowery motifs, are a very important source of the community income. In Fig. 1, we observe a detail of the flipchart produced by a team of girls, in which the drawings are similar to those placed in the textile production.

Finally, an important element is the use of language in the classroom. The teacher spoke in Spanish, and in the school, the language used was Spanish, but Spanish is not the mother tongue of the students, and their use of Spanish in oral and written form is awkward.

### 5.1.2 Results: Explaining the Diversity of Maize

In exploring their explanations for the different varieties of maize, students wrote things like "nobody had ever asked me" (Pedro-E1) and "the truth is I don't know" (Miguel-E2). They also produced tautological answers: "because there are different maizes" (Macrina-E1); "because there are many different *milpas* and different maizes" (Erika-E2).

At E1 school, tautological explanations prevailed; however, at E2, students incorporated some ideas related to genetic manipulation: "it could be about the genetic project as new individuals evolved and thus cause varieties" (Oliver-E2); and "I think it was a question of genetics, perhaps it was the cause of evolution or genetic alterations of the human being" (Miguel-E2). Some attributed the existence of variation to the process of pollination "as there is wind the pollen of the flower reaches another or there are insects that come to suck the pollen and sticks in their legs and if they suck another falls on the cob" (Agustina-E2).

Some students at E2 school expressed ideas that indicate a creationist worldview "so nothing else, they grew on earth and nature, so it was only that the maize came out different from the others" (Alberto-E2). According to González and Meinardi (2009), these ideas refer to an epistemological finalist obstacle, in the sense of explanations whose teleological character "lies particularly in the origin of individual variability which, according to the conceptions of the students, is oriented to the adaptation" (p. 1275).

The exploration indicates that these indigenous students do not have theoretical elements that allow them to explain the origin of the diversity of maize from a biological perspective. These students, according to curricular standards, should know basic ideas of natural selection and domestication. As do students in other contexts, they show finalist explanations and little understanding of the basic ideas of natural or artificial selection (Rosengren et al., 2012). They have, however, knowledge about sowing and selecting the best maize for the next season.

## 5.2 *Activities in Indigenous Groups, Second Stage*

In November 2016, two of us worked with one teacher in two groups (1c and 1d) of the first grade of secondary school in Highlands of Chiapas, in a different community called Ejido Candelaria. This is a "concentration" school, because it receives students from various surrounding communities. Students have to walk from five minutes to three hours to reach the school. The groups, of about 25 boys and girls (13 years old), were mainly Tsotsil, but there were also some Tzeltales and *mestizos*.[3] The classroom teacher, with 14 years of experience, has a degree in biology and a master's degree in the teaching of the natural sciences.

We carried out five activities of 50 min each in both groups. We proposed a trajectory of work, but after each session, the two researchers and the teacher discussed and agreed the next intervention. We used an adaptive methodology, where the ideas constructed by students and the questions and topics they brought to the classroom were considered to take decisions.

To elaborate the working trajectory, we considered the results of the exploratory study. In these, we identified students' extensive knowledge about *milpa* and maize cultivation, as well as the lack of understanding of antecedent concepts required to explain natural selection, for example, biological population, heritability of characteristics and ecological relations.

---

[3]*Mestizos* is the name that is used to identify non-indigenous people.

**Table 3** Activities carried out in the second visit to Mayan Highlands in Chiapas

| | |
|---|---|
| Session 1 | *This is what I know about my* milpa<br>Students made a sketch (to complete at home)<br>*What kind of maize do I know?*<br>Students bring maize from home to highlight diversity<br>Students made a scheme with varieties of maize, indicating characteristics and places each one is sown (they have to complete it at home) |
| Session 2 | *How can we explain the diversity of maize?*<br>Students discuss in a whole-group session<br>A video that introduces the idea of domestication is used<br>*How do I decide which ears to separate to sow in the next crop?*<br>Students have to investigate at home, asking parents and grandparents |
| Session 3 | *Artificial selection simulation*<br>Students simulate the artificial selection of maize |
| Session 4 | *Establishing relations with actual harvest*<br>Students discuss questions related to variety such as: "Why when I sow yellow corn seeds can I get yellow, *pinto* (cobs with two or more colours) or black corn in the harvest?"<br>*How does corn reproduce?*<br>Discussion of sexual reproduction of corn<br>Students draw how the corn seed receives information from both parents |
| Session 5 | *Regulation and metacognitive activities*<br>What did I learn?<br>What activity did I like most and why?<br>How did I feel in the activities? |

In italics the activities that were proposed to students

At this stage, students drew their *milpa*, made a comparative table of maize races, brought information from their communities around the selection of kernels, shared experiences around sowing and established the relation with meteorological conditions, among others. An overview of the main activities is presented in Table 3, although there were slight differences for each group.

In this second stage, we compiled students' productions, took photographs of the flipcharts and recorded and transcribed conversations between the teacher and students. In the following section, we present some examples.

### 5.2.1 Results of Second Stage

The activities were aimed to promote dialogical meaning making from a cultural perspective. We tried to identify the culture in which students' personal ideas are contextualised and then introduce a different cultural point of view, that is, the culture of school science in the context of indigenous students' knowledge (Aikenhead, 2001). This exploratory study allows us to identify traditional knowledge that can have a potential relation with scientific knowledge about artificial selection. We

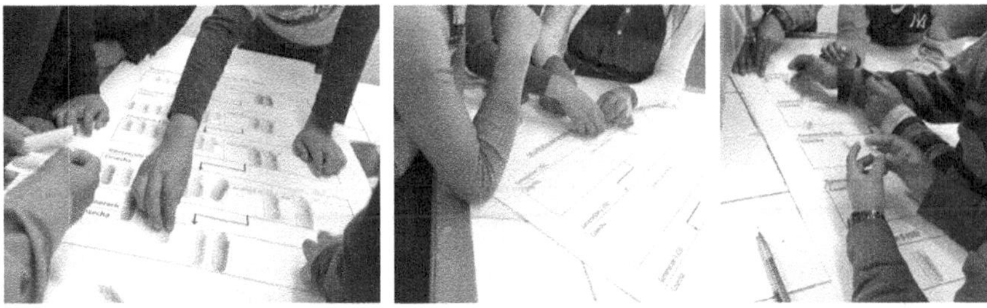

**Fig. 2** Performing a simulation of artificial selection of corn. The interest and collaboration work between students can be seen. *Photograph* Gómez Galindo A.A.

present a brief analysis of the most relevant points that were discussed, trying to emphasize the relations students established and that could be used in the classroom.

*Selection of grain for sowing in my milpa and the intentionality in artificial selection.*

Students know they select the grains for the next harvest from their production in *milpa*. To answer the question *What characteristics do we use to select?*, they turned to their community for information. Students asked their parents, grandparents and other community members. Later, they discussed this information in the school with their peers. Some ideas included that "big and strong" cobs and sobs with "no worms" are selected.

This activity allows us to bring community knowledge to school. The wisdom of indigenous people is distributed between different members of the community, according to the different functions they perform. We should consider the prevalence of oral tradition in indigenous groups, which implies that knowledge is not located in books, nor on the Internet, but in the senior members and the people in charge of the tasks in which this knowledge is used and recreated. In this case, the teacher can open spaces of communication between the curricular contents and the community knowledge.

On the other hand, the students also simulate five generations of maize harvest in which they choose a variant of a trait (larger size) as desirable. Actually, in *milpa*, during the harvest, the families select the best corncobs and use them for the next sowing, so students have to simulate the same process. Students were provided with several cards with corncobs representing the first generation of harvest, and they have to select the best ones (considering their size) for the next sowing. After selection, the teacher gives students the next harvest (several cards with images of corncobs). They repeat the same process for five generations (see Fig. 2). The size distribution of corncobs—in the cards—is carefully adjusted to obtain a change in size distribution of corncobs (the simulation was generated using some of the information in Table 1). After completing the simulation, students identify the increase in the size of the corncobs and reflect on the conditions that allowed these results, establishing relations of similarity between harvest, the simulation and the artificial selection process that takes place in the community.

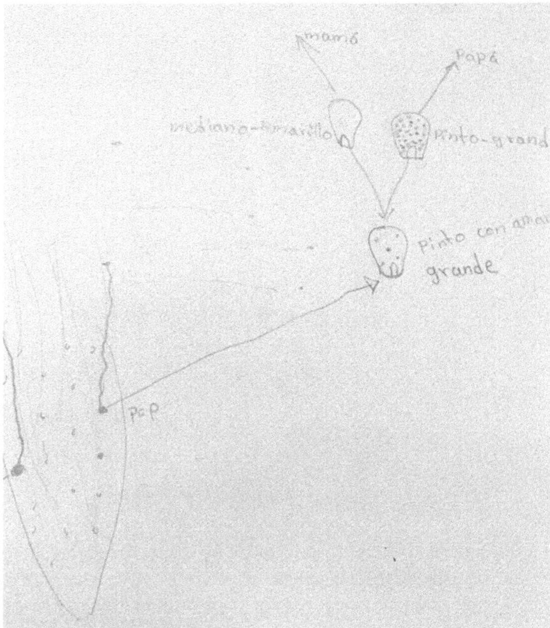


**Fig. 3** Representation of a student on cross-fertilization of maize and inheritance of characteristics of two parents. At the top left, *mamá* (mother) is medium and yellow, and at the right, *papá* (father) is big and *pinto*; students represent some characteristic of the progenitors that are heritable (*pinto* with yellow and big). At the bottom left, students represent the maize and the process of fertilization

The simulation was very motivating for the students; they interacted in teams, showed enthusiasm and interest (see Fig. 2) and talked about how it relates to their *milpa*. Students recognized that they could modify "how does it (maize) comes out", according to the decision of which grain to select.

*Diversity of maize in my milpa in relation to what I planted and the sexual reproduction of maize.*

Students know that the colour of ears in the harvest is not always the same as the colour of the grain they sowed: "if I sow black, it comes out black, *pinto* and white". Faced with the question of why this happens, we referred to human reproduction which they had studied in previous lessons. We also discussed the reproduction of maize and the inheritance of characters from parents in cross-fertilization. Students explained it through drawings (see Fig. 3).

In relation to the origin of heritable inter-individual variants, students recognize the heritability of characters and cross-pollination, but do not identify recombination and mutation as sources of variability. It should be noted that recombination is not part of their pre-existing knowledge in the curriculum, and mutation has not been studied either. The level of construction of students' explanations remains at the possibility of recognizing that there are characteristics inherited from the parents.

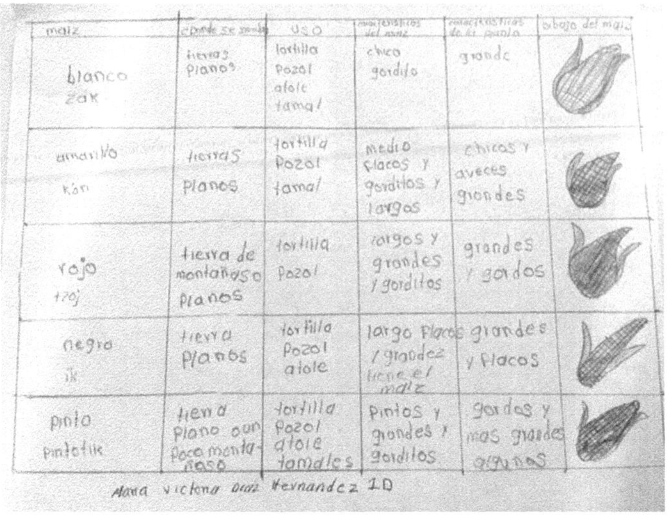

| maiz | dónde se siembra | uso | características del maiz | características de la planta | dibujo del maiz |
|---|---|---|---|---|---|
| blanco zak | tierras planos | tortilla pozol atole tamal | chico gordito | grande | |
| amarillo k'on | tierras planos | tortilla pozol tamal | medio flacos y garditos y largos | chicos y aveces grandes | |
| rojo tzoj | tierra de montañoso planos | tortilla pozol | largos y grandes y gorditos | grandes y gordos | |
| negro ik' | tierra planos | tortilla pozol atole | largo flacos / grandez tiene el maiz | grandes y flacos | |
| pinto pintotik | tierra plano aun poco montañoso | tortilla pozol atole tamales | pintos y grandes y gorditos | gordos y mas grandes algunos | |

Maria victoria Diaz Hernandez 1D

**Fig. 4** Table of types of maize and places to sow, made by team of students. In the columns left to right: name of maize in Spanish and Tsotsil; place where it is sown; what is it used for (different kinds of meals), phenotypical characteristics; type of corn kernel; and graphic representation

**Fig. 5** Some of the diversity of corncobs students brought to school. *Photograph* Gómez Galindo A.A.

*Diversity of my maize, where I sow it and what I use it for, and the short-term response strategy to climate change.*

Students brought to school community knowledge about the relationship between conditions of the ground where *milpa* is located (slope, average temperature, humidity) and the climate conditions expected for the year, e.g. a student says "[in this year] the rains have been delayed" and also the type or race of maize they sow. In Fig. 4, students represent the type—race—of maize, the place for sowing and its characteristics. Figure 5 shows some of the diversity of maize that students have in their *milpas*.

This allowed us to establish a dialogue about the changes that have been taking place year after year (students stated they had not heard about global warming) and

**Table 4** Extract of conversation between teacher and a girl, over hypothetical environmental conditions and types of maize that could be sown

| |
|---|
| Teacher: And, if the year comes with little water, which would you sow? |
| Daniela: The black, because it holds little water |
| Teacher: And if it comes with very little rain and the rainy season is late? |
| Daniela: Yellow because it is stronger |
| Teacher: And if there is a lot of rain and it rains a lot in the year, which one? |
| Daniela: None grows with much rain, it rots |
| Conversation with a student in group 1d in session 3 |

the importance of taking care of maize diversity. Students recognize the relationship between the different kind of maize and their resistance to drought and also the limits to the surviving of species (see conversation in Table 4).

After all activities had been undertaken, students were only beginning to relate the process of selecting maize for the next harvest to artificial selection. The simulation helps student focus on a characteristic of maize and how the choice has consequences for the size of the population. From this point on, it would be necessary to introduce other aspects that could be used in the construction of an artificial selection model that could then be related to natural selection.

## 6 Reflection

The activities were carried out with indigenous students looking to establish a dialogical perspective in the teaching of evolution. Even though we only had the opportunity to implement a few activities, our data indicate some key challenges for intercultural scientific education.

The logic used to introduce ideas is determined not only by the conceptual organization of the discipline or students' conceptions, but for the possibility of retrieving students' knowledge and establishing communication bridges between such knowledge and scientific (school) ideas.

The relation among conceptual knowledge and community knowledge is diverse: sometimes scientific ideas were used to begin an explanation of phenomena observed in the *milpa*; for example, when students sow yellow maize, they can get black maize or a combined maize (*pinto*). This can be explained by cross-pollination and biparental inheritance. At other times, disciplinary knowledge helped organize students' ideas and enabled them to reflect on the relevance of maintaining certain practices, for example, when students inquired in the community which are the optimal conditions to sow certain types of maize, which is related to maize diversity. This way of relating science and traditional knowledge could be important for empowering indigenous groups through recognition in school of the value of their knowledge and practices.

The knowledge about *milpa* and sowing is almost never considered in school. Students had never been asked to draw what is in their *milpa* or to discuss the different varieties of maize and their characteristics. Aikenhead and the teachers who implemented cross-cultural curricular units found something similar: "some students discovered that they already possessed some of this Aboriginal knowledge because it had been taught at home, but it was not highly valued as legitimate knowledge for school" (Aikenhead, 2001, p. 342). School often ignores students' wealth of knowledge, and this brings us to the problem of a scientific education that is completely decontextualized and has no meaning for students (Lave, 1996).

> To the extent that schooling negates the subjective, socioculturally constituted voices that students develop from their lived experience … and to the extent that teachers insist that dialogue can only occur on their terms, schooling becomes an instrument of power that serves to perpetuate the social class and racial inequities that are already inherent in society. (O'Loughlin 1992 in Aikenhead, 2001, p. 816)

The relation between traditional and scientific knowledge needs to be reassessed, taking into consideration not only what students know but also what is relevant for their culture. Research on multicultural education, equity and social justice (Atwater, 2011) can shed light on the way in which the relevance of knowledge is fundamental for students to learn and participate in their communities (Roth & Calabrese, 2004; Varelas, Martin, & Kane, 2012). We need to be aware that this knowledge is an integral part of a culture, inserted in their everyday lives, and that has been relevant for their own survival.

It is evident from this work that there is a strong emotional relationship of students with *milpa* and maize. This emotional aspect of learning science has been minimized in research and policy. In this sense, Kayumova and Tippins (2016, p. 568) state:

> From our situated experiences reforms, colleges of education, schools, and curriculum place not enough emphasis on affective and bodily dimensions of teaching and learning. Instead, the privilege seems to be given to reason, evidence, and rationalities, which continue to reinforce dominant ways of knowing and experiencing.

We also identified some obstacles shared by students elsewhere. Students do not fully understand the existence of phenotypical differences and that transmission of such differences is partially heritable because they are due to genetic differences. They also misunderstand the scientific idea of what a population is and do not recognize the variability inside populations.

In this experience, some of the students' ideas could be fruitful for establishing relations with ideas expressed in Table 1, in particular intentionality, possibility of directing changes through selection, heritability of characters, the relation between phenotype and environmental factors (nature of differential reproductive success).

We are well aware that there is still a long road to travel from where we got to our destination, but working with indigenous students and their teachers has allowed us to recognize the great opportunity for teaching natural selection to students who participate in daily practices involving artificial selection. Their knowledge of the environment and recognition of diversity brings new elements to be considered in a teaching sequence.

We recognize that the establishment of these spaces for dialogue is very complex because sometimes the instructional style of school science collides with the way indigenous students learn in their communities. There are also language obstacles in the classroom, given that oral language is the privileged way for knowledge construction in the classroom (Lemke, 1990). With indigenous students whose mother tongue is different from the language of instruction, this is even more relevant.

## 7 Conclusions and Perspectives

The intercultural dialogic perspective we have discussed in this chapter is relevant for indigenous students because it could be used to construct the tools needed to survive in this highly technological world without denying their own identity. Brandt (2008) has shown, through the story of a Navajo student, that fostering supportive peer relationships and establishing classroom environments that respect indigenous perspectives can make a significant difference in how students view science and are able to navigate between different models to make sense of natural phenomena.

An intercultural dialogic approach could also benefit non-indigenous students because there are links between biodiversity and cultural diversity (Maffi, 2007) that every student has the right to know. In the particular case of Mexico, where pluricultural composition is recognized in the constitution, there is also a long history of racism and discrimination against indigenous people. Science classrooms could become spaces to question prejudices about indigenous people being ignorant and dependant on government handouts (Varelas et al., 2012).

Our work proposes that, in order to learn significantly about evolution, specific contexts and cultures of students need to be considered. We need to go beyond the conceptual structure of evolution towards finding ways to increase the relevancy of this knowledge for students, so that they can use it to make sense of the world they inhabit.

For us, this work represents the first steps in the construction of an intercultural approach to evolution education. Domestication of maize could be considered as a paradigmatic case for teaching evolution not only for its biological but also for its cultural implications.

**Acknowledgements** Thanks to students in the Chiapas Highlands, their teachers, and the reviewers of this chapter. This work was supported by grants SEP/SEB 2013, No. 231425 y SEP/SEB 2014-01, No. 240192, CONACYT, México.

## References

Aikenhead, G. (2001). Integrating western and aboriginal sciences: Cross-cultural science teaching. *Research in Science Education, 31*, 337–355.

Aikenhead, G. (2006). *Science education for everyday life: Evidence based practice*. New York: Teachers College Press.

Aikenhead, G., & Michell, H. (2011). *Bridging cultures. Indigenous and scientific ways of knowing nature*. Canada: Pearson.

Álvarez-Buylla, E., Carreón, A., & San Vicente, A. (2011). *Haciendo milpa. La protección de las semillas y la agricultura campesina*. México: UNAM.

Atwater, M. (2011). Significant science education research on multicultural science education, equity, and social justice. *Journal of Research in Science Teaching, 49*(1), 1–5.

Bergstrom, C., & Dugatkin, L. (2012). *Evolution*. New York: Norton & Company.

Bishop, B., & Anderson, C. (1990). Students conceptions of natural selection and its role in evolution. *Journal of Research in Science Teaching, 27*(5), 415–427.

Brandt, C. (2008). Scientific discourse in the academy: A case study of an American Indian undergraduate. *Science Education, 92,* 825–847.

Burnett, D. (2009). Savage selection: Analogy and elision in on the origin of species. *Endeavour, 33*(4), 120–125.

Carrillo Trueba, C. (2010). *La milpa y la cosmovisión de los pueblos mesoamericanos* (p. 34). La Jornada: La Jornada del Campo.

CCA (Comisión para la Cooperación Ambiental). (2004). *Maíz y biodiversidad*. http://www3.cec.org/islandora/en/item/2152-maize-and-biodiversity-effects-transgenic-maize-in-mexico-key-findings-and-es.pdf. Accessed April 21, 2017.

Clement, J. (1993). Using bridging analogies and anchoring intuitions to deal with students' preconceptions in physics. *Journal of Research in Science Teaching, 30*(10), 1241–1257.

Chinn, P. (2007). Decolonizing methodologies and indigenous knowledge: The role of culture, place and personal experience in professional development. *Journal of Research in Science Teaching, 44*(9), 1247–1268.

Coley, J., & Tanner, K. (2012). Common origins of diverse misconceptions: Cognitive principles and the development of biology thinking. *CBE-Life Sciences Education, 11*(3), 209–215.

Dyer, G. A., López-Feldman, A., Yúnez-Naude, A., & Taylor, J. E. (2014). Genetic erosion in maize's center of origin. *Proceedings of the National Academy of Science, 111*(39), 14094–14099.

Evans, E. M. (2008). Conceptual change and evolutionary biology: A developmental analysis. In S. Vosniadou (Ed.), *International handbook of research on conceptual change* (pp. 263–294). New York: Routledge.

Fulton, T. M., Buckler, C. S., & Kissel, R. A. (2011). *The teacher-friendly guide to the evolution of maize*. New York: Paleontological Research Institution.

García Franco, A. (2015). La milpa como proyecto didáctico. In A. Gómez Galindo & M. Quintanilla (Eds.), *La enseñanza de las ciencias naturales basad en proyectos* (pp. 155–172). Chile: Bellaterra.

García Franco, A., & Lazos Ramírez, L. (2016). Designing educational material for scientific intercultural education: The harvest of milpa in Mexico as an example for dialogue. *Revista Brasileira de Pesquisa em Educao em Ciencias, 16*(3), 871–888.

González, L., & Meinardi, E. (2009). *El pensamiento finalista como obstáculo epistemológico para la enseñanza del modelo darwiniano. Enseñanza de las Ciencias, Número Extra*. http://ensciencias.uab.es/congreso09/numeroextra/art-1274-1276.pdf. Accessed April 11, 2017.

Gregory, R. (2009). Artificial selection and domestication: Modern lessons for Darwin's enduring analogy. *Evolution, Education and Outreach, 2*(1), 5–27.

Harrison, A., & Treagust, D. (2005). Teaching and learning with analogies: friend or foe? In P. Aubusson, A. Harrison, & S. Ritchie (Eds.), *Metaphor and analogy in science education*. Holland: Springer.

ICSU. (2002). *Science and traditional knowledge*. Report from the ICSU Study Group on Science and Traditional Knowledge.

Inagaki, K., & Hatano, G. (2006). Young children´s conception of the biological world. *Current Direction of Psychological Science, 15*(4), 177–181.

Jiménez Aleixandre, M. (1992). Thinking about theories or thinking with theories?: A classroom study with natural selection. *International Journal of Science Education, 14*(1), 51–61.

Jungwirth, E. (1975). The problem of teleology in biology as a problem of biology-teacher education. *Journal of Biological Education, 9*(6), 243–246.

Keown, D. (1988). Teaching evolution: Improved approaches of unprepared students. *The American Biology Teacher, 50*(7), 407–410.

Kampourakis, K. (2014). *Understanding evolution*. England: Cambridge University Press.

Kampourakis, K., & Zogza, V. (2009). Preliminary evolutionary explanations: A basic framework for conceptual change and explanatory coherence in evolution. *Science & Education, 18*(10), 1313–1340.

Kayumova, S., & Tippins, D. (2016). Toward re-thinking science education in terms of affective practices: Reflections from the field. *Cultural Studies of Science Education, 11*(3), 567–575.

Kelemen, D. (2012). Teleological minds. How natural intuitions about agency and purpose influence learning about evolution. In K. Rosengren, S. Brem, & G. Sinatra (Eds.), *Evolution challenges. Integrating research and practice in teaching and learning about evolution* (pp. 66–92). England: Oxford University Press.

Lave, J. (1996). La práctica del aprendizaje. In S. Chalkin & J. Lave (Eds.), *Estudiar las prácticas, perspectivas sobre actividad y contexto* (pp. 15–45). Buenos Aires: Amorrortu.

Lemke, J. (1990). *Talking science: Language, learning and values*. Norwood, NJ: Ablex.

Lucas, A. (1971). The teaching of "adaptation". *Journal of Biological Education, 5*(2), 86–90.

Maffi, L. (2007). Biocultural diversity and sustainability. In J. Pretty, A. Ball, T. Benton, et al. (Eds.), *Handbook of environment and society* (pp. 267–277). UK: Sage.

Matsuoka, Y., Vigouroux, Y., Goodman, M. M., Sánchez, G. J., Buckler, E., & Doebley, J. (2002). A single domestication for maize shown by multilocus microsatellite genotyping. *Proceedings of the National Academy of Sciences, 99,* 6080–6084.

Mayr, E. (1988). *Toward a new philosophy of biology: Observations of an evolutionist*. Cambridge: Harvard University Press.

Olivé, L. (2007). *La ciencia y la tecnología en la sociedad del conocimiento. Ética, política y epistemología*. México: Fondo de Cultura Económica.

Parrota, J. A., & Trosper, R. L. (2012). *Traditional forest-related knowledge*. The Netherlands: Springer.

Pérez Ruiz, M., & Argueta Villamar, A. (2011). Saberes indígenas y diálogo intercultural. *Cultura y Representaciones Sociales, 5*(10), 31–55.

Rosengren, K., Brem, S., & Sinatra, G. (Eds.). (2012). *Evolution challenges. Integrating research and practice in teaching and learning about evolution*. Oxford, England: Oxford University Press.

Roth, W. M., & Calabrese, A. (2004). *Rethinking scientific literacy*. New York: Routledge.

Settlage, J. (1994). Conceptions of natural selection: A snapshot of the sense-making process. *Journal of Research in Science Teaching, 31*(5), 449–457.

Sickel, A., & Friedrichsen, P. (2013). Examining the evolution education literature with focus on teachers: Major findings, goals for teacher preparation, and directions for future research. *Evolution, Education and Outreach, 6*(23). http://www.evolution-outreach.com/content/6/1/23/. Accessed March 17, 2017.

Sinatra, G., Brem, S., & Evans, M. (2008). Changing minds? Implications of conceptual change for teaching. Learning about biological evolution. *Evolution Education and Outreach, 1*(2), 189–195.

Smith, M. (2010). Current status of research in teaching and learning evolution: II. Pedagogical issues. *Science and Education, 19*(4–8), 523–538.

Varelas, M., Martin, D. B., & Kane, J. M. (2012). Content learning and identity construction (CLIC): A framework to strengthen African American students' mathematics and science learning in urban elementary schools. *Human Development, 55,* 319–339.

WGTE (Working Group on Teaching Evolution). (1998). *Teaching about evolution and the nature of science*. Washington, DC: National Academy Press.

**A. A. Gómez Galindo** is Professor in the Center for Research and Advanced Studies of the National Polytechnic Institute (Cinvestav) in Monterrey, México. She has a degree as teacher of early childhood education, is Marine Biologist and has a Ph.D. in didactics of science. She performs qualitative research on biology education, focusing on teaching–learning sequences for modelling by using multimodal representations and analogies. Her current projects include the analysis of dialogic perspectives for teaching evolution in cultural diversity contexts and developing of learning progressions for central models in biology to guide learning from kindergarten to middle school.

**Alejandra García Franco** is Chemical Engineer and has a Ph.D. in pedagogy from the National Autonomous University of Mexico (UNAM). She is Teacher-Researcher at the Autonomous Metropolitan University—Cuajimalpa. She is interested in intercultural scientific education and has collaborated for education projects with indigenous people in Mexico. She is also interested in chemistry learning and the design of teaching learning sequences for teacher training.

**Leonardo González Galli** has a Ph.D. in biological science and an M.Ed. in Biology Education of Secondary and Higher Education from Buenos Aires University. He is Research Assistant for the National Council of Scientific and Technological Research (CONICET) and Assistant Professor of Didactics of Biology at the CeFIEC Institute of Buenos Aires University. His current line of research is focused on the problems of learning and teaching models of evolutionary biology, a subject on which he has published numerous journal articles and book chapters; he has taught lectures and teacher training courses.

**José de la Cruz Torres Frías** has a Ph.D. in Education and is currently Postdoctoral Fellow in didactics of science in the Center for Research and Advanced Studies (Cinvestav) of the Polytechnic Institute Nacional-Unidad Monterrey, Mexico. He is Member of the National System of Researchers in Mexico. Current lines of research are training for research in higher education and postgraduate courses and in-service science teachers' training in primary and secondary school.

# Transforming a College Biology Course to Engage Students: Exploring Shifts in Evolution Knowledge and Mechanistic Reasoning

**Lisa O. Kenyon, Emily M. Walter and William L. Romine**

## 1 Instructional Vignette

### *1.1 Environmental Conditions Are Changing and the Birds Are Dying*

Students are about to embark on an online investigation to figure out what is happening to the birds. They clambered onto the only accessible rock to Daphne Major, accompanied by Drs. Peter and Rosemary Grant, 300 miles west of Ecuador. So began their engagement into synthesizing ground finch data from the Galapagos Islands. Students look at large data sets, determine patterns of evidence, and construct explanations about why some finches die and some survived. The students do not always agree. With their partners, students search through the database looking at environmental factors, food availability, predator–prey interactions, and morphometric traits such as weight, wing, beak, and leg length to find evidence that supports their claims about what happened on the Island.

As I walk around the room listening to their discussions, I hear comments such as "I think it has something to do with the rainfall, look [pointing to a graph], the rainfall decreased from 200 cm in wet 1973 to 25 cm in 1977. This was happening the same time the finch population was decreasing, look [pointing to another graph]

L. O. Kenyon (✉) · W. L. Romine
Departments of Biological Sciences and Teacher Education, Wright State University, 3640 Colonel Glenn Highway, Dayton, OH 45435-0001, USA
e-mail: lisa.kenyon@wright.edu

W. L. Romine
e-mail: romine.william@gmail.com

E. M. Walter
Science & Mathematics Education Center, California State University, Fresno 2555 E. San Ramon Ave, Fresno, CA 93740-8034, USA
e-mail: ewalter@csufresno.edu



U. Harms and M. J. Reiss (eds.), *Evolution Education Re-considered*,
https://doi.org/10.1007/978-3-030-14698-6_14

the population decreased from 60 in wet 1976 to 23 in wet 1977." Another group is arguing that they found evidence that the hawks are eating the finches. "Here, look [pointing at field notes in wet season 1976]. Gf71 was swept up by a hawk, dropped near the waterfront, and devoured."

Students ask me if they have the right answer, checking for my approval. I continually redirect their focus and ask them to think about their evidence; does it support their claim? How could the evidence rebut the claim? Once partners have constructed their explanations, I pair groups to work together converging on one explanation. Students are very chatty; all groups are busy evaluating, modifying, and defending ideas. Students arrive at a variety of explanations, some claim the results are gender-driven, with females out-surviving males, while others are convinced it has to do with beak size and eating the available harder seeds. Students tell me, "Wow, I never really understood natural selection, this makes sense to me know," "We never really learned about evolution in high school," "I never really understood this before because no one explained it this way," and "I totally get this, especially when thinking about how the finch population changed with respect to beak size."

## 2 Course Overview and Rationale

A central action of many post-secondary pedagogical initiatives is to encourage college and university instructors to adopt approaches based in research on how people learn (AAAS, 2011; Bransford, Brown, & Cocking, 2000). Despite this, efforts to transform the nature of post-secondary instruction have had limited success (e.g., AAAS, 2013), and as many as 70–90% of post-secondary instructors teach exclusively through lecture (Alters, 2005). In these settings, students learn to play the game of memorizing information, but have little or no meaningful learning. These challenges are compounded for topics like evolution, which are difficult to comprehend (e.g., Bishop & Anderson, 1990; Demastes, Good, & Peebles, 1996; Moore et al., 2002) and may be in conflict with students' worldviews.

There are a number of promising strategies for addressing the challenge of effective evolution instruction. The focus of our work in this project was a shift to providing more opportunities for students to build reasoning skills around content knowledge (Berland et al., 2016).

As such, it was our goal to transform an introductory, lecture-based biology course toward a more active learning environment built around science practices. We continued to follow a course content sequence of the typical biology textbook (Reece et al., 2014), but incorporated a different instructional framework based on the Next Generation Science Standards storyline K-12 model (Reiser, 2017). This framework provides a coherent sequence of lessons in which students generate questions by experiencing scientific phenomena. These questions then lead to investigations, situating students in contexts where they figure out problems while engaged in the science practices (NGSS Lead States, 2013).

We foregrounded the importance of evidence as the main objective, thus fostering the use of evidence in figuring out problems through explanatory thinking. We prompted students to answer "how and why" questions through mechanistic responses. Russ, Scherr, Hammer, and Mikeska (2008) define mechanism as a type of causal reasoning addressing how and why individual components of a phenomenon interact with one another over a period of time. More specifically, mechanisms represent non-teleological reasoning (Russ, Coffey, Hammer, & Hutchinson, 2009) and provide the rationale for why a phenomenon occurs. For instance, mechanistic models focus on several particular conditions: target detailed phenomena, identify initial conditions, identify entities, identify the organization of entities, and chain thoughts by working backwards or forwards to explain the situation. Other literature has interpreted mechanisms in similar ways as theoretical accounts that allow for causal explanations and testable predictions about the natural world (Darden & Craver, 2002; Machamer, Darden, & Craver, 2000). Our students collected empirical evidence and used the data to make sense of (a) mechanisms of evolutionary change, (b) body systems, (c) plant biology, and (d) ecology and then connected these phenomena back to the unifying theme of evolution.

## 3 Literature Review

Like all quality teaching, effective evolution instruction is based at least partly on understanding educational psychology. Much of the theory on how people learn has been translated into various evolution education interventions. At the root of these interventions is the principle that active, evidence-based learning has significant advantages over traditional, lecture-based approaches. In an exhaustive meta-analysis of STEM education research papers, average examination scores improved by about 6% in active learning sections, and students in classes with traditional lecturing were about 1.5 times more likely to fail than were students in classes with active learning (Freeman et al., 2014). A discussion of many of these pedagogical interventions as related to evolution can be found in Andersson and Wallin (2006) and Smith (2010). We summarize some of this literature herein as it fits with our study.

Our intervention model is grounded in two active learning precepts. First, *reasoning and critical thinking are key to building understanding of evolution* (Clough & Wood-Robinson, 1985; Lawson & Weser, 1990; Wandersee, 1985) and acceptance of evolution (Lawson & Worsnop, 1992). Second, *direct experience with phenomena is key to building understanding* of evolution (e.g., Nehm & Reilly, 2007).

### 3.1 Reasoning and Critical Thinking Is Key to Building Understanding of Evolution

Passmore and Stewart (2002) report on the design and implementation (but not the evaluation) of a nine-week elective high school course on evolution. The goal of this course was to "initiate students into the reasoning patterns of the discipline by engaging them in inquiry contexts that required them to develop, use, and extend Darwin's model of natural selection and to gain some experience with the significance of historical reconstructions" (p. 190). Students examined four real-world data-rich cases using the models of Paley, Lamarck, and Darwin, examining the phenomena to be explained, comparing underlying assumptions/beliefs, and comparing the explanatory power of each. This approach is noted as well grounded and worthy of replication and evaluation (Smith, 2010).

Evaluation of interventions built to improve students' reasoning and critical thinking skills around evolution was a focus of several early evolution education research studies. Much of this work has been done at the primary and secondary levels, but could still apply to a population of college science learners. Lawson and Thompson (1988) argued that formal reasoning skills enable students to modify prior beliefs (e.g., Posner, Strike, Hewson, & Gertzog, 1982), and therefore, the extent to which students hold non-scientific beliefs should be related to this skill. They examined seventh-grade students and found that after receiving instruction on genetics and natural selection, a sample of concrete operational (per Piaget) students held significantly more misconceptions than their formal operational peers. In the case of our study, this implies that if understanding evidence requires formal reasoning skills, it would seem necessary for the students to be formal operational; hence, instruction must be designed to promote its development in concrete operational students. Along these lines, Lawson and Worsnop (1992) noted that skill in reflective reasoning facilitated conceptual knowledge acquisition. They found that grade 10 students who were accomplished reflective (hypothetico-deductive) thinkers exhibited greater conceptual knowledge gains about evolution and natural selection than peers who were less skilled at reasoning.

Lawson and Weser (1990) found that college students who were less skilled at reasoning were more likely to hold non-scientific beliefs and were less likely to change those beliefs during instruction. They also discovered that students who were less skilled at reasoning were also less likely to be strongly committed to scientific beliefs. In other words, students who have poorly developed hypothetico-deductive reasoning skills may hold a correct scientific conception, but may not be strongly committed to that perception. Such students agree with an idea because they have been told that it is correct, rather than arriving at that idea themselves through an internal hypothetico-deductive dialogue around the evidence.

### 3.2 Direct Experience with Phenomena Is Key to Building Understanding

According to the NGSS (NGSS Lead States, 2013) natural phenomena are "observable events that occur in the universe and that we can use our science knowledge to explain or predict. The goal of building knowledge in science is to develop general ideas, based on evidence, that can explain or predict phenomena." Our intervention model is also supported on the premise that experience with evolutionary phenomena is beneficial to understanding. Scientific ideas are more likely to occur when students can experience phenomena directly (Alters, 2005). Live, eukaryotic organisms with a short generational time are best for observing evolution in action. Experiments using genetically modified foods, *Drosophila* (Coleman & Jensen, 2007; Plunkett & Yampolsky, 2010; Salata, 2002), *E. coli*, or cross-fertilization of plants (Sinatra et al., 2008) provide actual observations of natural, artificial, and/or sexual selection phenomena.

## 4 Instructional Intervention

As noted in our course design framework (Fig. 1), we integrated multiple activity structures to emphasize empirical reasoning skills and experience with evolutionary phenomena. We first set underlying concepts, anchored by evolution as a unifying concept, and focused on quality rather than quantity of the biological content. We explored questions about the evolutionary origin of animals and plants, their morphology and physiology, and the ecological interactions between organisms and ecosystems they inhabit. Throughout the course, evolution was the unifying theme (e.g., Coker, 2009).

Once we determined weekly topics and associated chapters, we planned for specific activities to teach each day. We selected resources that required students to examine data and use data as evidence to figure out scientific questions and/or make scientific explanations. Based on these criteria, we used Howard Hughes Medical Institute (HHMI) BioInteractive, a free online Web site. HHMI BioInteractive provides a variety of multimedia, apps, videos, interactives, and virtual laboratories that allow students to explore science through a scientific lens. Most media are coupled with student handouts for active learning exercises. These served as formative assessments for the course. We identified four short films, coupled with apps, interactives, and virtual laboratories as contexts that pushed on examining scientific evidence.

In addition to HHMI BioInteractive, we used another computer resource called *Gizmo*. *Gizmos* are online learning simulations that allow students to figure out concepts through making predictions, collecting data, interpreting graphs, and justifying conclusions. We used two *Gizmos* during the class to support student understanding about the digestive and circulatory systems. Other additional computer resources included a Web site called *BGuILE*, The Galapagos Finches, to examine both quan-

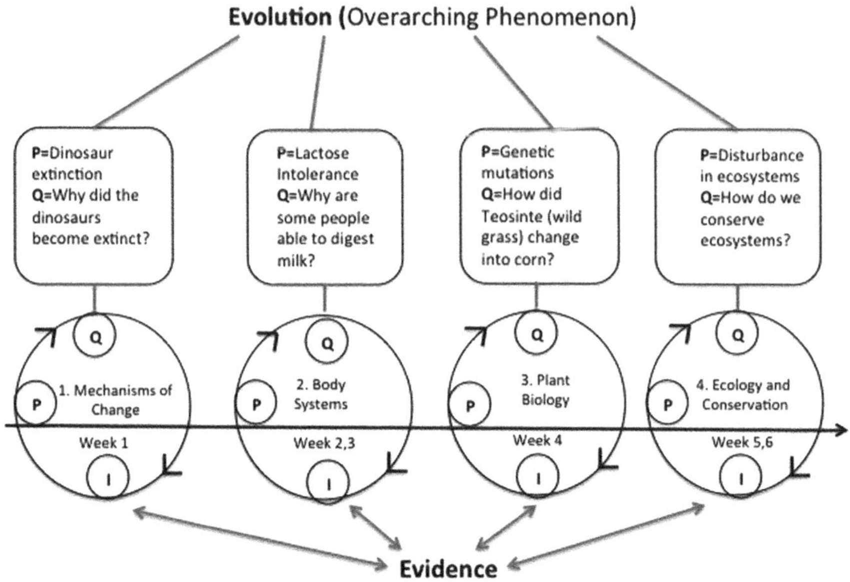


**Fig. 1** Course design framework. P = phenomenon, Q = questioning, I = investigation. Using HHMI video resources, each unit was anchored with a phenomenon, followed by a question to investigate using evidence. Evolution was the overarching phenomenon that was threaded throughout each unit and provided course coherence. Arguing with evidence was foregrounded as the key practice to figuring out the investigation and used to explain phenomena

titative and qualitative data and explore interrelationships between organisms and environmental influences on a finch population. And lastly, we used a computer simulation program called *NetLogo* to examine changes made to a community of organisms where students were tasked in predicting what happens when we add an unknown invader to this ecological system. In Table 1, we show examples of active learning activities aligned with the lecture material for the course. This table is not a complete list of the activities, but a summary of typical weeks. For each week, we would teach four lectures including 2–3 active learning days. Each activity would include one of the practices (modeling argumentation, or explanation). Students worked in groups during class, making sense of the activity practices or problem-based questions.

To achieve emphasis on empirical reasoning skills and experience with evolutionary phenomena, we developed instructional activities around the NGSS Science and Engineering Practices (SEPs; NGSS Lead States, 2013). While the NGSS was written for a primary and secondary (K-12) audience, the framework could be applied to college classrooms, as there is little reason to believe that high school students learn differently than early college students. Furthermore, both the DBER Report and Vision and Change give scientific practices and content equal importance (AAAS, 2011; NRC, 2012).

**Table 1** Examples of lecture material for the course aligned with active learning activities

| Lecture material | Active learning activities during course time |
|---|---|
| History of Life on Earth<br>Mechanisms of Evolution<br>Evolution of Population<br>Origin of Species | Activity 1: HHMI Film *The Day the Mesozoic Died*. Students worked in groups of four to examine the KT boundary trail of *evidence* for why the dinosaurs became extinct. Handouts used in this activity were provided by the HHMI Web site for this particular film<br>Activity 2: BGuILE, Beak of Finch (bird) Investigation. Students worked in pairs to examine a database to construct an explanation for why finches died and why some survived. They used their explanations to argue their claims using *evidence* as they formed consensus explanations with another group pair |
| Animal Form and Function<br>Animal Nutrition | Activity 3: Gizmo on the Digestive System. Students, working in pairs, use a computer simulation to learn the structure and function of the digestive system. They manipulate the addition and removal of structures to design the best functioning digestive system<br>Activity 4: HHMI film on *Got Lactase? The Co-evolution of Genes and Culture*. Students examine the *evidence* that explains genetic changes associated with the ability to digest lactose in milk, while tracing it to earlier civilizations where human populations started domesticating animals. Handouts used in this activity were provided by the HHMI Web site for this particular film |
| Circulation<br>Gas Exchange | Activity 5: Students construct scientific models on how and why blood moves through the body. Students share these models in class, participate in a Gizmo to gather *evidence* on circulation, and then revise their models |
| Evolution of Seed Plants<br>Plant Structure and Growth | Activity 6: HHMI film *Popped Secret: The Mysterious Origin of Corn*. Students, working in pairs, examine the *evidence* of how genetic changes were involved in the transformation of Teosinte (wild grass) into corn. Handouts used in this activity were provided by the HHMI Web site for this particular film |

A *scientific practice* represents social and scientific construction, evaluation, and communication of scientific knowledge (Duschl, Schweingruber, & Shouse, 2007). The goal is that students become well-grounded in scientific theory and thus able to form legitimate questions about the natural world around them and then use these practices to discover the answers to their questions. The eight NGSS SEPs include ideas such as *Developing and Using Models*, *Constructing Explanations*, *Planning and Carrying out Investigations*, *Analyzing and Interpreting Data,* and *Engaging in Argument from Evidence*. In the case of our intervention, we engaged students specifically in three NGSS SEPs: (a) developing and using models, (b) constructing explanations, and (c) engaging in an argument from evidence.

In a traditional college course, student engagement with SEPs likely would be relegated to the laboratory. This is a missed opportunity from our perspective. Therefore, our ancillary goal was to transform a lecture environment using technological tools for social sense making around the SEPs. We introduced "*untethering*," a process that begins with mobile device mirroring with a tablet (in this case, an iPad). The instructor uses the iPad as a tool to untether from the podium and walk around the room to engage students in the material and discussions. The use of the iPad device provided opportunities for leveraging inquiry-based apps that allowed the professor and students to share student work and ideas for whole class discussions. More students participated in peer-collaborative learning through the technology with this pedagogical initiative (Thinley, Reye, & Geva, 2014).

# 5 Research Methods

## *5.1 Paradigm*

We assume a post-positivist paradigm for this research study, reflecting a single, objective reality that is measurable by survey data. We therefore chose to ask research questions that fit a quantitative approach to approximate a single reality, i.e., "What do the students know?" Post-positivism as a paradigm challenges the traditional, positivist idea of an absolute truth (Phillips & Burbules, 2000) and recognizes that we cannot always know reality when studying behavior and actions of human subjects. Reality from a post-positivist view is based on cautious observation and measurement of the objective reality that exists "out there" in the world. In our case, exploring the thinking of individuals through survey data reflects our post-positivist paradigm (Creswell, 2009).

## *5.2 Context*

We examined 70 science majors in an introductory biology class at a research-intensive, open-enrollment university in the Midwest USA. The total minority student enrollment was 19.7% (10.4% African-American, 3.3% two or more races, 2.8% Asian American, 2.9% Hispanic American, 0.2% American Indian or Alaskan Native, and 0.1% Native Hawaiian or Pacific Islander), and the total international student enrollment was 11% (65 countries). As this class was an introductory course, its demographic distribution was consistent with the university as a whole. The course took place during a summer term when students met for six weeks, four days a week, for 1.5 h a session. Most students were science majors (many pre-medical profession), engineering, and computer science majors and had taken the prerequisite course on cells and genetic biology. Few, if any, students experienced active learning in their prior courses since attending the university. Typically, this summer session class was lecture only. These longer-time sessions allowed for this redesign opportunity to engage students in practices that support both cognitive and social approaches to the scientific discipline. The course professor had a graduate degree in biological sciences and, most influentially, had a doctorate in science education with an emphasis on curriculum design and instruction. The professor used her research and teaching philosophy to guide the design modifications in this course.

## *5.3 Instrumentation*

### 5.3.1 Knowledge of Natural Selection

We used the Conceptual Inventory of Natural Selection (CINS; Anderson, Fisher, & Norman, 2002) as a measure for knowledge of natural selection before and after instruction. The CINS was developed in response to previous instruments (Bishop & Anderson, 1990; Settlage & Odom, 1995) because the authors found the old instruments to be overly simplistic and abstract. Their solution to this was to develop an instrument that used actual evolutionary examples (e.g., Galapagos finches, Venezuelan guppies *Poecilia reticulata*, and Canary Island lizards). The 20-item CINS was therefore developed to measure non-science majors' understanding of natural selection. It was designed for each item to have one correct answer and three distracter answers based on common alternative conceptions about natural selection. The questions on the CINS target seven key concepts of natural selection (Mayr, 1982) and two additional key concepts (origin of variation and origin of species). Two questions target each key concept to enhance reliability. In the context of this study, selection refers to: causes of phenotypic variation (e.g., mutation, recombination, sexual reproduction); heritability of phenotypic variation; the over-reproductive capacity of individuals; limited environmental resources or carrying capacity; competition or limited survival potential; selective survival based on heritable traits; and changes in the frequency of individuals with certain heritable traits (Mayr, 1982, pp. 479–80).

Prior research with the CINS demonstrates validity and reliability sufficient for group or temporal comparisons (reliability > 0.7). In this study, we used the CINS as a measure of a single latent variable (knowledge of natural selection) in line with the validity analysis of Anderson et al. (2002). When used in this way, the CINS demonstrated satisfactory reliability (Rasch reliability = 0.75) in our sample of college students and all items demonstrated satisfactory weighted mean squares fit with the Rasch validity model (Wright & Stone, 1979).

### 5.3.2 Mechanistic Reasoning

We used a single constructed response assessment item to solicit mechanistic reasoning (Krist, Schwarz, & Reiser, 2018). In response to the prompt, students were asked to specify the factors they believed contributed to increased percentage of elephants without tusks and were subsequently asked to *explain their reasoning* behind their response. Students responded to the following prompt:

- African elephants are known for their large tusks, which the animals use for digging and defense. These tusks are valuable to people because of their ivory, which can be used in jewelry and decorations. Poachers hunt and kill elephants for their tusks, often before elephants are able to reproduce. Some elephants never grow tusks. In 1930, 1% of adult elephants didn't have tusks. In some areas today, up to 38% of adult elephants don't have tusks.
- How and why is the percentage of elephants without tusks higher today than it was in 1930?

### 5.3.3 Student Assessment of Learning Gains

At the end of the semester, we administered an online survey called "Student Assessment of their Learning Gains" (SALG) to measure students' self-reported learning gains and other progress toward course learning outcomes. The survey consisted of a variety of constructs including content understanding, skills, attitudes, class activities, class resources, and student support. Student responses were reported to the instructor after the course was completed. Aligning with our research questions, we report student survey data regarding content understanding and increase in skills.

# 6 Research Questions

## *6.1 How Did Students' Understanding of Natural Selection Change During the Course?*

We sought interpretations of how concepts about natural selection changed, which involved: (1) statistical significance of gains; (2) changes in conceptions implied by the gains; and (3) students' assessment of their learning gains. First, we were interested in whether knowledge of natural selection improved. To aid interpretation, Rasch logit measures on the CINS were first rescaled onto a range of 0–20 (the range for the original CINS scale). Change in the mean measure before and after the class was evaluated at the 0.05 alpha level using a paired t test. Since the distributions were not normal, standard errors, confidence intervals, and $p$ values were derived from a bootstrap distribution based on 10,000 simple random draws with replacement from the data. We used the percentile method to generate a 95% confidence interval for gains (0.025 and 0.975 quantiles of the bootstrap distribution) (Banjanovic & Osborne, 2016). The standardized mean gain (Cohen's D) was used as a measure of practical significance. Cohen's (1988) guidelines were used to qualify the size of the effect from the standardized mean difference.

Second, we constructed a Wright map of student and item Rasch measures along the common CINS scale (Fig. 2). The Wright map is a plot of student and item measures along a common scale and allows one to predict concepts that individual students have mastered based on their relative location along the scale (Boone, 2016).

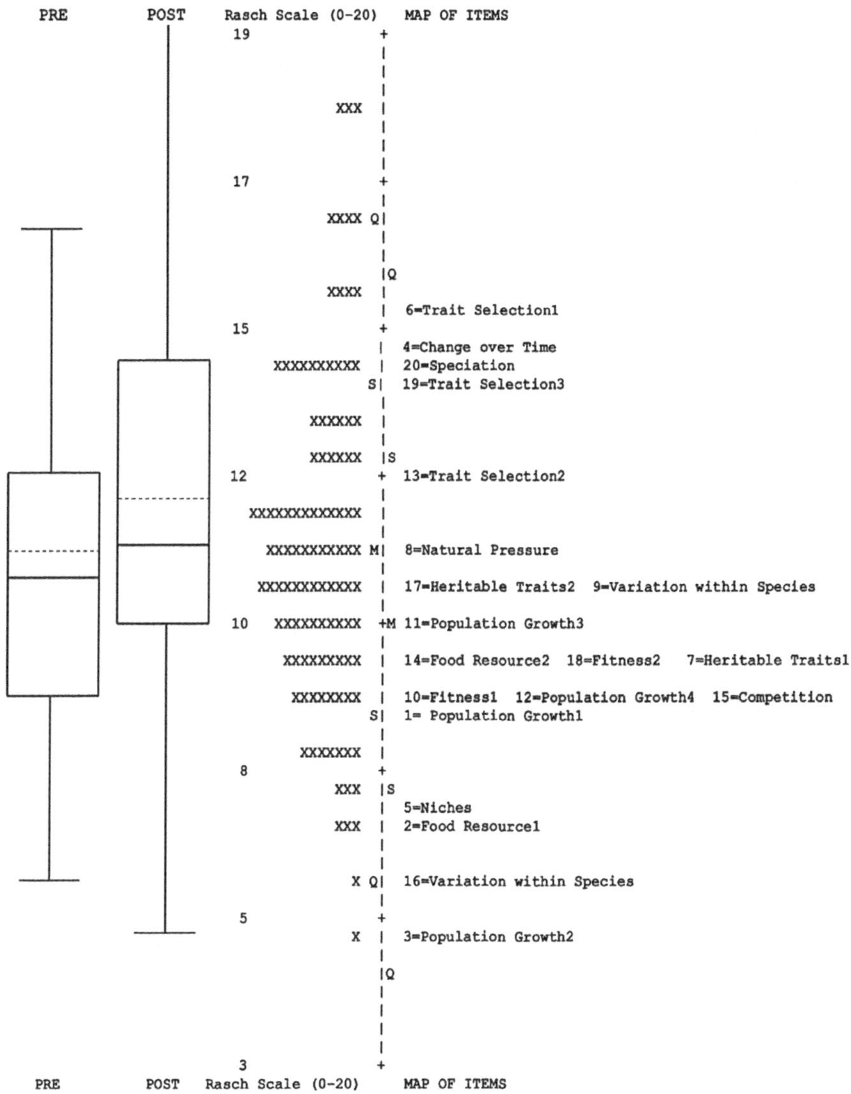


**Fig. 2** Wright map of student and item measures along the CINS scale (0–20). Students who are positioned below the location of the item are predicted to get that item incorrect, indicating non-mastery of that concept. The x's show the total distribution of student measures (pre and post together). The box plots indicate the distribution of students' measures before and after instruction

Specifically, if a student's ability location sits above the item's difficulty location on the scale, then that student is predicted to have mastered the concept associated with that item. From the Wright map, we were able to deduce concepts that were comparatively easy or difficult for students, and how mastery of particular concepts changed between the beginning and end of instruction. All analyses were carried out under the assumption that interpretation of measures on the CINS did not change between the pre- and posttest.

## 6.2 How Did Students' Mechanistic Reasoning Around Natural Selection Change?

We scored the mechanistic reasoning prompt on two levels. For level 1, we coded students' responses based on factors they believed caused more elephants without tusks, "within variation in a trait or genes exists within a population of organisms" (1a in Table 2) and "humans caused a change in the environment which selected for elephants without tusks" (1b in Table 2). We also were interested in whether students provided inaccurate alternative explanations (1c in Table 2). For level 2, after students described the factors, we asked them to explain their reasoning. If students reasoned that variation in traits or genes (1a) happens because organisms reproduce and pass on genes or traits to offspring, then they were deemed to show appropriate reasoning (2a in Table 2). Similarly, appropriate reasoning around natural selection (1b) involved explanation that human hunting affected the elephant population over time (2b in Table 2).

We also documented responses containing reasoning behind misconceptions about how other processes may have caused the elephant population to change (2c in Table 2). We hypothesized that effective instruction would increase the proportion of students who specified the correct factors causing the change in the elephant population (increase in 1a and 2a), and correct reasoning around how these factors caused the change (increase in 2a and 2b). Since all six of these tasks were scored dichotomously (and hence the distributions were not normal), standard errors, confidence intervals, and $p$ values were again derived from a bootstrap distribution based on 10,000 simple random draws with replacement from the data. We used the percentile method to generate 95% confidence intervals for gains (0.025 and 0.975 quantiles of the bootstrap distribution) (Banjanovic & Osborne, 2016). The standardized mean gain (Cohen's D) was used as a measure of practical significance of student gains or losses before and after the unit.

**Table 2** Changes in students' concept understanding and mechanistic reasoning around natural selection through the course

| Construct | Pre | Post | $P$ value | 95% $CI_{Gain}$ | $D_{gain}$ |
|---|---|---|---|---|---|
| Natural selection[a] | 10.83 (2.49) | 11.83 (3.06) | 0.011 | 0.29 to 1.74 | 0.39* |
| Variation 1a[b] | 0.22 (0.42) | 0.32 (0.47) | 0.31 | −0.08 to 0.28 | 0.15 |
| Selection 1b[b] | 0.80 (0.40) | 0.78 (0.42) | 0.79 | −0.16 to 0.12 | −0.04 |
| Inaccurate 1c[b] | 0.58 (0.50) | 0.38 (0.49) | 0.026 | −0.36 to −0.04 | −0.33* |
| Correct reasoning 2a[b] | 0.06 (0.24) | 0.24 (0.43) | 0.017 | 0.04 to 0.32 | 0.34* |
| Correct explanation 2b[b] | 0.54 (0.50) | 0.44 (0.50) | 0.214 | −0.26 to 0.04 | −0.18 |
| Incorrect reasoning 2c[b] | 0.44 (0.50) | 0.26 (0.44) | 0.049 | −0.34 to 0.00 | −0.29* |

*Effect size is significant at 0.05 alpha level
[a]Paired t test for difference in means ($N = 46$) reported as Mean (SD). Cohen's D ($D_{gain}$) used as an effect size measure. Cohen's D ($D_{gain}$) was used as an effect size measure. Since this distribution is not normal, 95% confidence intervals and 2-tailed $p$ values were derived using the percentile method (0.025 and 0.975 quantiles of the bootstrap distribution) using 10,000 simple random draws with replacement from the data
[b]Paired t test for difference in means ($N = 50$) reported as Mean (SD). Cohen's D ($D_{gain}$) was used as an effect size measure. Since these distributions are not normal, 95% confidence intervals and 2-tailed p values were derived using the percentile method (0.025 and 0.975 quantiles of the bootstrap distribution) using 10,000 simple random draws with replacement from the data

## 7 Findings and Discussion

In this section, we present our findings and relate them to the literature. We focus on data from the CINS and our mechanistic reasoning prompt. We conclude with pedagogical implications and a reflection on the course (including SALG comments from students), as well as implications for future research and faculty development.

### *7.1 Overall Shifts in Natural Selection Knowledge*

Students' conceptions about evolution were more sophisticated by the end of the course (bootstrap 95% CI = 0.29–1.74, $p = 0.011$, $D_{gain} = 0.39$). A standardized mean difference of 0.39 is equivalent to moving from the 50th percentile to the 65th percentile in average performance on the CINS. We found that this gain was accompanied by a proportional decrease (bootstrap 95% CI = −0.36 to −0.04, $p = 0.026$, $D_{gain} = 0.33$) in inaccurate alternative explanations (Inaccurate 1c in Table 2). Both these findings concord with Nehm and Reilly (2007), who also investigated science majors' natural selection knowledge and alternative conceptions in an active learning setting. Nehm and Reilly quantified individual use of key concepts versus alternative conceptions into a single composite measure called the natural selection performance quotient (NSPQ). A passing NSPQ was 65 (out of 100), a score calibrated to require employment of at least four of Mayr's seven key concepts (1982). As in our study, knowledge of natural selection was low prior to instruction (62; failing). Post-course, Nehm and Reilly documented a significant knowledge increase in their active learning group (from 62 to 79).

## 7.2 *Item-Level Shifts in Natural Selection Knowledge*

In Fig. 2, we document differences in mastery of concepts before and after instruction based on our CINS data. As it is common for students to show extremes on one side or another, we were most interested in the middle of the distribution; namely how the first, second (median), and third quartiles of the distributions changed, and the subsequent inferences we can draw with respect to concept mastery.

The median (50th percentile) of the student measure distribution (solid line in the middle of each box) shifted from 10.5 to 11.1. Item 8 sits between these two measures, indicating that a median student did not understand the role of the environment in selecting for certain beak types in Darwin's finches before the class, but that they gained this understanding by the end of the class. The first quartile (25th percentile) of the distribution shifted from 8.6 to 9.9 between the pre- and posttests. Items 7, 14, and 18 sit between these levels, indicating that students at the second quartile obtained mastery of the concepts of (a) heritability, (b) competition for resources, and (c) fitness. Furthermore, mastery of item 7 indicates that the course helped these students abandon the Lamarckian misconception that change occurs due to a need or desire. This was replaced by the understanding that genes are a driver of evolution. The students also understood the biological definition of "fitness" by the end of the class (item 18) and expressed understanding that resource limitations exist (item 14).

The third quartile (75th percentile) of the distribution shifted from 12.4 to 14.1, indicating mastery of items 19 and 20. These items relate to selection of traits and speciation. Item 19 indicates that the unit may have helped students replace Lamarckian misconceptions of within-species phenotypic variation with the understanding that random genetic mutations are the initial driver of variation, and item 20 indicates that students were then able to apply this idea toward scientifically accepted understanding of how speciation occurs.

## 7.3 *Changes in Students' Mechanistic Reasoning*

We originally hypothesized that effective instruction would increase the proportion of students with correct reasoning around how natural selection caused changes (items 2a and 2b). Students' ability to qualify the variability in genes and traits in a population did not change through the course (Item 2b), but their ability to explain the reasoning behind this—that organisms pass genes and traits to offspring (item 2a)—did increase significantly (bootstrap 95% CI = 0.04 to 0.32, $p = 0.017$, $D_{gain} = 0.34$). Only 6% of students could express this reasoning clearly at the beginning of the course; this increased to 24% of the students by the end of the course. While this is not the level of reasoning mastery we would like to see in our science major students, we see this as a step in the right direction.

## *7.4 Pedagogical Implications and Reflections on the Course*

Our goals were to present students with phenomena and engage them in using evidence to explain and reason through the phenomena. Our data indicate that the course was effective in transforming students' conceptions about evolution. In particular, we found that the introductory investigation of how and why the dinosaurs died was pivotal in changing the climate and mindset of the course. We showed this HHMI video on the first day and students worked in small groups figuring out the evidence of what happened to the dinosaurs. It was clear that the students were excited and felt that this course was going to be "different" than other courses. For example, on the SALG survey student comments included:

- "The class activities were awesome! They helped me so much to understand the material—I would even go home and talk about what I learned to others because I found it very interesting and exciting!"
- "I liked looking back and applying what we learned at the start of the course and building up and growing upon the idea of evolution and branching from there."

When students learn about evolution, it is often through direct instruction, personal experiences, and or in bits and pieces, which may lead to misconceptions and incomplete understandings (Coker, 2009; Gil-Perez & Carrascosa, 1990; Sinclair, Pendarvis, & Baldwin, 1997). Decontextualized experiences may explain in part why students experience difficulty when learning evolution. We attribute our relative success to the framing of the course and the relevancy of using the situated context for learning about the overarching phenomenon of evolution. We chose the history of life on Earth as an entry point to thinking about evolution, particularly understanding the adaptive radiation of mammals. The film about dinosaur extinction allowed for this content to become more interesting. Most students have heard about the dinosaur extinction, dating back to their preschool and elementary years, but this topic rarely appears in their upper science classes. This context opened up the space for bringing in different student ideas of what happened, which led to a variety of questions to investigate. Thus, students had a personal interest in the topic.

Our focus on foregrounding evidence was key for facilitating student buy-in and understanding and explaining evolution. The course message was not about finding the right answer, but rather examining the evidence to figure out the most convincing claim. Changing the language of the classroom environment to include attention to audience and persuasion with evidence contributed to a positive change in student thinking and learning. When students were asked what skills they learned from participation in this course, they commented, "I learned how to learn," "Being able to analyze different pieces of evidence and putting that together, like what explained the K-T extinction" and "One of the main skills that I have gained as a result of this class is looking at both sides of an argument. I feel like I tend to always pick one side but never really look on the other side of the argument. This class really challenged me to analyze both sides of the argument and actually find evidence to 'support the claim'."

The film also provided a look into the personal side of science, where scientists agreed or disagreed with one another. Finally, we used technology as a tool throughout the course to provide opportunities for student learning. Student comments on the SALG included, "The technology was extremely helpful, and how she tied in many different forms of learning. I do not necessarily learn well from just being lectured at, but rather we watched videos and did interactive gizmos, which extremely helped." and "I really appreciated the great efforts in use of technology in the class, it was great having such a forward thinking professor."

## *7.5 Research Implications*

We encourage other researchers to explore mechanistic reasoning behind evolutionary concepts. We document gains in reasoning (Table 1) and strong indications that the science practice approach positively influenced student understandings of natural selection. In particular, we are interested in exploring mechanistic reasoning more comprehensively—and exploring not only reasoning behind not natural selection mechanisms, but also those related to speciation. This approach addresses the call for students to have a complete scientific understanding of evolution. They should learn examples of natural selection and speciation on both microevolutionary and macroevolutionary scales (Catley, 2006). Evolution across long timescales may be particularly important as knowledge of macroevolution has been reported to be significantly correlated with acceptance of evolution for both biology (Nadelson & Southerland, 2010) and non-science majors (Romine, Walter, Bosse, & Todd, 2016; Walter, 2013; Walter, Halverson, & Boyce, 2013).

## *7.6 Implications for Faculty Development*

Instructors often create lessons, select readings, and design assessments in the same way they always have (Wilson, 2010), calling on their experiences as learners to inform how they teach (Tobin, Tippins, & Gallard, 1994). In this way, instructors can perpetuate ineffective and antiquated lecture norms as they operate under the belief that teaching occurs by transmitting knowledge (DeHaan, 2005).

We deviate from lecture-only approaches in our intervention, as we are using phenomena to help students engage in authentic science practices. Unlike other studies that incorporate principles of inquiry in this manner (e.g., Demastes, Settlage, & Good, 1995; Robbins & Roy, 2007), our intervention occurs in a large enrollment lecture hall, not a laboratory classroom. Since the SEPs model how scientists understand and practice in their own work, implementation of a practice-based teaching approach may provide an easier pedagogical transition for faculty new to active learning strategies. For example, we postulate that a faculty member unsure on how to implement an approach like "problem-based learning" may feel more comfortable with guiding students to "build an argument from evidence." In this way, our study could be used as a bridge between these two worlds: how instructors teach and how students learn.

## References


Abraham, J. K., Meir, E., Perry, J., Herron, J. C., Maruca, S., & Stal, D. (2009). Addressing undergraduate student misconceptions about natural selection with an interactive simulated laboratory. *Evolution: Education and Outreach, 2,* 393–404.

Alters, B. T. (2005). *Teaching biological evolution in higher education: Methodological, religious, and nonreligious issues*. Sudbury, MA: Jones and Bartlett Publishers.

American Association for the Advancement of Science [AAAS]. (2011). *Vision and change in undergraduate biology education: A call to action.* Washington, DC: Author.

American Association for the Advancement of Science [AAAS]. (2013). *Measuring STEM teaching practices: A report from a national meeting on the measurement of undergraduate Science, Technology, Engineering, and Mathematics (STEM) teaching*. Washington, DC: Author.

Anderson, D. L., Fisher, K. M., & Norman, G. J. (2002). Development and evaluation of the conceptual inventory of natural selection. *Journal of Research in Science Teaching, 39,* 952–978.

Andersson, B., & Wallin, A. (2006). On developing content-oriented theories taking biological evolution as an example. *International Journal of Science Education, 28,* 673–695.

Banjanovic, E. S., & Osborne, J. W. (2016). Confidence intervals for effect sizes: Applying bootstrap resampling. *Practical Assessment, Research & Evaluation, 21*(5), 1–20.

Berland, L. K., Schwartz, C. V., Krist, C., Kenyon, L., Lo, A. S., & Reiser, B. J. (2016). Epistemologies in practice: Making scientific practices meaningful for students. *Journal of Research in Science Teaching, 53,* 1082–1112.

Bishop, B. A., & Anderson, C. W. (1990). Student conceptions of natural selection and its role in evolution. *Journal of Research in Science Teaching, 27,* 415–427.

Boone, W. J. (2016). Rasch analysis for instrument development: Why, when, and how? *CBE-Life Sciences Education, 15*(4), rm4.

Bransford, J. D., Brown, A. L., & Cocking, R. R. (2000). *How people learn.* Washington, DC: National Academies Press.

Campbell, N., & Reece, J. (2012). *Biology* (10th ed.). San Francisco, CA: Pearson Benjamin Cummings.

Catley, K. M. (2006). Darwin's missing link—A novel paradigm for evolution education. *Science Education, 90,* 767–783.

Clough, E., & Wood-Robinson, C. (1985). How secondary students interpret instances of biological adaptation. *Journal of Biological Education, 19*(2), 125–130.

Cohen, J. (1988). *Statistical power analysis for the behavioral sciences* (2nd ed.). Hillsdale, NJ: Lawrence Erlbaum Associates.

Coker, J. S. (2009). Reinventing life: Introductory biology for a rapidly evolving world. *The American Biology Teacher, 71,* 281–284.

Coleman, S. W., & Jensen, M. S. (2007). Male mating success: Preference or prowess? Investigating sexual selection in the laboratory using *Drosophila melanogaster. The American Biology Teacher, 69,* 351–358.

Creswell, J. W. (2009). *Research design: Qualitative, quantitative, and mixed methods approaches* (3rd ed.). Los Angeles, CA: SAGE.

Darden, L., & Craver, C. (2002). Strategies in the interfield discovery of the mechanism of protein synthesis. *Studies in History and Philosophy of Science Part C: Studies in History and Philosophy of Biological and Biomedical Sciences, 33*(1), 1–28.

DeHaan, R. (2005). The impending revolution in undergraduate science education. *Journal of Science Education and Technology, 14,* 253–269.

Demastes, S. S., Good, R. G., & Peebles, P. (1996). Students' conceptual ecologies and the process of conceptual change in evolution. *Science Education, 79,* 637–666.

Demastes, S. S., Settlage, J., & Good, R. (1995). Students' conceptions of natural selection and its role in evolution: Cases of replication and comparison. *Journal of Research in Science Teaching, 32,* 535–550.

Detterman, D. K., & Sternberg, R. J. (1993). *Transfer on trial: Intelligence, cognition, and instruction.* Westport, CT: Ablex Publishing.

Duschl, R. A., Schweingruber, H. A., & Shouse, A. W. (Eds.). (2007). *Taking science to school: Learning and teaching science in grades K-8.* Washington, DC: National Academies Press.

Evans, J. H., & Evans, M. S. (2008). Religion and science: Beyond the epistemological conflict narrative. *Annual Review of Sociology, 34,* 87–105.

Freeman, S., Eddy, S. L., McDonough, M., Smith, M. K., Okoroafor, N., Jordt, H., et al. (2014). Active learning increases student performance in science, engineering, and mathematics. *Proceedings of the National Academy of Sciences, 111,* 8410–8415.

Gil-Perez, D., & Carrascosa, J. (1990). What to do about science "misconceptions". *Science Education, 74,* 531–540.

Krist, C., Schwarz, C. V., & Reiser, B. J. (2018). Identifying essential epistemic heuristics for guiding mechanistic reasoning in science learning. *Journal of the Learning Sciences.* https://doi.org/10.1080/10508406.2018.1510404.

Lawson, A. E., & Thompson, L. D. (1988). Formal reasoning ability and misconceptions concerning genetics and natural selection. *Journal of Research in Science Teaching, 25,* 733–746.

Lawson, A. E., & Weser, J. (1990). The rejection of nonscientific beliefs about life: Effects of instruction and reasoning skills. *Journal of Research in Science Teaching, 27,* 589–606.

Lawson, A. E., & Worsnop, W. A. (1992). Learning about evolution and rejecting a belief in special creation: Effects of reflective reasoning skill, prior knowledge, prior belief and religious commitment. *Journal of Research in Science Teaching, 29*(2), 143–166.

Lemke, J. L. (1990). *Talking science: Language, learning, and values.* Norwood, NJ: Ablex Publishing.

Machamer, P., Darden, L., & Craver, C. F. (2000). Thinking about mechanisms. *Philosophy of Science,* 1–25.

Mayr, E. (1982). *The growth of biological thought.* Cambridge, MA: Harvard University Press.

McKeachie, W. J., Lin, Y. G., & Strayer, J. (2002). Creationist vs. evolutionary beliefs: Effects on learning biology. *The American Biology Teacher, 64*(3), 189–192.

Moore, R., Mitchell, G., Bally, R., Inglis, M., Day, J., & Jacobs, D. (2002). Undergraduates' understanding of evolution: Ascriptions of agency as a problem for student learning. *Journal of Biological Education, 36,* 65–71.

Nadelson, L. S., & Southerland, S. A. (2010). Examining the interaction of acceptance and understanding: How does the relationship change with a focus on macroevolution. *Evolution: Education and Outreach, 4,* 82–88.

National Research Council [NRC]. (2012). *Discipline-based education research: Understanding and improving learning in undergraduate science and engineering*. Washington, DC: National Academies Press.

Nehm, R. H., & Reilly, L. (2007). Biology majors' knowledge and misconceptions of natural selection. *BioScience, 57,* 263–272.

Nehm, R. H., & Schonfeld, I. S. (2007). Does increasing biology teacher knowledge of evolution and the nature of science lead to greater preference for the teaching of evolution in schools? *Journal of Science Teacher Education, 18,* 699–723.

Nelson, C. E. (2008). Teaching evolution (and all of biology) more effectively: Strategies for engagement, critical reasoning, and confronting misconceptions. *Integrative and Comparative Biology, 48,* 213–225.

NGSS Lead States. (2013). *Next generation science standards: For states, by states*. Washington, DC: The National Academies Press.

Passmore, C., & Stewart, J. (2002). A modeling approach to teaching evolutionary biology in high schools. *Journal of Research in Science Teaching, 39,* 185–204.

Phillips, D. C., & Burbules, N. C. (2000). *Post-positivism and educational research*. Lanham, MD: Rowman & Littlefield.

Plunkett, A. D., & Yampolsky, L. Y. (2010). When a fly has to fly to reproduce: Selection against conditional recessive alleles in *Drosophila. The American Biology Teacher, 72*(1), 12–15.

Posner, G. J., Strike, K. A., Hewson, P. W., & Gertzog, W. A. (1982). Accommodation of a scientific conception: Toward a theory of conceptual change. *Science Education, 66,* 211–227.

Reece, J. B., Urry, L. A., Cain, M. L., Wasserman, S. A., Minorsky, P. V., Jackson, R., et al. (2014). *Campbell biology* (10th ed.). Boston, MA: Pearson.

Reiser, B. J. (2017). *Developing coherent storylines to support three-dimensional science learning*. Presented at Rhode Island Science Teachers Association (RISTA) Conference.

Robbins, J. R., & Roy, P. (2007). The natural selection: Identifying & correcting non-science student preconceptions through an inquiry-based, critical approach to evolution. *The American Biology Teacher, 69,* 460–466.

Romine, W. L., Walter, E. M., Bosse, E., & Todd, A. (2016). Understanding patterns of evolution acceptance—A new implementation of the Measure of Acceptance of the Theory of Evolution (MATE) with midwestern college students. *Journal of Research in Science Teaching, 54*(5), 642–671.

Russ, R. S., Coffey, J. E., Hammer, D., & Hutchison, P. (2009). Making classroom assessment more accountable to scientific reasoning: A case for attending to mechanistic thinking. *Science Education, 93*(5), 875–891.

Russ, R. S., Scherr, R. E., Hammer, D., & Mikeska, J. (2008). Recognizing mechanistic reasoning in student scientific inquiry: A framework for discourse analysis developed from philosophy of science. *Science Education, 92,* 499–525.

Salata, M. (2002). Evolution lab with *Drosophila. Bioscene, 28*(2), 3–6.

Settlage, J., & Odom, A. L. (1995, April). *Natural selection conceptions assessment: Development of the two-tier test "Understanding Biological Change"*. Paper presentation at the National Association of Research in Science Teaching Annual Meeting, San Francisco, CA.

Sinatra, G. M., Brem, S. K., & Evans, E. M. (2008). Changing minds? Implications of conceptual change for teaching and learning about biological evolution. *Evolution: Education and Outreach, 1,* 189–195.

Sinclair, A., Pendarvis, M. P., & Baldwin, B. (1997). The relationship between college zoology students' beliefs about evolutionary theory and religion. *Journal of Research and Development in Education, 30,* 118–125.

Smith, M. U. (2010). Current status of research in teaching and learning evolution: II. Pedagogical issues. *Science & Education, 19,* 539–571.

Thinley, P., Geva, S., & Reye, J. (2014). Tablets (iPad) for M-learning in the context of social constructivism to institute an effective learning environment. *International Journal of Interactive Mobile Technologies, 8*(1), 16–20.
Tobin, K. G., Tippins, D. J., & Gallard, A. J. (1994). Research on instructional strategies for teaching science. In D. L. Gabel (Ed.), *Handbook of research on science teaching and learning* (pp. 45–93). New York, NY: Macmillan.
Walter, E. M. (2013). *The influence of pedagogical content knowledge (PCK) for teaching macroevolution on student outcomes in a general education biology course* (Doctoral dissertation). Retrieved from ProQuest Dissertations and Theses (#3576041).
Walter, E. M., Halverson, K. L., & Boyce, C.-J. (2013). Investigating the relationship between college students' acceptance of evolution and tree thinking understanding. *Evolution: Education and Outreach, 6,* 26.
Wandersee, J. (1985). Are there too many terms to learn in biology? *The American Biology Teacher, 47,* 346–347.
Wilson, R. (2010, September). Why teaching is not priority no. 1. *Chronicle of Higher Education, 57,* A1–A8.
Wright, B. D., & Stone, M. H. (1979). *Best test design.* Chicago, IL: University of Chicago Social Research.

**Dr. Lisa Kenyon** is Associate Professor in the Department of Biological Sciences in the College of Science and Mathematics and the Department of Teacher Education in the College of Education and Human Services at Wright State University. Her research focuses on engaging K-12 students and teachers in scientific practices such as explanation, argumentation, and scientific modeling—specifically, examining how students use their epistemologies of science to support these practices. Other research interests include curriculum design, project-based inquiry, and teacher professional development.

**Emily M. Walter** is Assistant Professor of Biology at California State University—Fresno. At Fresno State, she is Director of the Science and Mathematics Education Center, a group of faculty and students that collaborate across STEM departments and with the School of Education to support STEM education initiatives and research at all grade levels. Her areas of expertise include evolution education and exploring the cognitive, social, and organizational factors that influence faculty adoption and sustained use of evidence-based instructional practices.

**William Romine** is Associate Professor in Wright State University's Biological Sciences department, with joint appointments in Teacher Education and Computer Science. He directs the *Data Science for Education* laboratory at Wright State where he and his students focus on educational measurement using traditional tests and surveys, social media, and sensors. He and his students use these measures to construct dynamic models for both learning and affective change related to science.

# Improving Student Understanding of Randomness and Probability to Support Learning About Evolution

**Ute Harms and Daniela Fiedler**

## 1 Introduction

The theory of evolution is widely considered to be one of the most important and groundbreaking theories in the history of science; it essentially underpins all modern biology, from ecology through to medicine. Charles Darwin's theory of evolution—developed simultaneously by Alfred Russel Wallace—explains the fundamental mechanism by which all life is related and descended from a common ancestor. However, our understanding of evolutionary biology did not end with Darwin and Wallace; researchers continued, and continue, to make remarkable strides in the years thereafter. These include, for instance, integrating genetic concepts with those from evolutionary biology ('the modern synthesis') and the population genetics which emerged from this, developing the 'neutral theory of evolution,' and conducting research into social interactions and cooperation mechanisms.

Since the theory of evolution was first presented more than 150 years ago, results from across the life sciences have verified and enhanced elements in the theory. The multitude of common applications provides several examples of practical societal applications of evolutionary aspects, including ones connected to antibiotic resistance, the emergence of new diseases and biodiversity management, as well as responses and adaptations to climate change (Meagher, 2007). Other examples relate to genetically engineered organisms, cancer, resistance to HIV treatment, and modern biotechnology (e.g., energy-, nano-, and gene technology). Therefore, a meaningful understanding of evolutionary theory is essential for many areas of individual, social, and scientific life. Still, a growing number of adults either lack such an understanding

U. Harms (✉) · D. Fiedler
IPN - Leibniz Institute for Science and Mathematics Education,
Biology Education, Olshausenstr. 62, 24118 Kiel, Germany
e-mail: harms@ipn.uni-kiel.de

D. Fiedler
e-mail: fiedler@ipn.uni-kiel.de



U. Harms and M. J. Reiss (eds.), *Evolution Education Re-considered*,
https://doi.org/10.1007/978-3-030-14698-6_15

of the processes of evolution or resist accepting evolutionary theory as the best scientific explanation of all processes related to life (Berkman & Plutzer, 2011; Miller, Scott, & Okamoto, 2006).

## 2 Evolution, Learning, and Threshold Concepts

Science education research has shown that the theory of evolution can present severe problems for learners (e.g., Bishop & Anderson, 1990; Gregory, 2009; Kampourakis & Zogza, 2008). In fact, many students and teachers lack a naturalistic scientific worldview (Evans et al., 2010). The segregation of evolution from other biological topics in textbooks and teaching is only one source of the problem (Nehm et al., 2009). One suggestion to overcome learners' difficulties is emphasizing the relevance of evolution in students' lives (Hillis, 2007). Another aspect of receiving comparably little attention is the role that language and discourse practices play in the formulation of mental models of evolution (Lemke, 1990; Pinker, 2007). In fact, language, complexity, dynamics, and dimensionality are key threshold concepts in biology. Many words in science, like adaptation and fitness, also appear in everyday language with slightly different meanings (Sinatra, Brem, & Evans, 2008). In addition, reasoning about evolution, natural selection, and adaptation is often characterized by *Teleology* (the idea that changes that occur are purposeful), *Anthropomorphism* (using human actions as a model for other species), *Intentionality* (organisms or Nature acting with intention), and *Essentialism* (each species has a unifying essence and differences within species are considered as anomalies or not considered at all) (e.g., Coley & Tanner, 2012; Mayr, 1982; Sinatra et al., 2008). This makes students confused and risks introducing misconceptions and misunderstandings (e.g., Coley & Tanner, 2015; Nehm & Schonfeld, 2008). Gregory (2009) showed that students tend to think of natural selection and evolution in terms of an 'event' rather than as a process, while development is considered as an 'either–or' happening where all the fit creatures survive, and the non-fit creatures die. In light of this, many of the points where students' reasoning goes astray could be considered due to the limitations of language. Thus, Nehm and Reilly (2007) have suggested targeting misconceptions and core concepts as tools for explaining particular evolutionary scenarios, which is in line with well-established conceptual change theories in science education (Strike & Posner, 1992).

Current science education research also indicates learning difficulties with those evolutionary concepts that are strongly related to abstract concepts like randomness and probability, so-called threshold concepts (Garvin-Doxas & Klymkowsky, 2008; Mead & Scott, 2010; Ross et al., 2010). Threshold concepts are described as conceptual gateways that, once passed, open up a new way of thinking and are distinguished from 'key' or 'core' concepts, as they are more than mere building blocks toward understanding within a discipline (Meyer & Land, 2003). Thus, evolutionary theory involves a conglomerate of various threshold concepts such as randomness, probability, spatial scale, temporal scale, and complexity (Ross et al., 2010). Tibell and Harms (2017) concluded that complete understanding of evolutionary theory might require the understanding of these more general abstract concepts. Students particularly struggle with the importance and nature of randomness in evolution (Garvin-Doxas & Klymkowsky, 2008; Robson & Burns, 2011).

Randomness is often difficult to understand because of its different meanings in diverse contexts. In everyday usage, the term 'random' means that a phenomenon is without order, predictability or pattern (Wagner, 2012). Nevertheless, scientists (and mathematicians) use the term to suggest unpredictability, but not intended purposelessness (Buiatti & Longo, 2013; Mead & Scott, 2010). The evolutionary notion of randomness is often quite specific: The word randomness refers to events (e.g., mutations, genetic drift) that are independent of an organism's needs and the directionality provided by natural selection in the process of adaptation (Mead & Scott, 2010; Millstein, 2000). In this sense, mutations in particular are said to be random. It cannot be predicted precisely where and when a mutation will appear and they are not directed to individuals' adaptations (Heams, 2014). In contrast, natural selection itself can be described as a probabilistic process, since the process of selection can be defined as the probabilities of individuals with different traits in a given population surviving and reproducing in a specific environment (Tibell & Harms, 2017). Still, students tend to struggle with both probability and the notion of randomness in the evolutionary context (Brumby, 1979; Gregory, 2009; Robson & Burns, 2011). However, a clear understanding of randomness and probability is essential both for understanding evolution and for molecular and cellular biology (Lenormand, Roze, & Rousset, 2009).

Based on this, the first study of this chapter will focus on the impact of understanding randomness and probability to comprehend evolution. The study starts from the hypothesis that the misconceptions identified do not arise from the biological principles (i.e., variation, selection, and reproduction) themselves, that together merge to form the theory of evolution, but from underlying general abstract concepts such as randomness and probability (Tibell & Harms, 2017). From an empirical point of view, we must determine how to assess these constructs to answer this particular question, and whether the understanding of randomness and probability in evolution is the same as in mathematics or everyday contexts. Despite the wide variety of instruments measuring the knowledge of evolutionary theory, there is a lack of tools for measuring threshold concepts like randomness and probability. The development of such an instrument is in focus of the second study presented in this chapter.

## 3 The Impact of Understanding Randomness and Probability for the Comprehension of Evolution

Inadequate comprehension of underlying abstract threshold concepts such as randomness or probability is regarded as a learning difficulty for deeper understanding of evolutionary theory. As a preliminary step, we conducted an explorative study to obtain evidence on whether an understanding of the threshold concepts of randomness and probability is connected to and facilitates an understanding of the theory of evolution. For this, we investigated the relationship between the understanding of randomness and probability and the understanding of evolution. In addition, we explored the influence of three different learning conditions to foster students' understanding of randomness and probability.

The presented explorative study was carried out as a quasi-experimental intervention study in a pretest–posttest design with three treatment groups. A total of 20 German university students participated in the study (four males, 16 females; age: $M = 26.7$ years; $SD = 3.3$ years). They were studying toward a Master of Education degree with a focus on both STEM and non-STEM subjects. These participants were divided randomly into three different treatment groups: animation ($n = 8$), text ($n = 6$), and mathematical tasks ($n = 6$), that will be explained below.

In the first treatment (i.e., animation), students were given the animation 'evolving lines'[1] that focuses on the visualization of randomness (BBC & Open University, 2011). The moderator asks volunteers to draw a straight line on a tablet PC. Although people can accurately draw such lines with a special pen, mistakes will generally occur. The line represents the DNA, while drawn mistakes represent errors that occur during DNA replication, so-called mutations. Thus, the line will change when the line is drawn once by each of 200 volunteers, who see only the previous line and not the original straight line. As every mistake is copied, the random changes in 'DNA' accumulate within the 200 generations. Furthermore, the moderator describes speciation by duplicating the line on a new tablet PC after 25 generations and again after 175 generations, while replication continues from these branching points. In the end, three distinct lines have emerged that simulate a family tree. Finally, the moderator explains that both random changes in DNA and natural selection are relevant for an evolutionary change. Without random genetic changes, all organisms would be the same, and natural selection would not occur. In biology, visualizations such as animations play an important role, particularly at the molecular level (Kozma, 2000). Moreover, research indicates that animations can facilitate the learning of dynamic processes (Ploetzner & Lowe, 2004). For this reason, we wanted to explore if randomness in evolution can be made tangible by using the selected animation.

In the second treatment (i.e., text), students were given a text that explains randomness by focusing on scientific and everyday events (542 words). The text starts with a dialog of two students and their teacher on the newspaper topic 'Lightning strike in the Bavarian tent during Kieler Week'. The participants in the text discuss the probabilities of this event by referring to subjective and objective randomness, also using physical and biological examples (quantum mechanics and mutations, respectively; cf. Spektrum, 1999). In the end, a small glossary briefly explains the biological and physical terms used in the text (7 terms, 175 words). We decided to work with a typical text-based instruction on randomness because (a) textbooks are

[1]The original animation is no longer available, but copies may be found on the Internet. The playing time was 5 min and 28 s with English spoken language.

still core teaching resources, and (b) learning sessions are often organized around text-based instructions (McDonald, 2016).

In the last treatment (i.e., mathematical tasks), students have to work through four mathematical tasks in the field of probability calculus at a difficulty appropriate for 16 years old (10th graders). Two of the tasks focus on evolutionary processes such as mutations and genetic drift, while the other two tasks refer to more general events (i.e., picking a defective electrical fuse and picking three times the same marble). Each time, the solution could have been determined by either using a tree diagram or calculating probabilities. Mathematical descriptions of randomness and probability are often key elements of evolutionary processes (Buiatti & Longo, 2013; Wagner, 2012). Thus, we chose the mathematical tasks to explore the potential influence on students' understanding of randomness and probability.

Before and after the intervention, students were asked to complete several test instruments. The pretest included students' demographic data, their preconceptions of the concept of randomness, and their understanding of randomness, probability, and evolution. The posttests consisted of a general cognitive ability test and the tests for students' understanding of randomness, probability, and evolution.

To measure students' preconceptions of the concept randomness, we used three open-response items and twelve single-choice items, in which students have to order the presented events as either 'random' or 'non-random' (Table 1; adapted and modified from Döhrmann, 2004). The items were used to get an impression of students' general understanding of the threshold concept of randomness but were not meant to be included in the statistical analyses. Students' open responses were coded using qualitative content analysis (e.g., Kuckartz, 2012). For the single-choice items, the reliability coefficient computed by Cronbach's alpha was 0.86.

To assess students' understanding of randomness and probability, we used ten multiple-choice items focusing on mathematical equations in an everyday and evolutionary context. The items were developed by researchers of the *EvoVis*[2] project group and translated into German. Due to negative discrimination indices, two items were excluded, and Cronbach's alpha of the remaining items was 0.62 for the pretest and 0.67 for the posttest. The order of the items varied between the pretest and the posttest to reduce memory effects.

We also used the Conceptual Inventory of Natural Selection (CINS; Anderson, Fisher, & Norman, 2002) to assess students' understanding of evolution through natural selection. This test consists of 20 multiple-choice items that focus on key ideas related to natural selection and common misconceptions. Cronbach's alpha was 0.87 and 0.88 for the pretest and posttest, respectively. The order of the items varied between the pretest (finches, guppies, salamanders) and the posttest (salamanders, finches, guppies) to reduce memory effects.

To assess whether general cognitive abilities may have an influence on students' understanding of randomness and probability (and evolution), we used a verbal (25 items), a quantitative (20 items), and a nonverbal (25 items) subscale of the KFT

---

[2]*EvoVis: Challenging Threshold Concepts in Life Science—enhancing understanding of evolution by visualization* is a Swedish-German cooperation project funded by the Swedish Research Council.

**Table 1** Open-response questions and single-choice items to measure students' preconceptions of the concept of randomness

| **Preconceptions of the concept of *randomness*** |
|---|
| **Open-response questions** |
| 1. What comes into your mind when you think of the term *chance*? |
| 2. Name five events that you consider to be random. |
| 3. What words do you think can be used to describe the term *random?* |
| **Single-choice items** |
| *Mark which of the following events do you think are random*: |
| 1. Two former classmates meet on the plane to the island Mallorca. |
| 2. To roll a six with a die. |
| 3. It will rain tomorrow. |
| 4. Mutations occur during the replication process of DNA. |
| 5. When throwing a coin eight times, the result will be HTHTHTHT (H = Head; T = Tail). |
| 6. Winning the lottery. |
| 7. At the pregnancy checkup, your gynecologist tells you that you are having a girl. |
| 8. The weather forecast announces rain for tomorrow, and it is actually raining on the day after. |
| 9. When throwing a coin eight times, the result will be HTTHTHHT (H = Head; T = Tail). |
| 10. In a storm, seeds are caught from a plant of the mainland and then spread to an island where the plant did not exist previously. |
| 11. At the next appointment, your dentist will find a hole in your left, upper canine. |
| 12. To throw a six three times in a row. |

4–12 + R (Heller & Perleth, 2000), the German version of the Cognitive Abilities Tests (CAT; Thorndike & Hagen, 1971). Each scale consists of items that present a word (verbal), a row of numbers (quantitative) or a pair of meaningfully related drawings connected with another single drawing (nonverbal), to which the appropriate counterpart (word/synonym, number, or figure, respectively) has to be selected from five answer options. Cronbach's alpha was 0.09 for the verbal scale, 0.79 for the quantitative scale, and 0.90 for the nonverbal scale. Due to the extreme low reliability of the verbal scale, we excluded the variable 'verbal abilities' from further analyses.

The results of the open-ended items showed that students often explained the terms random or randomness as something happening unexpectedly, something that cannot be planned or is not controllable. Furthermore, some students also indicated the predictability of random events by calculating their probability. Two non-STEM students explained that randomness does not exist. In contrast, random events were rated as being either negative or positive and, from a philosophical point of view, connected with destiny. On this basis, students' preconceptions of the term randomness are very basic and connected to its everyday usage.

To investigate whether and to what extent understanding of randomness and probability is related to the understanding of evolution, general cognitive abilities, and

**Table 2** Summary of the regression analyses for the variables explaining understanding of evolution ($N = 20$)

| Predictor | *B* | *SE B* | $\beta$ | $R^2$ |
|---|---|---|---|---|
| Understanding randomness and probability | 1.80 | 0.69 | 0.63* | 0.707 |
| Quantitative abilities | −0.48 | 0.33 | −0.42 | |
| Nonverbal abilities | 0.27 | 0.16 | 0.35 | |
| Mathematics grade | 0.09 | 0.28 | 0.07 | |
| STEM/non-STEM[a] | −3.78 | 1.46 | −0.45* | |

*B* unstandardized regression coefficients, *SE B* standard error of *B*, $\beta$ standardized regression coefficient

*$p < 0.05$, [a]Students studying either STEM or non-STEM subjects

**Table 3** Descriptive statistics (mean with standard deviation in brackets) and effect sizes of pretest to posttest scores for each intervention

| | Understanding randomness | | Effect size |
|---|---|---|---|
| | Pretest | Posttest | Cohen's *d* |
| Animation | 6.38 (1.69) | 6.25 (1.98) | −0.07 |
| Text | 6.50 (1.05) | 6.67 (1.03) | 0.16 |
| Mathematical tasks | 5.50 (1.64) | 6.00 (1.67) | 0.30 |

demographic variables (e.g., last high school grade in biology or mathematics), correlation and regression analyses were performed using pretest data. Findings revealed that understanding of randomness and probability was significantly positively related to the understanding of evolution ($r = 0.74$, $p < 0.001$), quantitative abilities ($r = 0.71$, $p < 0.001$), and students' last grade in mathematics ($r = 0.74$, $p = 0.001$). The last two effects could probably be explained by the design of the test instrument for understanding randomness and probability since the items included a lot of numbers and mathematical reasoning. Nevertheless, the results of the multiple regression analyses (forced entry) indicated that understanding of randomness and probability showed the highest positive effect on students' understanding of evolution (Table 2). Altogether, the included variables accounted for 71% of the variance: $F(5, 14) = 6.75$, $p = 0.002$.

Finally, we investigated the effect of the three treatments (animation, text, and mathematical tasks) used for influencing the understanding of randomness and probability. Due to the small sample size, groups lack a representativeness of the larger population, and calculating inference statistics would be inappropriate. Therefore, we focused on descriptive statistics and calculated effect sizes of pretest to posttest scores in each group. These findings indicated no effect for the animation group, a small effect for the text group, and a medium effect for the mathematical tasks group (Table 3). The animation showed the least effect, which could be due to the abstractedness of the subjects and students' inability to transfer an abstract construct into their model of the real world (Scalise et al., 2011).

## 4 A Way to Measure Students' Understanding of Randomness and Probability

For a more thorough investigation of the relationship between the understanding of threshold concepts such as randomness and probability and the understanding of evolution, an instrument was needed that generates valid and reliable inferences to confirm our hypothesis.[3]

In a pilot study, we revised the primary test instrument (see Sect. 3) to measure students' understanding of randomness and probability (hereafter, RaPro) and decided to add more items on both random and probabilistic phenomena. This resulted in a set of 28 items (26 multiple-choice and two open-response items) focusing on students' conceptual knowledge of randomness ($n = 10$) and probability ($n = 18$). Additionally, we used three open-response items of Nehm and Reilly (2007) to assess students' use of randomness and probability in evolution as well as to measure their evolutionary explanations. These three items (i.e., antibiotic-resistant bacteria, fast running cheetahs, and blind cave salamanders) focus on trait gain and loss in species.

The pilot paper-based version was administered to a group of 48 German university students (79% female; age: $M = 22.7$ years; $SD = 2.7$ years). We calculated internal reliability by using Cronbach's alpha as well as item difficulty and item discrimination indices of the scales for randomness and probability. Based on these analyses, we excluded four items from the subscale randomness and ten items from the subscale probability. The remaining item set consistently showed satisfactory item characteristics (difficulty: 0.15–0.85; discrimination: 0.21–0.41). The internal consistencies of the subscales randomness ($\alpha = 0.52$; $n = 6$) and probability ($\alpha = 0.59$; $n = 8$) were lower than desired but still acceptable (Cronbach, 1951; Taber 2017). On average, students achieved 7.13 points ($SD = 2.17$) in the RaPro.

For the open-response items, we developed a coding schema for students' use of randomness and probability in evolution as well as for their understanding of evolution. The use of randomness and probability in evolution (hereafter, OpenRaPro) was measured using three categories: (1) randomness (naming random processes), (2) probability of survival (e.g., different traits cause different probabilities to survive), and (3) probability of reproduction (e.g., the probability that advantageous traits are inherited to the offspring). For the understanding of evolution (hereafter, OpenEvo), we measured eight key concepts: (1) origin of variation, (2) individual variation, (3) different survival potential, (4) limiting resources, (5) competition, (6) changes in gene pool, (7) inheritance of the trait, and (8) reproductive success (Anderson et al., 2002; Nehm & Reilly, 2007). The scoring rubrics were used to quantify the presence or absence of the respective categories and concepts in students' written explanations. The internal reliability measured by Cohen's Kappa was $\kappa = 0.84$ for the OpenRaPro, and $\kappa = 0.76$ for the OpenEvo. Students' answers to the evolutionary open-response items showed a low usage of randomness and prob-

[3]This study is part of the *EvoVis* project.

ability (OpenRaPro: $M = 0.92$, $SD = 0.93$) and a moderate usage of evolutionary key concepts (OpenEvo: $M = 4.69$, $SD = 2.09$).

Finally, correlation analyses were performed to investigate possible relationships between the test instruments' scores. Students' OpenRaPro scores showed a significant positive correlation to the OpenEvo scores ($r = 0.65$, $p < 0.001$), while there was a nonsignificant positive correlation between the RaPro and the OpenEvo scores ($r = 0.11$, $p = 0.490$). The design of the test instrument might be an explanation for these findings. The open-response items (OpenEvo/OpenRaPro) focused on students' explanations of evolutionary changes, while the multiple-choice items of the RaPro instrument focused on randomness and probability in a mathematical as well as everyday context.

We then decided to revise the developmental process of the instruments by adding more closed-response items that specifically focus on random and probabilistic processes of evolution. Additionally, we wanted to explore the empirical structure of students' understanding of randomness and probability in the contexts of evolution and mathematics. To date, there is no empirical evidence about students' conceptual structures regarding randomness and probability in biological contexts, and their connections (if any) to conceptual structures in mathematics contexts. Still, mathematical explanations of randomness or probability often serve as the explanatory basis for random processes in biology (Buiatti & Longo, 2013; Wagner, 2012), and the explorative study outlined above (see Sect. 3) indicated a connection.

For this purpose, we designed the 'Randomness and Probability test in the context of Evolution' (RaProEvo, 21 items) and its sister instrument the 'Randomness and Probability test in the context of Mathematics' (RaProMath, 33 items; Fiedler, Tröbst, & Harms, 2017). The results revealed that the two test instruments RaProEvo and RaProMath measure separate competencies. Furthermore, evidence of the instruments' reliability measures was promising, while experts and criterion-related indications confirmed their validity measures. Nevertheless, the developed instruments were neither intended to be summative evaluation tools nor to assess every aspect of randomness and probability exhaustively. Still, we hope that our instruments will facilitate efforts to design more tools to assess students' conceptual knowledge of randomness and probability.

## 5 Conclusion

Even though the results of the first explorative study should be interpreted with caution, it provides preliminary insight into the importance of the comprehension of abstract underlying threshold concepts for the understanding of a complex biological theory (i.e., evolution). As indicated by the correlation and regression analyses, students' understanding of randomness and probability is highly connected to their understanding of evolution. Indeed, focusing on underlying abstract concepts might also be relevant for other scientific topics that have been shown to be challenging to understand meaningfully by students such as the concept of energy (e.g., Opitz,

Blankenstein, & Harms, 2017). In this context, it might be interesting for the scientific community to discuss it as a relevant perspective on learning problems in science that, so far, have been rather neglected.

By means of the developed RaProEvo, we provide an instrument that measures students' understanding of randomness and probability in evolutionary context. In addition, focusing on the understanding of threshold concepts in learning science could lead to the development of new instructional methods that could help overcome learning obstacles revealed by science education research in the last decade and thus warrants attention.

## References


Anderson, D. L., Fisher, K. M., & Norman, G. J. (2002). Development and evaluation of the conceptual inventory of natural selection. *Journal of Research in Science Teaching, 39*(10), 952–978. https://doi.org/10.1002/tea.10053.

BBC & Open University (Producers) (2011). Bang goes the theory – evolving lines [Video clip]. Retrieved from http://www.bbc.co.uk/programmes/p00wwvfs.

Berkman, M. B., & Plutzer, E. (2011). Defeating creationism in the courtroom, but not in the classroom. *Science, 331*(6016), 404–405. https://doi.org/10.1126/science.1198902.

Bishop, B. A., & Anderson, C. W. (1990). Student conceptions of natural selection and its role in evolution. *Journal of Research in Science Teaching, 27*(5), 415–427. https://doi.org/10.1002/tea.3660270503.

Brumby, M. (1979). Problems in learning the concept of natural selection. *Journal of Biological Education, 13*(2), 119–122. https://doi.org/10.1080/00219266.1979.9654240.

Buiatti, M., & Longo, G. (2013). Randomness and multilevel interactions in biology. *Theory in Biosciences, 132*(3), 139–158. https://doi.org/10.1007/s12064-013-0179-2.

Coley, J. D., & Tanner, K. D. (2012). Common origins of diverse misconceptions: Cognitive principles and the development of biology thinking. *CBE-Life Sciences Education, 11*(3), 209–215. https://doi.org/10.1187/cbe.12-06-0074.

Coley, J. D., & Tanner, K. D. (2015). Relations between intuitive biological thinking and biological misconceptions in biology majors and nonmajors. *CBE-Life Sciences Education, 14*(1), ar8. https://doi.org/10.1187/cbe.14-06-0094.

Cronbach, L. J. (1951). Coefficient alpha and the internal structure of tests. *Psychometrika, 16*(3), 297–334. https://doi.org/10.1007/BF02310555.

Döhrmann, M. (2004). *Zufall, Aktien und Mathematik: Vorschläge für einen aktuellen und realitätsbezogenen Stochastikunterricht [Chance, stocks, and math: Suggestions for a current and real-world teaching of stochastics]*. Hildesheim, Germany: Franzbecker.

Evans, E. M., Spiegel, A. N., Gram, W., Frazier, B. N., Tare, M., Thompson, S., et al. (2010). A conceptual guide to natural history museum visitors' understanding of evolution. *Journal of Research in Science Teaching, 47*(3), 326–353. https://doi.org/10.1002/tea.20337.

Fiedler, D., Tröbst, S., & Harms, U. (2017). University students' conceptual knowledge of randomness and probability in the contexts of evolution and mathematics. *CBE-Life Sciences Education, 16*(2), ar38. https://doi.org/10.1187/cbe.16-07-0230.

Garvin-Doxas, K., & Klymkowsky, M. W. (2008). Understanding randomness and its impact on student learning: lessons learned from building the Biology Concept Inventory (BCI). *CBE-Life Sciences Education, 7*(2), 227–233. https://doi.org/10.1187/cbe.07-08-0063.

Gregory, T. R. (2009). Understanding natural selection: Essential concepts and common misconceptions. *Evolution: Education and Outreach, 2*(2), 156–175. https://doi.org/10.1007/s12052-009-0128-1.
Heams, T. (2014). Randomness in biology. *Mathematical Structures in Computer Science, 24*(03), e240308. https://doi.org/10.1017/S096012951200076X.
Heller, K. A., & Perleth, C. (2000). *KFT 4-12 + R: kognitiver Fähigkeitstest für 4. bis 12. Klassen, Revision [Cognitive abilities test for grades 4 to 12]*. Beltz Test: Göttingen, Germany.
Hillis, D. M. (2007). Making evolution relevant and exciting to biology students. *Evolution, 61*(6), 1261–1264. https://doi.org/10.1111/j.1558-5646.2007.00126.x.
Kampourakis, K., & Zogza, V. (2008). Students' intuitive explanations of the causes of homologies and adaptations. *Science & Education, 17*(1), 27–47. https://doi.org/10.1007/s11191-007-9075-9.
Kozma, R. (2000). Reflections on the state of educational technology research and development. *Educational Technology Research and Development, 48*(1), 5–15. https://doi.org/10.1007/Bf02313481.
Kuckartz, U. (2012). *Qualitative Inhaltsanalyse. Methoden, Praxis, Computer unterstützung [Qualitative content analysis. Methods, practice, computer-assistance]*. Beltz Juventa: Weinheim, Germany.
Lemke, J. L. (1990). *Talking science: Language, learning, and values*. Norwood, NJ: Ablex Publishing Corporation.
Lenormand, T., Roze, D., & Rousset, F. (2009). Stochasticity in evolution. *Trends in Ecology & Evolution, 24*(3), 157–165. https://doi.org/10.1016/j.tree.2008.09.014.
Mayr, E. (1982). *The growth of biological thought: Diversity, evolution, and inheritance*. Cambridge, MA: Harvard University Press.
McDonald, C. V. (2016). Evaluating junior secondary science textbook usage in Australian schools. *Research in Science Education, 46*(4), 481–509. https://doi.org/10.1007/s11165-015-9468-8.
Mead, L. S., & Scott, E. C. (2010). Problem concepts in evolution part II: Cause and chance. *Evolution: Education and Outreach, 3*(2), 261–264. https://doi.org/10.1007/s12052-010-0231-3.
Meagher, T. R. (2007). Is evolutionary biology strategic science? *Evolution, 61*(1), 239–244. https://doi.org/10.1111/j.1558-5646.2007.00041.x.
Meyer, J. H., & Land, R. (2003). Threshold concepts and troublesome knowledge: Linkages to ways of thinking and practising within the disciplines. In C. Rust (Ed.), *Improving student learning: Theory and practice ten years on* (pp. 412–424). Oxford, United Kingdom: Oxford Centre for Staff and Learning Development (OCSLD).
Miller, J. D., Scott, E. C., & Okamoto, S. (2006). Public acceptance of evolution. *Science, 313*(5788), 765–766. https://doi.org/10.1126/science.1126746.
Millstein, R. L. (2000). Chance and macroevolution. *Philosophy of Science, 67*(4), 603–624. https://doi.org/10.1086/392857.
Nehm, R. H., Poole, T. M., Lyford, M. E., Hoskins, S. G., Carruth, L., Ewers, B. E., & Colberg, P. J. (2009). Does the segregation of evolution in biology textbooks and introductory courses reinforce students' faulty mental models of biology and evolution? *Evolution: Education and Outreach, 2*(3), 527–532. https://doi.org/10.1007/s12052-008-0100-5.
Nehm, R. H., & Reilly, L. (2007). Biology majors' knowledge and misconceptions of natural selection. *BioScience, 57*(3), 263–272. https://doi.org/10.1641/b570311.
Nehm, R. H., & Schonfeld, I. S. (2008). Measuring knowledge of natural selection: A comparison of the CINS, an open-response instrument, and an oral interview. *Journal of Research in Science Teaching, 45*(10), 1131–1160. https://doi.org/10.1002/tea.20251.

Opitz, S. T., Blankenstein, A., & Harms, U. (2017). Student conceptions about energy in biological contexts. *Journal of Biological Education, 51*(4), 427–440. https://doi.org/10.1080/00219266.2016.1257504.

Pinker, S. (2007). *The stuff of thought: Language as a window into human nature*. London, United Kingdom: Viking Penguin.

Ploetzner, R., & Lowe, R. (2004). Dynamic visualisations and learning. *Learning and Instruction, 14*(3), 235–240. https://doi.org/10.1016/j.learninstruc.2004.06.001.

Robson, R. L., & Burns, S. (2011). Gain in student understanding of the role of random variation in evolution following teaching intervention based on Luria-Delbruck experiment. *Journal of Microbiology & Biology Education: JMBE, 12*(1), 3–7. https://doi.org/10.1128/jmbe.v12i1.272.

Ross, P. M., Taylor, C. E., Hughes, C., Whitaker, N., Lutze-Mann, L., Kofod, M., et al. (2010). Threshold concepts in learning biology and evolution. *Biology International, 47,* 47–52.

Scalise, K., Timms, M., Moorjani, A., Clark, L., Holtermann, K., & Irvin, P. S. (2011). Student learning in science simulations: Design features that promote learning gains. *Journal of Research in Science Teaching, 48*(9), 1050–1078. https://doi.org/10.1002/tea.20437.

Sinatra, G. M., Brem, S. K., & Evans, E. M. (2008). Changing minds? Implications of conceptual change for teaching and learning about biological evolution. *Evolution: Education and Outreach, 1*(2), 189–195. https://doi.org/10.1007/s12052-008-0037-8.

Spektrum (1999). *Zufall in der Biologie [Chance in biology]*. Retrieved from http://www.spektrum.de/lexikon/biologie/zufall-in-der-biologie/72005.

Strike, K. A., & Posner, G. J. (1992). A revisionist theory of conceptual change. In R. A. Duschl & R. J. Hamilton (Eds.), *Philosophy of science, cognitive psychology, and educational theory and practice* (pp. 147–176). Albany, NY: State University of New York Press.

Taber, K. S. (2017). The use of cronbach's alpha when developing and reporting research instruments in science education. *Research in Science Education* (pp. 1–24). https://doi.org/10.1007/s11165-016-9602-2.

Tibell, L. A., & Harms, U. (2017). Biological principles and threshold concepts for understanding natural selection. *Science & Education, 26*(7–9), 953–973. https://doi.org/10.1007/s11191-017-9935-x.

Thorndike, R. L., & Hagen, E. (1971). *Cognitive abilities test*. Boston, MA: Houghton-Mifflin.

Wagner, A. (2012). The role of randomness in Darwinian evolution. *Philosophy of Science, 79*(1), 95–119. https://doi.org/10.1086/663239.

**Ute Harms** is Director at the IPN - Leibniz Institute for science and mathematics education, Full Professor for Biology Education at the University of Kiel (Germany) since 2007, and Fellow of the Royal Society of Biology (Great Britain). She has a Ph.D. in cell biology and has worked as a high school teacher for several years. In 2000, she got her first Professorship for Biology Education at the Ludwig-Maximilians-University in Munich (Germany). From 2006 to 2007, she held a chair in Biology Education at the University of Bremen. Her main research interests are conceptual learning in biology and in science, focusing on evolution and energy, biology teacher education, biology-related competitions and transfer of contemporary topics in the life sciences to the public.

**Daniela Fiedler** is currently a Postdoctoral Researcher in Biology Education at the IPN - Leibniz Institute for Science and Mathematics Education. She studies Biology at Giessen University, owns a Master of Science with focus on Animal Ecology and Environmental Science, and finished her Doctoral Thesis in Fall 2018. From 2014 on, she is part of the Swedish-German cooperation project *EvoVis: Challenging Threshold Concepts in Life Science—enhancing understanding of evolution by visualization* funded by the Swedish Research Council. Her main research interests are teaching and learning of biology focusing on evolution.

# Evolution Learning and Creationism: Thinking in Informal Learning Environments

**Jorge Groß, Kerstin Kremer and Julia Arnold**

## 1 Evolution and Creation—Two Ends of a Continuum

Teaching evolution is very challenging: the biological topic is complex and everyday conceptions which are not in accordance with scientific conceptions are widely spread and can hinder a deeper understanding of science. It is not only the members of creationist movements who can hold creationists views, but everyday conceptions are also influenced by biblical explanations (Blancke, Hjermitslev, Braeckman, & Kjærgaard, 2013). In this chapter, we speak about views and conceptions in terms of concepts and explanations about the natural world that can, but do not necessarily have to, be associated with religious faith (e.g. Cobern, 1994; Kutschera, 2008). Concerning these explanations, one can use Scott's (2009) description of worldviews, where evolution and creationism are the two ends of a continuum with many variations in between. Accordingly, worldviews, or philosophical conceptions, can reach from literal creationism to atheistic evolutionism and they have implications for the understanding of and acceptance in scientific—especially evolutionary—explanations of the natural world.

Scott's descriptions (2009) can be summarized as follows: at the creationism-end of the continuum, the strictest biblical literalism can be found, which assumes that the earth is of young age. For example, in *flat earth creationism*, people believe

J. Groß (✉)
Science Education, University of Bamberg, Markusplatz 3-Noddack-Haus, 96047 Bamberg, Germany
e-mail: jorge.gross@uni-bamberg.de

K. Kremer
Leibniz University of Hanover, Hanover, Germany
e-mail: kremer@idn.uni-hannover.de

J. Arnold
University of Applied Sciences and Arts Northwestern Switzerland (FHNW), School of Education, Basel, Switzerland
e-mail: Julia.arnold@fhnw.ch



U. Harms and M. J. Reiss (eds.), *Evolution Education Re-considered*,
https://doi.org/10.1007/978-3-030-14698-6_16

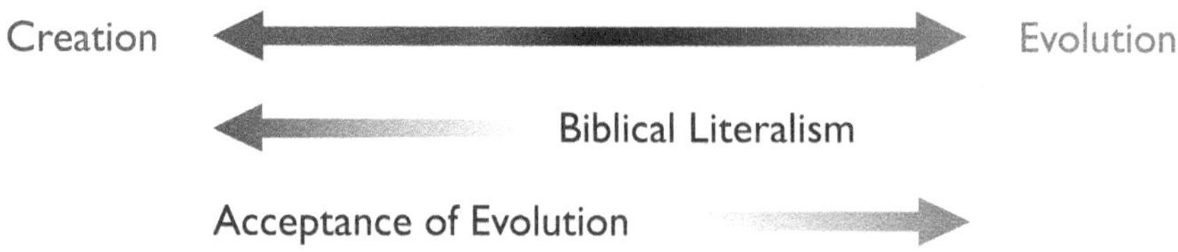


**Fig. 1** Creation-evolution continuum (from a scientific viewpoint)

that the earth is flat and has been created by God only six thousand years ago and within six days. *Geocentrists* may accept the Earth as being spherical, but deny the sun as the centre of the solar system, while *young earth creationism* accepts the earth being spherical and the heliocentric worldview but states that the earth was created relatively recently. Views at the creationism end of the continuum all have the conception of 'special creation' in common, which means that God created the earth and all its living beings as separate 'kinds' in their existing forms (relatively recently). This biblical literalism explains geological features by biblical events (e.g. the Grand Canyon was formed by Noah's Flood) and therefore denies many of the scientific findings and explanations of modern physics, astronomy, geology and biology (see Fig. 1). This is different in *old earth creationism*. While people holding this view accept most modern scientific explanations, like, for example, the earth being older than six thousand years, they see God as the acting agent and, therefore, refer to many explanations presented in the Bible. For example, *gap creationism* explains the age of the earth by a first creation long before the creation within six days and a 'gap' in between (corresponding to the two versions in *Genesis*), while *day-age creationism* interprets the days of creation as rather long periods of time (millions of years). But most creationists today belong to the group of *progressive creationism*. In this form, most physical and geological science is undisputed, but biological evolution is mostly denied. This is in contrast to *evolutionary creationism* and *theistic evolutionism*, which are very close to each other. In these variations, evolution is accepted, but God is seen as the driving force behind all physical, geological or biological events. This also holds true for the *intelligent design movement*, which is special, because it can be described as a form of creationism, but members of this movement themselves describe it as a science without any religious content or goals; that is why, intelligent design cannot be assigned a special range within this continuum, according to Scott (2009). At the other end of the continuum stands *agnostic evolution*, which describes the acceptance of science and evolution without making any statements about God.

Since it is the goal of science education to develop scientifically literate people, creationism and its 'alternative explanations' pose a challenge to fostering both the acceptance of science, and the nature of science (NOS) (Working Group on Teaching Evolution, National Academy of Sciences, 1998). In biology and in other sciences, evolution is a key concept, explaining the world and life on earth (as is shown throughout this book), as well as being an example of teaching the nature of science

(Rudolph & Stewart, 1998). Here, informal learning environments can be a way to address students' conceptions.

## 2 Evolution Learning in Informal Learning Environments

We know from research on learners' conceptions on evolution and creationism that students often consider assertions about the two sides of the continuum (evolution and creationism) to be directly contradictory to each other and that they declare scientific assertions as objective truth and faith as a subjective belief (Kattmann, 2016). Furthermore, they try to construct subjective theories to overcome the perceived contradiction (Kattmann, 2016).

Many studies have shown that students start evolution education with firmly established everyday conceptions of evolution that often do not change even after instruction (Bishop & Anderson, 1990; Bizzo, 2007; Settlage, 1994; Sinclair & Pendarvis, 1997). Students often arrive to instruction with a worldview that may be at odds with mainstream science (Lee, 1999) and the exposure to science is not enough to foster the self-reflection of everyday conceptions, or the acceptance and understanding of scientific explanations. Here, informal learning settings can provide even novice learners with valuable insights and authentic learning opportunities that add to the evolution learning opportunities in the classroom.

When dealing with evolution, informal learning (e.g. in natural history museums, zoos or science centres), conceptions of creation affect the learning of evolution concepts and it is also the case that conceptions of creation can be affected by this conceptual understanding. Informal learning environments offer specific learning possibilities for evolution learning, e.g. authentic objects in natural history museums or the presentation of designed objects for self-regulated learning. In the following, we present two case studies, focusing on the interplay between creationist conceptions and evolution understanding in informal learning environments.

## 3 Case Study 1: Who Am I and Where Do I Come from?

The first study deals with the conflicted topic of the emergence of humankind, which has been debated ever since the thought of evolution and its mechanisms emerged.

### *3.1 Subject of the Study*

To explain the theory of evolution and to foster the dialogue between science and society, teaching programmes have been organized in out-of-school learning environments. But empirical studies have shown that the goals of these educational initiatives are often not accomplished (Groß & Gropengießer, 2008). Human evolution is not understood by means of variation and selection. Rather, evolution is (mis)understood as an intended process from a single-celled organism to humans as the top, end point of creation. We assume that these difficulties are not only due to the complexity of the topic, but also to its educational implementation. The goal of our project was to design an exhibition in order to facilitate evolutionary biology learning for a wider public. This piece of the study relates to a station focused on human evolution. The topic of human evolution is particularly suitable, because it is at the centre of creationist-evolution positions. From a creationist perspective, it is unthinkable that humans and apes should have common ancestors with monkeys.

## 3.2 Theoretical Framework

As a theoretical background for our studies, we used the Model of Educational Reconstruction (Duit, Gropengiesser, & Kattmann, 2005) with a focus on learners' conceptions regarding evolution. In this model, students' conceptions are not understood as misconceptions, but rather as learning opportunities. On the basis of our previous studies, we know that particularly complex topics (such as the theory of evolution) are not understood, because students lack experience in this abstract field: students do not think in terms of populations, natural selection and long time periods. Rather, it has been shown that students—whenever they lack experience—use metaphors. From this, we see evolution in terms of evolving into a 'higher species', because we, too, have had the embodied experience of our own development into an upright position, from child to adult. Here, there is a disagreement with scientific ideas. For instance, Darwin wrote: "It is absurd to talk of one animal being *higher* than another" (Wyhe, 2002). In order to investigate the genesis of these metaphors, we analyse the specific use of metaphors.

The embodiment explanation of metaphors by George Lakoff, Mark Johnson and his colleagues (Gallese & Lakoff, 2005; Lakoff & Johnson, 1999, 2003) offers the most conclusive theory concerning them. Metaphors are not flashy rhetorical tools; rather, they provide a discreet structure of thinking. Humans can understand things such as up, down, behind and in front because they have made bodily or social experiences in the source domain. These conceptions are expressed by means of language, and affect the actions of people. In their theory of experiential realism, a metaphor is explained by two kinds of theoretical approaches.

Firstly, the processes of metaphorical understanding are motivated by bodily experience. Additionally, they can involve enduring conceptual structures such as kinesthetic image schemas or basic-level structures (Lakoff & Johnson, 1980). An easy example is the metaphorical structure of 'High is good and low is bad'. This example includes the embodied concept of 'high' and 'low' which arise from direct experience with our bodies. It can also be used in a metaphorical context if we speak about 'highly developed species' or 'upper class society' (Groß & Gropengießer, 2008; Lakoff & Johnson, 2003). The example indicates that the concept of 'high and low' is directly linked to 'good and bad' because of our recurring experience with objects and our body, e.g. as we grow up. Although scientists know that species cannot be higher or lesser developed, this metaphorical structure is often used in educational contexts.

Secondly, metaphors are multi-modal representations. Metaphorical understanding operates by using embodied simulations of the terms being compared. In this case, metaphors are processed via simulations that draw on sensorimotor encodings stored in modality-specific areas of the brain (e.g. Barsalou, 2005). The following will show how this theoretical framework was used to develop the exhibition.

### 3.2.1 The Exhibition 'Evolution Creates Diversity'

Educational programmes and exhibitions on evolution have been organized in out-of-school learning environments to foster the dialogue between science and society. Nevertheless, empirical studies show that their educational impact is often rather marginal. For example, we have evidence that a complexly designed exhibition in the Science Centre about variation was not suitable for promoting something to learn but rather reinforced existing everyday conceptions (Groß & Gropengießer, 2006). We assume that these difficulties are due to the educational implementation. These programmes and exhibitions do not—or at least not at a reasonable level—consider the learners' conceptions, even though everyday conceptions are essential premises for successful learning.

The goal of our project is to engage learners' interests and to encourage evolutionary biology learning for a wider public. In cooperation with Ulrich Kattmann and Annette Scheersoi, an exhibition was designed. Ulrich Kattmann is a specialist for teaching the theory of evolution and Annette Scheersoi for learning in informal learning environments. In our exhibition, evolution is introduced to the visitors in the context of everyday life: the various exhibits are integrated and displayed throughout an IKEA store (Groß, Kattmann, & Scheersoi, 2009). IKEA stores are built with a one-way path for visitors. The design of the exhibition was specifically integrated into these IKEA concepts: a progression in evolution and time corresponded to the progression of the branched shopping path at the store. All exhibits were predominantly built with IKEA furniture, in particular transparent SAMLA boxes could be used. For example, the creatures or the pinball game were installed in modified SAMLA boxes (see Figs. 2 and 4). Other exhibits were in or on furniture of the store. Everyday conceptions regarding evolution are used as starting points for the presentation of the biological topics. Assorted media (interactive media, short texts, original artifacts etc.) offer numerous entry points to the exhibition's theme. Visitors are encouraged to become engaged—touching and playing with the objects is intended ('hands-on'). Biological information is introduced to the visitors step by step ('minds-on') and presented in a 'multi-level-form' (each level successively provides more information). In order to achieve a learning success among the visitors in the sense of the conceptual change theory, the exhibition was supplemented with various interventions: In addition to the possibility of self-directed learning, IKEA

**Fig. 2** Station A at the exhibition 'evolution creates diversity' (picture: gewerk design)

had a lecture series and an extensive learning booklet with exercises that deepened the topics of the stations.

The exhibition is composed of eleven stations (see Table 1) with different media equipment (see Fig. 2) offering the visitors numerous entry points into the concept of evolution and possibilities for participation. To check which stations were used by the visitors and to provide an incentive for the use of as many stations as possible, each visitor was given a stamp pass when entering the store. The stamp pass could only be stamped out at the respective station on the one-way shopping path. Full stamp passes also participated in a raffle.

The exhibition's concept is based on the results of various empirical studies concerning the analysis of scientific and everyday conceptions of the theory of evolution. These studies focus on a self-instructive learning path, e.g. as demonstrated by the gorilla path at Hanover Zoo. The path was specifically constructed to exemplify important steps in human evolution, but educational success was not achieved. The intended educational goal—to show that humans and other great apes have the same origin—was not accomplished (see Groß & Gropengießer, 2008). Instead, it can be concluded that students did not understand the scientific content of evolution in the exhibits, but rather fell back on their everyday concepts. Therefore, these studies were helpful to identify typical everyday conceptions on the topic of evolution. For example, the theory of selection was presented as one (ideal) type of organism is left over, survives, and therefore, the ever-present variability is ignored. Nothing new develops (conception of stability) or evolution processes are regarded as target-oriented changes intended by organisms: "First, humans walked hunched and a little bit on legs and then they rose higher and higher … walked still a little bit stooped and someday they walked in a completely upright position, just like we do today." (Nils, 19 years). These studies indicate that the interplay of variation and selection

**Table 1** Exhibition is composed of eleven stations

| Number | Claim | Goal | Content |
|---|---|---|---|
| Station A | Diversity of living | Amazed by the incredible diversity | Various animals and plants as silhouettes at the IKEA store |
| Station B | Variation and selection | All species are in constant change | PC game with different coloured snails |
| Station C | Versatile wood | Woods and veneers are products of evolution | A tree displaying different types of woods |
| Station D | Value of diversity | We live on diversity | Podium with treasure chest of biology |
| Station E | Life of the dinosaurs | In the past, diversity was different | Meat and herbivores. Dinosaurs in comparison |
| Station F | Earth's history | Life has a long history. It goes back to a common source 3.5 billion years ago | Earth history as a timeline with 'highlights' and disasters |
| Station G | Flowers and insects | Flowers and pollinators are in constant competition | Game: which animal pollinates which plant? |
| Station H | Darwin's world tour | A naturalist pays attention to every little thing | interactive world map with '*The Beagle*' |
| Station I | Many faces | All humans belong to a kind | Photo wall with integrated screens for visitor pictures |
| Station J | Ancestors | Humans and great apes have a common ancestor | Interactive pinball game |
| Station K | Sounding variation | Without evolution, the world would be silent | Vocalizations are assigned to animals |

processes is generally not understood (Lewis & Kattmann, 2004). In fact, the central concept in all interviews was that humankind was descended from a recent monkey (like a chimpanzee) and had not developed from a common ancestor. It is common to all students' conceptions that humans are the most developed species, they are always 'on top'. Figure 3 shows the four empirically determined student conceptions of human evolution.

The theory of experiential realism explains why people regard themselves always as 'higher' evolved in comparison to other mammals of the order Primates. We found two central schemata: the 'start-path-goal schema' and the 'high is good and low is bad' schema (Lakoff & Johnson, 1999). Both schemata are not only based on the choice of words, but in the way students think about human evolution and perceive change at all: transformation is not only thought of as a gradual gain in height, but also as an evolutionary development. Progress in evolution is associated with progress in

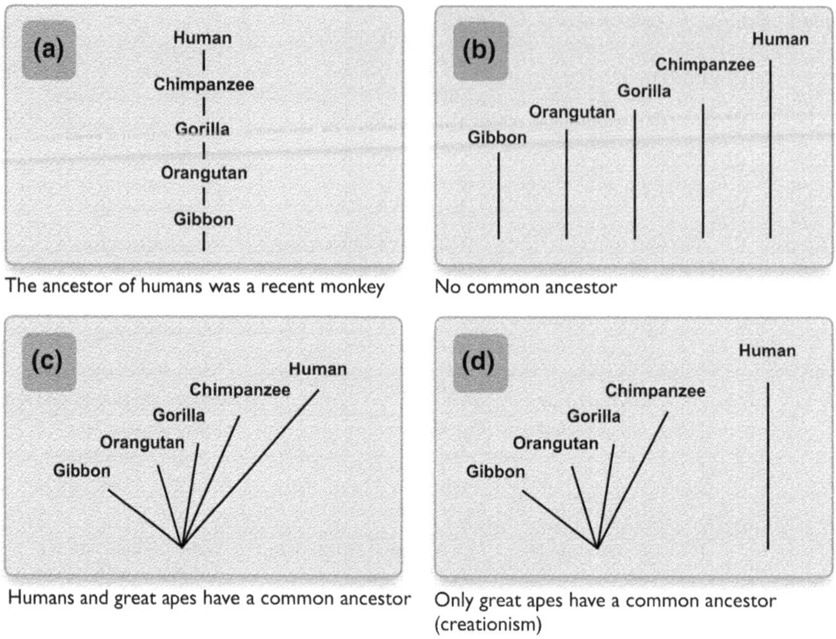


**Fig. 3** Students' conceptions of human evolution

movement. The idea of higher, therefore, coincides with the notion of further, in the sense of better.

Based on these findings, we developed the pinball game (see Fig. 4). The game is built like a classic pinball machine, but it deals with human evolution. In this interactive game, the pinball corresponds with an entity on the path of human evolution. In this way, there are several branching possibilities: extinction corresponds to the loss of the ball. The random principle of evolution corresponds to the random course at points of branching in the pinball game. Environmental influences correspond to movements by the player. But the central idea was the inversion of the timeline: the pinball starts at the beginning of the timeline (20 million years BCE), and leads either to the extinction (lost pinball at Neanderthal man) or to a recent primate (chimpanzee or human). In contrast to existing approaches, the phylogenetic tree was both rotated and didactically reduced to the central species (chimpanzees, Lucy, humans, Neanderthals). Because of the inversion, the everyday conception of 'higher is better—lower is less' can no longer be thought and can lead to conflict. This central idea is empirically investigated in the following.

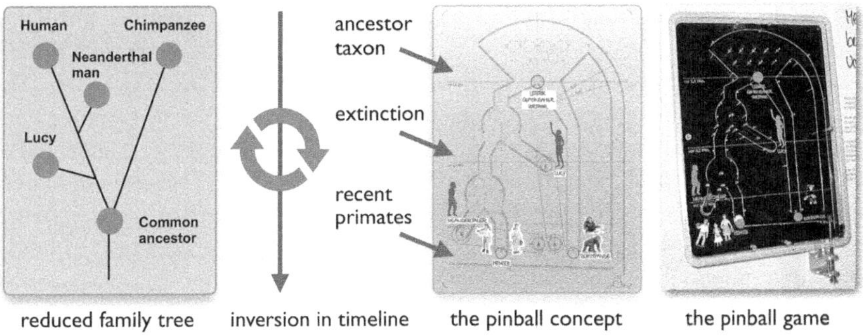


**Fig. 4** Development of the pinball game by inversion the timeline (station J 'Ancestors')

## 3.3 Research Questions

Our research questions were deduced from the above theoretical framework:

- How should an exhibition be designed in order to promote an understanding of human evolution for the wider public?
- To what extent can a learning environment taking into account students' conceptions promote the understanding of human evolution?

## 3.4 Research Methods

To obtain students' conceptions about evolution, we used semi-structured interviews as a qualitative research tool, also applying the method of the retrospective query on learning processes (Paul, Lederman, & Groß, 2016). A total of 53 individual interviews were conducted during the exhibition 'Evolution creates diversity'. For that purpose, all participating students were contacted concerning their willingness to be interviewed. From the positive responses, all students from 10 to 18 years were randomly chosen for the interviews to produce an approximately equal distribution of interviewees with respect to location, age and gender. Just picking people is surely not scientifically random. But regardless of age and gender, visitors ($N = 498$) were randomly approached on different days and times at the end of the store to measure if they had seen and used the station J (pinball game). All personalized data were made anonymous. Two different researchers conducted the interviews. An interview lasted for about 30 min and started 30–60 min after the visit to the exhibition. We used a structured guideline to align the 53 interviews for reproducibility. The interviews were conducted by Franziska Perau and associated researchers (Perau, 2010).

The interview guideline integrated two methodological approaches: firstly, problem-orientation, open and half-open questions to collect the current conceptions about human evolution, and, secondly, retrospective questions on the individual learning process. The strengths of this method of retrospective questions thus lie in the clarification of the individual learning processes, while at the same time linking these to the subjects' named causes of these learning processes (Paul et al., 2016). The guideline started with open questions about evolution and the pinball game such as: "Please tell me in two or three sentences, what did you do during the visit at the IKEA store?". In the second section, subjects were asked to think and reflect on their conceptions about human evolution from their point of view. Appropriate questions were, for instance: "What is the relationship between chimpanzees and humans?". Finally, the causes of possible conceptual changes were requested. For this purpose, we asked questions such as: "Why did you change your conceptions about evolution?". The interviews were captured using a voice recorder. The interrelationship between questions and answers was validated by three different researchers based on qualitative content analysis (Mayring, 2010).

## *3.5 Results*

The results of these studies indicate that our goals could be attained to awaken the visitor's interest, and to clarify misunderstandings about the theory of evolution. Interest development was supported by different factors: for example, surprising moments caught the attention of visitors (e.g. original artefacts presented in IKEA furniture), and they became aware of connections between the biological topics presented and their individual lives (Groß et al., 2009). Such connections were starting points for further engagement with the objects, and meaning-making processes were initiated.

One of the most effective exhibits was the pinball game: visitors tend to assume that recent apes are the ancestors of humankind (Groß & Gropengießer, 2008). To challenge this everyday conception, we designed a pinball game similar to a phylogenetic tree. In total, we counted 50,862 games in the IKEA store, at an average of 706 games per day. The findings indicate three outcomes: (a) Most (adult) IKEA clients passed by, but 58% of students or children used the pinball game; (b) If visitors stopped, some changed their conceptions about evolution; and (c) Learners generally do not read the text information (see Table 2). However, when considering the data, it is important to note that some visitors were only shopping at IKEA. Additionally, the study pointed out that there is much work to show that while people often hold teleological views of evolution, reading phylogenies and other types of evolutionary tree is very complex. There is also the problem of understanding what the tree diagram is trying to show. Many people have the notion of the anagenetic model of evolution, with cladogenesis being less well understood (e.g. Novick, Stull, & Catley, 2012).

The interview data suggest that people understand—based on the pinball game—that chimpanzees and humans have a common ancestor and here we have an outcome of the pinball game experience. A typical statement comes from Lena (11 years-old, secondary school) and is representative of the interviewed students. After playing the pinball game, she answered: "Chimps and humans have the same ancestor". If the given text information is read, students are able to describe the

**Table 2** Hidden participant observation, station J 'Ancestors' in percentage ($N = 498$)

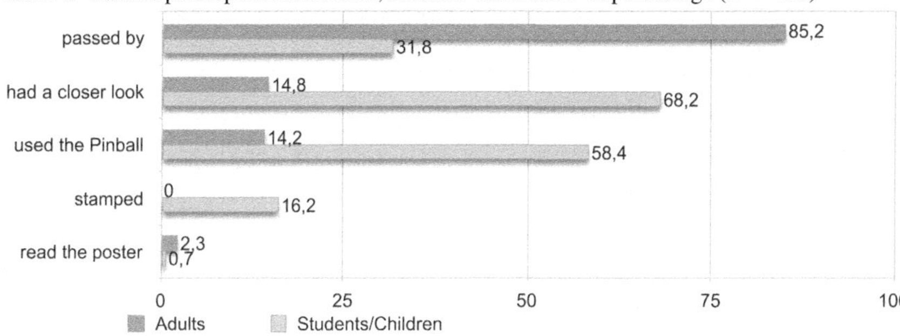


learning outcomes: "First, I thought human beings are a direct descendant of apes". Although students gained an idea of the branching of the presented species as a result of the intervention, learners like Lena still stick to everyday conceptions: "There are apes, which evolved to human beings and some remained apes". This was the name given to an everyday conception that only humans have undergone an evolution, but recent great apes stopped evolving a long time ago. In this sense, some students had a learning success in terms of the frame of branching, but just no conception of the equal evolution of recent humans and great apes.

## 3.6 *Conclusion from Case Study 1*

The simplifications in the learning environment were positive. Students transfer the random principle in the pinball game metaphorically to the process of evolution. This analogy seems to fall easily to students, since the random principle between pinball and environmental influences (in the game of the pinball players) greatly simplifies the complex interaction between populations and selection in the process of evolution. In addition, the rotation of the timeline does not allow any teleological thinking in the sense of 'higher' evolution: all recent primates are in the same timeline. This seems to be more scientific thinking than the students had before (see Figs. 3a–c). There was, however, no conceptual change in the argumentation of students with creationist conceptions. They remain with their idea of separate creation (as seen in Fig. 3d). Even though everyday conceptions are questioned, some of the goals of the exhibits are still too complex and explanatory texts are not read comprehensively. The wide range of attitude profiles high school children have towards evolution indicates that a simple acceptance/non-acceptance polarity is inadequate (Konnemann, Asshoff, & Hammann, 2016).

# 4 Case Study 2: How Did Our Earth and All Living Species Evolve?

The second study deals with an outreach programme in the natural history museum at the Ottoneum in Kassel in the form of a guided tour through the museum's permanent exhibition and a workshop on fossils. In an evaluative study, we examined in how far knowledge gain through the activity is related to creationist conceptions in novice learners.

## *4.1 Subject of the Study*

Natural history museums usually advertise educational outreach programmes for school groups of all ages from novice learners to advanced learners with a deeper knowledge of evolution and natural history. These exhibitions with their various learning opportunities are ideal places for research on the impact of evolution instruction on students' knowledge gains (Evans et al., 2010) and accompanying attitudes (Konnemann et al., 2016). This study aims at analyzing the relationship of an education programme in a natural history museum on novice students' (5th graders) reasoning about creationism. At this age, no systematic knowledge about natural history or evolution principles could have been gained and students' knowledge and attitudes are mostly affected by religious beliefs or everyday experiences (Astley & Francis, 2010).

According to Konnemann, Nick, Brinkmann, Asshoff and Hammann (2013), creationism is operationalized as a subjective, non-rational cognitive system of conceptions (Southerland & Sinatra, 2003) with the potential to influence the acceptance of a person towards evolution. According to Astley and Francis (2010), creationist conceptions are constituted by a lack of acceptance of evolution combined with conceptions of biblical literalism. For this study with a novice population of learners, we only focus on the second part of this definition, because we assume that novice learners are not able to report their acceptance of evolution adequately due to a lack of consistent knowledge or understanding about evolution.

Studies on the relationship of evolution understanding and evolution attitudes come to controversial results. While some studies—mostly relying on the MATE instrument on evolution acceptance (Rutledge & Warden, 1999)—report no significant relationships (Brem, Ranney, & Schindel, 2003), other authors do report relationships (Kim & Nehm, 2011). Studies exist that have investigated the relationship of evolution understanding and evolution conceptions or acceptance; however, the body of literature that explicitly addresses the effects of evolution instruction on creationist conceptions is still rare to date (Eve, Losh, & Nzekwe, 2010).

Therefore, the goal of this study was to track knowledge and creationist conceptions development of novice learners over a one-day education programme in a natural history museum. The relevant research questions guiding the investigation are the following:

1. Is there a long-term impact of a one-day education programme in a natural history museum on novice students' evolution knowledge?
2. Is there also a lasting impact on novice students' conceptions of creationism?
3. Is there a relationship between the creationist conceptions and the knowledge learned about natural history and evolution?

## 4.2 Design and Methods

The study was carried out with 42 5th grade students (24 girls, 18 boys). Thirty-six students reported that they had never learned about evolution or natural history in their biology lessons before. Twenty-two students reported that they had never visited a natural history museum before. Students visited the museum together with their biology teachers as a one-day school trip. At the museum, they participated in an education programme consisting of a guided tour and a fossil imprint activity. The students were instructed in two groups. Both groups were introduced to the same standardized programme parts, which are briefly described in the following.

*Guided Tour* Museum educators instructed a tour around the exhibition. They started with a timetable on the characteristics of geological eras, from the Precambrian to the Permian era, to give students a preview about the history of the earth in general. After this introduction, the tour around the exhibits of the museum started with a reconstruction of life during the Permian period, which was characterized by a hot and arid climate, rare vegetation and the predominance of reptiles which were best adjusted to the environmental conditions. From here on, ongoing earth ages were visualized in the museum by dioramas or scenes, and by original exhibits of fossils or bones found in the region of Northern Hesse (Middle Germany), the region where the museum is situated. Thereby, the museum educators' guides introduced students to the earth ages of their home region, explained the adaptation of life to the environmental conditions, and at the same time explained from which regional evidence (fossils, bones) how people know today what life on earth would have looked like a long time ago.

*Fossil Activity* The fossil activity was a hands-on activity for explaining the principles of fossilization and for visualizing the work of a palaeontologist when finding and specifying a fossil. The steps of work for the students were: making an imprint; filling the imprint with material; gaining a copy; specification. Students worked in small groups and created their own copies of a given fossil of a primeval lizard.

In order to track students' learning over time, they were asked to respond to a questionnaire on three occasions (see Table 3). Students answered questions about socio-demographic variables, responded to knowledge questions referring to the education programme and responded to test items measuring their conceptions about

**Table 3** Design of study 2

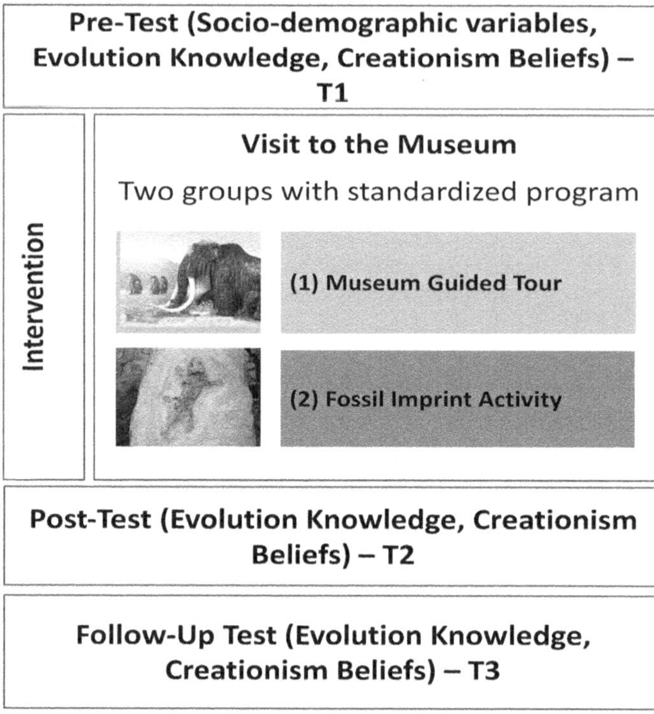


creationism. Data analyses were conducted based on the Rasch partial credit model using Winsteps (Linacre, 2011). Rasch analysis generates linear data. Measures are reported in logits. Further analyses were conducted using SPSS. For the initial analyses, data from the different measurement times were stacked. This procedure makes it possible to produce person measures for each test time within one frame of reference that can be used for further analyses.

*Evolution Knowledge* The test to evaluate knowledge progression consisted of twelve items constructed with reference to the education programme (closed and open-ended). The instrument was developed to cover the content of the outreach programme, items were evaluated by the two museum guides. Rasch analyses revealed that the item difficulties cover the whole range of the student sample. From this finding, we conclude suitability of the knowledge scale to assess knowledge gain in this particular setting. Item example: *How do scientists think dinosaurs became extinct?* Answer possibilities (right ones underlined): meteorite impact/illnesses/volcanic eruption/flooding. In total, students could reach 19 points from the knowledge test. Rasch analyses revealed test reliability (real person reliability) to be 0.67.

*Creationist Conceptions* The scale on creationist conceptions was used with reference to an instrument of Konnemann et al. (2013). These authors put together items from formerly used scales by Astley and Francis (2010) and Klose (2011) to study creationist conceptions and evolution acceptance. These studies also refer to validation aspects of the scales. By this means, Konnemann et al. (2013) report a reliable scale consisting of ten items about creationist conceptions and lack of evolution acceptance. In this study, due to the formulated research questions, only the five of the creationism items focusing on biblical literalism were analysed as one creationism scale. Item example: *When I observe nature, I belief that behind all life there exists a divine plan of creation.* Students rated the items on a 4-point Likert-type scale. For the analyses, all items were coded in a way that a higher rating on the Likert-type scale refers to greater biblical literalism. Test reliability (real person reliability) was found to be 0.72.

## 4.3 Analyses and Findings

To answer questions 1 and 2, the mean values of the person measures of evolution knowledge and creationist conceptions are compared over the three testing times (T1, T2 and T3; Table 4).

*Knowledge* To test if the means of test times differed significantly, Friedman's ANOVA was used. The knowledge about evolution did change significantly following the intervention: $\chi^2(2) = 48.161$, $p < .001$. It appeared that the knowledge about evolution did change significantly from T1 (Mdn = −.54) to T2 (Mdn = −1.19), $T = 9$, $p < .001$, $r = .6$ and that the change is still significant from T1 to T3 (Mdn = −1.01), $T = 27$, $p < .001$, $r = .55$.

*Creationist Conceptions*. Creationist conceptions did change significantly due to the education programme and stayed stable afterwards, $\chi^2(2) = 11.476$, $p < .01$. It appeared that conceptions did change significantly from T1 (Mdn = 1.3) to T2 (Mdn = 2.36), $T = 38.5$, $p < .001$, $r = .39$ and that the change is still significant from T1 to T3 (Mdn = 1.78), $T = 92.5$, $p < .01$, $r = .35$.

In order to answer question three, regression analyses determined that creationist conceptions (dependent variable) could be predicted from knowledge (independent variable). The results show that evolution knowledge significantly predicted creationism conceptions: $\beta = -.376$, $p < .001$. Knowledge also explained a significant proportion of variance in creationist conceptions, $r^2 = .141$, $F(1, 122) = 20.04$, $p < .001$, which means that 14.1% of the variance in creationist conceptions is predicted by the knowledge gain over the time.

## 4.4 Dialogues from the Guided Tour

The transcripts of the dialogues between guide and students from the guided tour were analysed, in order to find features of interactions. We could extract specific

**Table 4** Knowledge and conceptions in pre-test (T1), post-test (T2) and follow-up (T3)

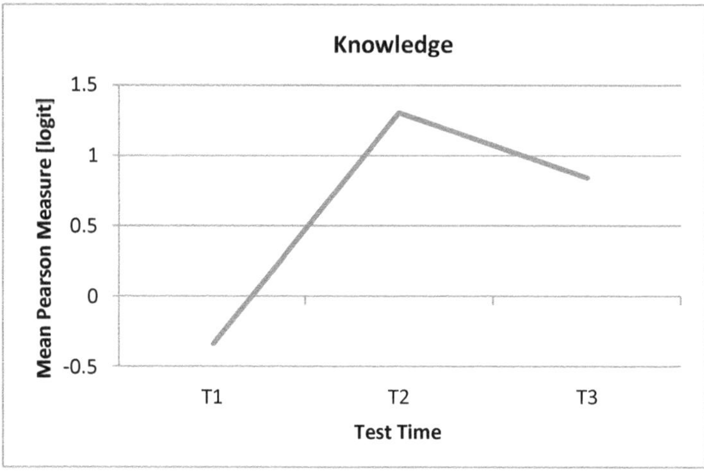


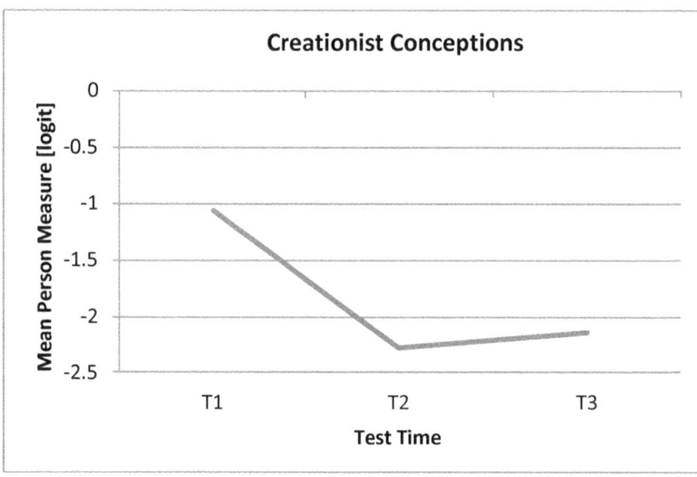


features from the dialogues that we assume to be relevant for the understanding of the exhibition: incorporation of prior knowledge, evidence and everyday experience.

The following example comes from the dialogue in front of the Permian Scenery in the museum.

**Guide**: During the Permian Era it became very dry. Amphibians had no more good opportunities to lay their eggs into the water, because water became scarce. From a part of amphibians developed reptiles. Which recent animals belong to the reptiles today?

**Students**: Snakes, crocodiles. […] Lizards, turtles.
**Guide**: Where do reptiles lay down their eggs?
**Student**: Into the sand.
**Guide**: Yes, into the dry and hot sand, and there the eggs get hatched. […] So the eggs are protected from dehydration by the sun. This adaptation to life ashore took place in the Permian Era.

The guide works with the students' prior knowledge about recent reptiles. He uses this knowledge to connect it to the impression of the scenery and the environmental change. The following excerpts show how students' everyday experiences were used to understand fossilization.

**Guide**: Triassic, Limestone. Now here was not a sea, anymore. Obviously, it became a desert. The animals left marks in the sand. Do you leave marks, when you walk in the sand?
**Student**: No, they are quickly gone with the wind.
**Guide**: Exactly, so it must have been wet, perhaps it rained, or the desert was wet. The animals walked in the mud and the footprints got dry and permanent. […] Other sediments overlaid the prints, it got compressed and rigid. […]

## *4.5 Conclusion from Case Study 2*

From the results, we conclude that a knowledge gain on evolution and natural history initiated by the here described education programme in a natural history museum was able to lead novice learners simultaneously reduce their creationist conceptions. We conclude from the dialogues of the guided tour that the way of explaining scientific evidence in connection with prior knowledge from everyday life is responsible for the change in conceptions. Students who mostly never before visited a natural history museum gained fruitful new explanations for the natural history of their home area. The conception change and knowledge gain are stable over time. Providing younger students with meaningful learning experiences about the progress of evolution seems to influence the formation of their conceptions about creationism as an explanation of how earth and life came into being.

Certainly, the tentativeness of the results has to be considered. The sample is not representative. Further research should include video analyses and interview studies subsequent to the guided tour to systematically analyse how the conception change proceeds during the museum visit, from which learning experiences it comes, which kind of instruction is most effective, and to what extent prior knowledge, everyday conceptions and prior learning experiences influence the knowledge gain and conception change.

## 5 Evolution Learning in Informal Settings

Concluding from these two studies, one can see that informal settings can be fruitful opportunities for learning evolution and even for changing creationist conceptions in novice learners. *Study one* showed that people can learn the concept of 'common ancestors' and overcome common inadequate conceptions. The results show on the one hand that our goals could be attained to awaken the visitor's interest, to clarify misunderstandings about the theory of evolution and to foster the dialogue between science and society. On the other hand, these results reveal learning boundaries in informal environments concerning the theory of evolution. However, not all exhibits proved to be as fruitful as the pinball game. Our results reveal learning boundaries concerning the theory of evolution (repetition): even though everyday conceptions are questioned, some of the exhibits are still too complex and explanatory texts are not read. In addition, *study two* showed that acquiring knowledge about evolution can come along with change in conceptions. In contrast to study one, in which no conceptions associated with creationism changed, study two indicates changed conceptions towards creationism and evolution in a novice learner population. One of the most significant differences between the studies was staffing, which was crucial to reflect non-rational creationism conceptions. The specific and unstaffed context in study one makes it harder to achieve a conceptual change. The playful approach of the pinball game seemed to us not scaffolded enough to change creationist ideas. It should be noted that all students were young people who still have a naive view of creationism, combined with rather little, if any, evolution knowledge. Here, staffing seems to be a more fruitful factor when reacting to creationist ideas. However, from both studies, it became clear that learning opportunities in informal contexts have to be designed with great care. Our results hint at how educational programmes on evolution should be designed to overcome existing difficulties in informal learning environments. This includes theoretical underpinnings as well as methodological implementation ('hands-on' and 'minds-on'). A future goal of research in evolution learning in informal settings should be to identify general design factors for effective evolution learning, specified to learning objectives and learner characteristics. Furthermore, science and religion represent two systems that can help people to organize their understanding of the world around them. When the two systems introduce two opposing explanations for the same phenomenon, there is a competition for 'explanatory space' (Preston & Epley, 2009) and conflict might well result with one explanation diminishing the perceived value of the other.

**Acknowledgements** We thank Franziska Perau for collecting data and Annette Scheersoi and Ulrich Kattmann for cooperation in study 1. We thank Anne Ahrens from Kassel University for collecting data and the Naturkundemuseum im Ottoneum Kassel for providing the museum infrastructure for study 2.

## References


Astley, J., & Francis, L. J. (2010). Promoting positive attitudes toward science and religion among sixth-form pupils: Dealing with scientism and creationism. *British Journal of Religious Education, 32*(3), 189–200.

Barsalou, L. W. (2005). Abstraction as dynamic interpretation in perceptual symbol systems. In L. Gershkoff-Stowe, & D. Rakison (Eds.), *Building object categories* (pp. 389–431). Carnegie Symposium Series. Majwah, NJ: Erlbaum.

Bishop, B. A., & Anderson, C. W. (1990). Student conceptions of natural selection and its role in evolution. *Journal of Research in Science Teaching, 27*(5), 415–427.

Bizzo, N. M. V. (2007). From down house Landlord to Brazilian high school students: What has happened to evolutionary knowledge on the way? *Journal of Research in Science Teaching, 31*(5), 537–556.

Blancke, S., Hjermitslev, H. H., Braeckman, J., & Kjærgaard, P. C. (2013). Creationism in Europe: Facts, gaps, and prospects. *Journal of the American Academy of Religion, 81*(4), 996–1028.

Brem, S. K., Ranney, M., & Schindel, J. (2003). Perceived consequences of evolution: College students perceive negative personal and social impact in evolutionary theory. *Science Education, 87*(2), 181–206.

Cobern, W. W. (1994). Comments and criticism. Point: Belief, understanding, and the teaching of evolution. *Journal of Research in Science Teaching, 31*(5), 583–590.

Duit, R., Gropengiesser, H., & Kattmann, U. (2005). Towards science education research that is relevant for improving practice: The model of educational reconstruction. In H. E. Fischer (Ed.), *Developing standards in research on science education* (pp. 1–9). London: Taylor & Francis.

Evans, E. M., Spiegel, A. N., Gram, W., Frazier, B. N., Tare, M., Thompson, S., et al. (2010). A conceptual guide to natural history museum visitors' understanding of evolution. *Journal of Research in Science Teaching, 47*(3), 326–353.

Eve, R. A., Losh, S. C., & Nzekwe, B. (2010). Lessons from the social psychology of evolution warfare: Good science is not enough. *Evolution: Education & Outreach, 3*, 183–192.

Gallese †, V., & Lakoff, G. (2005). The brain's concepts: The role of the sensory-motor system in conceptual knowledge. *Cognitive Neuropsychology*, 22(3–4), 455–479.

Groß, J., & Gropengießer, H. (2008). Warum Humanevolution so schwierig zu verstehen ist [Why human evolution is difficult to understand]. In Harms, U., & Sandmann, A. (Hrsg.), *Lehr- und Lernforschung in der Biologiedidaktik*. Band 3, Studienverlag, Innsbruck, 105–121.

Groß, J., & Gropengießer, H. (2006). Uniqueness and variation: The unexpected outcomes of free-choice learning. In *Proceedings of the VIth ERIDOB conference*, London.

Groß, J., Kattmann, U., & Scheersoi, A. (2009). *Darwin visiting IKEA*. Paper presented at the European Science Education Research Association (ESERA) 2009 Conference, Istanbul, Turkey, August 31, 2009–September 4, 2009.

Kim, Y. S., & Nehm, R. H. (2011). A cross-cultural comparison of Korean and American science teachers views of evolution and the nature of science. *International Journal of Science Education, 33*(2), 197–227.

Klose, B. (2011). Kreationismus, Wissenschaftsgläubigkeit und Werthaltung Jugendlicher [Creationism, scientism and value-attitudes of adolescents]. In H. Bayrhuber, A. Faber, & R. Leinfelder (Eds.), *Darwin und kein Ende? Kontroversen zu Evolution und Schöpfung* (pp. 146–151). Seelze: Klett.

Konnemann, C., Asshoff, R., & Hammann, M. (2016). Insights into the diversity of attitudes concerning evolution and creation: A multidimensional approach. *Science Education, 100*(4), 673–705.

Konnemann, C., Nick, M., Brinkmann, S., Asshoff, R., & Hammann, M. (2013). Entwicklung, Erprobung und Validierung von Erhebungsinstrumenten zur Erfassung von Kreationismus und Szientismus bei deutschen SchülerInnen [Development, use and validation of instruments to measure creationism and scientism with German students]. In U. Harms, & F. X. Bogner (Eds.), Lehr- und Lernforschung in der Biologiedidaktik, Band 5. Innsbruck: Studienverlag.

Kutschera, U. (2008). Creationism in Germany and its possible cause. *Evolution: Education and Outreach, 1*(1), 84–86.

Lakoff, G., & Johnson, M. (1999). *Philosophy in the flesh*. New York: Basic-Books.
Lakoff, G., & Johnson, M. (2003). *Metaphors we live by*. London: University of Chicago Press.
Lee, O. (1999). Science knowledge, world views, and information sources in social and cultural contexts: Making sense after a natural disaster. *American Educational Research Journal, 36*(2), 187–219.
Lewis, J., & Kattmann, U. (2004). Traits, genes, particles and information: Re-visiting students' understandings of genetics. *International Journal of Science Education, 26,* 195–206.
Linacre, J. M. (2011). *Winsteps® Rasch measurement computer program user's guide*. Beaverton, Oregon: Winsteps.com.
Mayring, P. (2010). *Qualitative Inhaltsanalyse* [Qualitative content analysis.]. Weinheim: Beltz Verlag.
Novick, L. R., Stull, A. T., & Catley, K. M. (2012). Reading phylogenetic trees: The effects of tree orientation and text processing on comprehension. *BioScience, 62*(8), 757–764.
Paul, J., Lederman, N. G., & Groß, J. (2016). Learning experimentation through science fairs. *International Journal of Science Education, IJSE, 38*(15), 2367–2387.
Perau, F. (2010). *Vorstellungen und Vorstellungsveränderungen zur Station „Vorfahren" der Ausstellung „Evolution schafft Vielfalt" im Kontext der Didaktischen Rekonstruktion*. [Conceptions and conceptual changes in the station "ancestors" at the exhibition "Evolution creates diversity"]. Masterarbeit der Leibniz Universität Hannover.
Preston, J., & Epley, N. (2009). Science and god: An automatic opposition between ultimate explanation. *Journal of Experimental Social Psychology, 45*(1), 238–241.
Rutledge, M. L., & Warden, M. A. (1999). The development and validation of the measure of acceptance of the theory of evolution instrument. *School Science and Mathematics, 99*(1), 13–18.
Scott, E. (2009). *Evolution vs. creationism: An introduction*. Berkeley, California: University of California Press.
Settlage, J. (1994). Conceptions of natural selection: A snapshot of the sense-making process. *Journal of Research in Science Teaching, 31*(5), 449–457.
Sinclair, A., & Pendarvis, M. P. (1997). Evolution vs. conservative religious beliefs. *Journal of College Science Teaching*; Washington, *27*(3), 167–170.
Wyhe, J. V. (Ed.) (2002). *The Complete work of Charles Darwin Online. Notebook on Transmutation. 1837*. Retrieved January, 2018 from http://darwin-online.org.uk/content/frameset?itemID=CUL-DAR121.-&viewtype=side&pageseq=1.

**Jorge Groß** graduated in biology and studied biology and chemistry for Science Education at the University of Hanover. He has taught at a secondary school in Hanover and in parallel finished his master's degree in communication in Kassel. He gained his Ph.D. in science education at the University of Hanover in 2006. Since 2012, he has been working as Professor of Sciences Education at the Otto-Friedrich-University of Bamberg and since 2013 as Director of the Institute for Research and Development of Subject-Related Teaching (EE-feU). Amongst others, he is a laureate of the Ars Legendi Award for Excellence in Mathematics and Science 2017. His research focuses on the analysis of students' conceptions and learning processes in science education.

**Kerstin Kremer** studied biology and chemistry to become a teacher. After attending teacher training and working three years as a school teacher, she completed her Ph.D. in biology education at the University of Kassel, Germany. Subsequently, she was a post-doctoral researcher at the University of Kassel and then took over professorships for Science Education at the Technical Universities of Munich and Aachen and for Biology Education at the Leibniz Institute for Science and Mathematics Education in Kiel, Germany. Now she is a Professor for biology education at the Leibniz University of Hanover, Germany. Research foci are the understanding of scientific inquiry and nature of science, sustainability education and informal learning.

**Julia Arnold** studied biology and English for secondary education. After her Ph.D. in biology education at Kassel University, she has been a post-doctoral researcher at RWTH Aachen University and at the Leibniz Institute for Science and Mathematics Education in Kiel, Germany. Now she is Head of the research group *Biology Education* at the Centre for Science and Technology Education, University of Applied Sciences and Arts Northwestern Switzerland—School of Education in Basel, Switzerland. Her main research interests are in health education, evolution learning, inquiry learning, and procedural understanding. She is particularly interested in the interplay between knowledge, beliefs and motivational factors.

# Participating in an Object-Based Learning Project to Support the Teaching and Learning of Biological Evolution: A Case Study at the Grant Museum of Zoology

**Jo Nicholl and Paul Davies**

## 1 Teaching and Learning About Evolution

Evolution is a unifying theme when teaching biology. However, both subject knowledge (SK) and pedagogic knowledge (PK) of evolution have been shown to be limited in teachers, as well as having the potential to conflict with personal worldviews (Taber, 2017). It has been shown that common misconceptions and alternative explanations surrounding biological evolution are maintained after formal education ceases (Nehm & Reilly, 2007; Yates & Marek, 2015), and even at degree level, biology students have been shown to leave university with major misconceptions of evolution and language-related misunderstandings (Burke da Silva, 2012; Smith, 2010). Unfortunately, trainee teachers are no exception to these findings (Crawford, Zembal-Saul, Munford, & Friedrichsen, 2005; Sanders & Ngxola, 2009). One difficulty in understanding evolution is because teachers (and museum exhibits) cannot directly transmit evolutionary knowledge to students, as this is not enough to overcome the conceptual and cognitive barriers, and when evolutionary explanations are learned 'by rote', these are not likely to include learning the *significance* of the concept and transforming the idea into 'personal culture' (Falchetti, 2012). Falchetti (2012) identifies some of the main hurdles towards understanding evolution as (i) comprehending the large time scales involved, (ii) understanding that the environment plays an *active* role in the process and (iii) classifying the biodiversity of life and the different levels of organisation, including the origin of different species.

It is not just the nature of the subject content that makes teaching about evolution challenging but also teachers' perceptions about their abilities to teach and their

J. Nicholl (✉)
UCL Institute of Education, London, England, UK
e-mail: j.nicholl@ucl.ac.uk

P. Davies
Queen's College, London, England, UK
e-mail: paul.davies@ucl.ac.uk



U. Harms and M. J. Reiss (eds.), *Evolution Education Re-considered*,
https://doi.org/10.1007/978-3-030-14698-6_17

students to learn. This notion, often described as teacher belief or self-efficacy, is hard to both describe and quantify (Parjares, 1992). Interwoven with this idea is the concept of teacher confidence, an important concept because it is about knowing how well one can successfully complete a task (Hargreaves & Fullan, 2012). Confidence can drive the development of self-efficacy, especially when teachers are given opportunities to reflect on their changing knowledge and skills and is a powerful driver in teacher development (Bandura, 1997).

Much of the current literature concentrates on improving evolution knowledge through professional development programmes, such as the knowledge enhancement course aimed at primary school teachers offered by a UK-based STEM education organisation (STEM, 2016). The research presented in this chapter looks specifically at the use of *objects within an object-based learning experience* to promote the understanding of evolution. The work was carried out in collaboration with the Grant Museum of Zoology, University College London and involved pre-service teachers (PSTs) studying for a postgraduate certificate in Science Education at a London university. All students volunteered to be part of the research, often identifying their lack of evolution knowledge as one of the main reasons for partaking in the workshops. The research is part of a larger project, but this chapter focuses on how using objects and being involved in an object-based learning project in a museum setting can develop teachers' SK of evolution, including their perceived confidence in the SK, and how their pedagogical practices may change as a result of involvement with such a project.

## 1.1 Using Museums to Teach Evolution

Public museums hold rare objects that are not commonly seen in daily life (Braund & Reiss, 2006). The specimens and artefacts that are housed in museums have long had an educational purpose (Pye, 2016). For example, The Natural History Museum, London was established in 1887 after the natural history collection of the British Museum was transferred there. It was here that there was a shift towards teaching the public about science rather than them being displayed just for academics (Davies & Nicholl, 2017). Open to the general public, these objects triggered an interest in the natural world, informing visitors of the life that has existed on the planet over the past 3.8 billion years.

Objects still remain at the very centre of a museum experience. These objects come in many different forms and, for natural history collections, they may come as a preserved specimen in a jar, fossil, bone remnants or a stuffed animal. They provide an immersive experience for the visitor where they are physically surrounded by evolutionary evidence. Consequently, natural history museums remain powerful vehicles to deliver knowledge to the public on evolution, presenting information in a way that stimulates interest, motivation and inquiry (Falchetti, 2012). The objects displayed are at the core of each museum's identity. Their collections provide a way of thinking deeply about the complexity of evolution, for example, comparing the features of fossils of prehistoric organisms presents evidence about evolution. Museum collections, in particular fossils, have long been the main source to evolutionary thinking, and using objects to help reconstruct the history of the Earth was something that Charles Darwin did himself. These objects can also be used as a stimulus for conversations surrounding key evolutionary ideas, such as deep time or natural selection (Gay, 2012).

While animal specimens and fossils are the most obvious and common objects used to help understand evolution in museums, sometimes other objects are used. For example, Darwin's notes and descriptions of specimens have been used at the Zoological Museum of Rome to showcase the flexibility of his interpretations (Falchetti, 2012). Most museum objects have taken centre stage for visitors to *see* rather than touch, where they are often displayed behind bars or in dimly lit cabinets. The main objection towards touching such objects comes from trying to protect and preserve the collections. There has been some attempt to go beyond the visual aid that collections provide. The Museum of Natural History of the University of Florence created cast skulls of humans and australopithecines so students could touch the skulls to explore the place of *Homo sapiens* (Dominici & Cioppi, 2012), while also having real specimens on display. Although their study touches on the handling of objects, it is more focused on bringing a framework of evolutionary theory together with academics and curators.

Real specimens are at the heart of the Grant Museum, where part of its philosophy is towards being able to touch and interact with specimens where possible. Although not all specimens can be handled, many of the skeletons can be touched and organised school workshops also have access to collections they can handle that would not be available to unscheduled visitors. The specimens at the museum range from full skeletons to specimens preserved in fluid. The museum houses a fascinating teaching collection of around 68,000 specimens, including a dodo and quagga, and is now the last university zoological museum in London.

### *1.2 Object-Based Learning*

Object-based learning is a more recent pedagogy associated with learning through touching objects (Chatterjee, 2011). Pioneered in university museums, interacting with such objects provides a stimulating and sensory experience that can be compared to that of a toddler exploring their world by receiving tactile feedback from an object (Hauf & Paulus, 2011). Thus, touch can be perceived to be the 'ultimate sense' to aid in building a 'complete representation of the world' (Critchley, 2008; Giachritsis, 2008). Touching objects also provide a focal point for the initiation of discussions and teamwork (Chatterjee, 2011) and promotes communication, curiosity and inquiry (Were, 2008). Reading Museum carried out a 10-month study evaluating the use of their loan boxes to schools and found that being able to actually handle and see the objects, in reality, did aid both the learning and how well the information associated with that object was retained (McAlpine, 2002).

Touching objects has also been linked to inducing emotional responses (Solway, Camic, Thomson, & Chatterjee, 2015). Emotive responses to handling objects can evoke memories, helping the elderly recall moments from youth and encourage them to talk (Kavanagh, 2000). They can assist in physical and mental health problems, encouraging young people to express their emotions (Martin & Jones, 2009). Not surprisingly, such tactile stimulation of a 3D object has been shown to maximise learning (Gallace & Spence, 2008).

Learning about evolution involves the understanding of many inter-related topics that fall under the umbrella of 'science'. You need to have an understanding of geology, genetics, ecosytems, the nature of science, the history of science. The list is extensive and ideas are complex. It is not as simple as learning how the kidney works. At present, work related to object-based learning has been mostly carried out in higher education (Chatterjee, 2015). The use of object-based learning has not been explored in teaching and understanding evolution.

A major aspect of understanding evolution involves observing living things. For example, the way the teeth of animals in a particular lineage change can be related to changes in diet. Similarly, examination of fossil material can shed light on the speed of evolutionary process such as dwarfing (Davies & Lister, 2001). Therefore, touching and interacting with these objects collaboratively may support the lateral thinking needed to help understand such a complex idea. While schools may have access to suitable materials for the learning of evolution, museums have been shown to provide authentic experiences for students that cannot normally be found in the formal classroom (e.g., Diamond & Evans, 2007). The Grant Museum, with its vast numbers of specimens (including cranial and post-cranial skeletal material and preserved specimens), is such a setting and may have the ability to represent such challenging ideas related to evolution in a way that promotes discussion and provides a more memorable learning experience.

Our study explored how the use of objects in the museum and involvement in an object-based learning project supports pre-service science teachers in both their SK and PK of biological evolution. To do this, we considered two main questions:

1. How does participation in an object-based learning experience develop pre-service teachers' subject knowledge of biological evolution of pre-service teachers?, and
2. How do the pedagogies of pre-service teachers change as a result of involvement in such a project?

We considered there to be multiple elements involved in the object-based learning experience that underpinned our research. Not only were the objects a central component of the project, but other factors, as shown in Fig. 1, were also regarded as key to the project and its outcomes. For example, since the project was carried out in a zoology museum, we deemed this to be of significance when considering and answering our research questions.

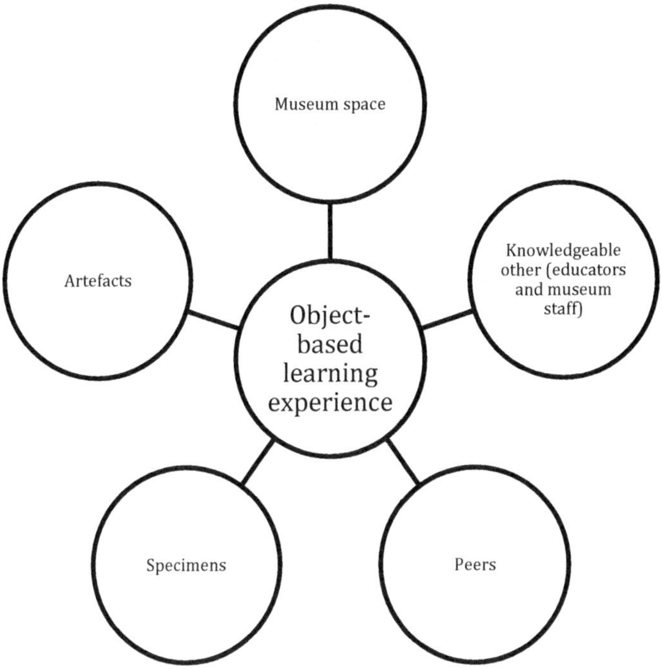


**Fig. 1** Our conceptualisation of the key elements involved in the object-based learning project

## 2 Methodology and Methods

The focus of the research was varied and, in order to capture as much valid data as possible, and collection of data through both qualitative and quantitative approaches was most appropriate; thus, we took a mixed methods approach (Cohen, Manion, & Morrison, 2013). This research was carried out as a case study, where we identified the work we carried out in the museum as the case. Defining a case study is not necessarily straightforward, with different authors taking varied perspectives on the scope of a case study, and its boundaries and application to other settings (Thomas, 2011). We chose to identify the workshops at the museum as the case because (i) The workshops were clearly bounded within the museum—both physically and in terms of the commitment that the participants made to the project, (ii) the museum was fully integrated into the project design, with the specimens used and the museum educators being all part of the museum and (iii) working in the museum was an essential aspect of the experiential focus of the project.

The Grant Museum of Zoology is a special place. The museum is the last of its kind in the University of London and a rarity within the UK. We were keen to bring together a range of experts in the project, with varying levels of experience, from different fields in order to bring richness to the project that is not always seen in research of this type. The team of five project members consisted of experts in high school and primary school biology education, museum education and a museum team member with extensive experience of using of objects in learning in various contexts. Collectively, the team designed and ran the project with the two authors of this chapter taking a specific lead in the data collection and analysis, including the recording of field notes.

## 2.1 Methods

The participants in this project were drawn from PSTs who were either training to teach science in high school or be the science specialist in primary school. All PSTs were studying on a one-year programme that involved a mixture of attending lectures at the university and experience of working in a range of schools in London. The PSTs were notified about the project by email and asked to volunteer. A meeting was held to outline the project and answer any queries. The PSTs then signed up to participate, in total 14 people. The participants were fully voluntary and in no way did their involvements on the project affect their progress on their teacher education programme. Table 1 shows the characteristics of the participants. As the data show, the majority of PSTs had a background in biology. However, there is a great range of subjects that come under the umbrella of 'biology' with only a small number of students have taken general biology courses. This matters because the study of biology at university has become more and more specialised with students graduating with very specific knowledge, for example about microbiology or genetics, but potentially without a more holistic appreciation of the science of biology, and thus a narrower understanding of evolution.

The mixed methods approach we took in this research allowed for the capture of a variety of data. Approaches of this type allow for triangulation of data and can potentially yield both highly valid and reliable data (Cohen et al., 2013).

In order to assess the best methods to approach our research questions, we considered each question in turn and identified what data would be necessary for each question to be answered. For question 1, we designed an identical (paired) questionnaire which was carried out at the start and end of the project. The questionnaire contained 17 items, asking a variety of questions that had a mixture of both open and closed responses. Many of these questions addressed common misconceptions. The questions were developed from a mixture of research into school students' (Kampourakis & Zogza, 2007, 2008, 2009; Shtulman, 2006; Spindler & Doherty, 2009), biology undergraduates' (Anderson, Fisher, & Norman, 2002; Baum & Smith, 2013; Jensen & Finley, 1996; Nehm & Reilly, 2007; Silva, Araújo, Gibram, & Carvalho, 2014) and science/biology teachers' (Kampourakis, 2014; Sa'adah, Hidayat, & Sudargo, 2017; Yates & Marek, 2015) understanding of evolution. In order to validate the questionnaire, we piloted it with PSTs following a programme that prepared them to teach science in high school ($n = 96$). This group did not include any PSTs involved in the project. Through this process, the project team was able to refine and develop the questionnaire items.

**Table 1** Characteristics of the project participants

| Participant number | Gender (F or M) | Age (years) | Subject specialism (as undergraduate) | Teaching experience |
|---|---|---|---|---|
| 1 | F | 24 | Biology | None |
| 2 | F | 24 | Biology | None |
| 3 | F | 25 | Biology | 'Summer camps' |
| 4 | F | 27 | Biology | None |
| 5 | F | 28 | Biology | None |
| 6 | F | 28 | Chemistry | School technician |
| 7 | F | 30 | Biology | None |
| 8 | F | 30 | Biology | None |
| 9 | F | 43 | Medicine | Supply teaching |
| 10 | M | 24 | Biology | None |
| 11 | M | 24 | Biology | None |
| 12 | M | 26 | Physics | None |
| 13 | M | 27 | Biology | School technician |
| 14 | M | 29 | Biology | Supply teaching |

Table 2 shows the questions that were asked in the questionnaire. Questions 1–4 were open response and were designed to probe the participants' understanding of the general, underlining ideas in evolution and the evidence that supported these ideas. These open response questions were developed to explore the thinking behind the participants' answers by asking them to explain their reasons for certain decisions they made in answering the questions. For example, one question focused on the ability to 'read' and interpret cladograms (from Baum, Smith, & Donovan, 2005) and asked the participant to identify the relationships between various groups depicted on the cladograms and to explain the reason(s) for their choice.

**Table 2** Pre- and post-questionnaire items and response type

| Question number | Question | Response type |
|---|---|---|
| 1 | What do you understand by the term 'biological evolution'? | Open |
| 2 | Make a list of any types of evidence that you can think of that supports the theory of evolution | Open |
| 3<br>a<br>b<br>c<br>d | What do you think these biological terms mean?<br>Adaptation<br>Homologous structure<br>Analogous structure<br>Last common ancestor | Open<br>Open<br>Open<br>Open |
| 4 | Cladogram interpretation (see Fig. 2) | Open |
| 5 | Evolution cannot be considered a reliable explanation because evolution is only a theory | Likert (1–5) with confidence scale (1–5)[a] |
| 6 | The scientific methods used to determine the age of fossils and the earth are reliable | Likert (1–5) with confidence scale (1–5)[a] |
| 7 | The earth is old enough for evolution to have occurred | Likert (1–5) with confidence scale (1–5)[a] |
| 8 | Evolution always results in improvement | Likert (1–5) with confidence scale (1–5)[a] |
| 9 | Members of a species evolve because of an inner need to evolve | Likert (1–5) with confidence scale (1–5)[a] |
| 10 | Traits acquired during the lifetime of an organism—such as large muscles produced by body building—will not be passed along to offspring | Likert (1–5) with confidence scale (1–5)[a] |
| 11 | New traits within a population appear at random | Likert (1–5) with confidence scale (1–5)[a] |
| 12 | Individual organisms adapt to their environments | Likert (1–5) with confidence scale (1–5)[a] |
| 13 | There exists a large amount of evidence supporting the theory of evolution | Likert (1–5) with confidence scale (1–5)[a] |
| 14 | 'Survival of the fittest' means basically that 'only the strong survive' | Likert (1–5) with confidence scale (1–5)[a] |

[a]Likert scale anchors: 1 = strongly disagree, 2 = disagree, 3 = neither disagree or agree, 4 = agree, 5 = strongly disagree. Students could also choose 'I don't know' or 'undecided'

Confidence scale: 1 = very unconfident, 2 = unconfident, 3 = somewhat confident, 4 = confident, 5 = very confident. Note that students were asked to leave the confidence box unanswered if they chose 'don't know' or 'undecided' for the answer

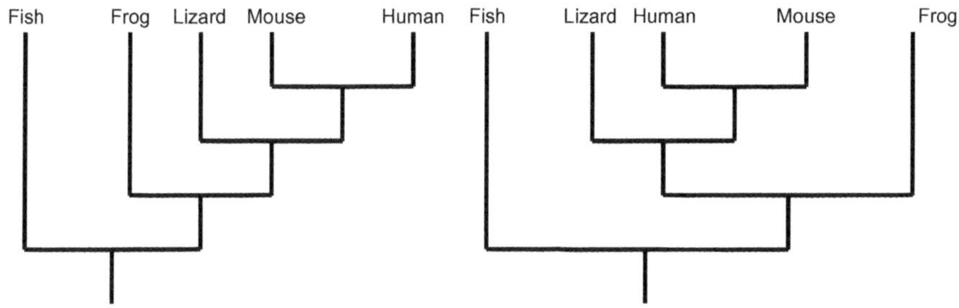


**Fig. 2** Cladogram used in questionnaire item 4. Participants were shown the figure and asked: on the basis of the tree on the left, is the frog more closely related to the fish or the human? Does the tree on the right change your mind? Why? (Based on Baum, Smith, & Donovan, 2005.)

Question 5–14 were Likert-style in design, giving the participants limited choice in their responses. These questions probed common misconceptions about evolution. At the end of each question, the participants were asked to rate the confidence of their response on a scale of 1–5. This was done to gain a sense of how the PSTs felt they had changed in terms of their own self-assuredness about their knowledge. Whilst not directly correlated with an increased ability to teach, nor an increase in *actual* subject knowledge, this measure was an important indicator of how the PSTs were developing as early-stage teachers and how they viewed their own abilities in teaching biological evolution. Confidence was also an important idea in terms of the field notes we gathered with it appearing in PSTs' discussions both in terms of language used and their expression.

Following the pre-workshop questionnaire, participants carried out a series of workshops introducing them to a number of object-based approaches that the museum used when teaching primary and high school students (Activity group A). Most activities involved objects that could be handled or touched. Table 3 provides details about each activity. During the activities, field notes were made, depending on what the participants discussed while touching/examining the objects. These were made as surreptitiously as possible as not to interrupt the flow of discussion. At the end of each activity, there was a general discussion that explored the participants' responses to the activities and their views about how these could be used with the students they were teaching in school, in order to capture any developments the workshops were having on their PK. These field notes and end-of-activity discussion notes involved a total of seven hours of observations and were used as data collection methods to contribute to answering both research questions 1 and 2. Table 3 identifies what research question each data collection strategy is linked to.

During the design phase of the project (Activity group B), participants spent time in the museum exploring the specimens and talking to the museum educators about activities they could develop if they brought their own students to the museum. After these discussions, captured through field notes, the participants wrote detailed plans to explain their activities. These were then presented to the project group, with each presentation ending with a conversation about their choices in the design and

**Table 3** Design of the workshop elements and data collection strategies

| Workshop number | Activity group | Activity title | Details of activity | Data collection strategy | Research question data collection related to |
|---|---|---|---|---|---|
| 1 | A | Pre-workshop questionnaire | Questionnaire completed individually | Mixture of open response and Likert-type questions | 1 |
| | | Animal limb bingo | A game involving participants working in pairs to explore the museum to match animal limb photographs to specimens | None | None |
| | | Skull adaptation | Hands-on examination of various vertebrate skulls (lion, bear, cow, beaver etc.) with discussion around adaptation | Field notes and end-of-activity discussion | 1 and 2 |
| | | Homologous and analogous structures | Exploration of the museum to find specimens to support ideas about homology and analogy as evidence for evolution | Field notes and end-of-activity discussion | 1 and 2 |
| 2 | A | Evolution-related object | Students were asked to bring in an object from home they felt was associated with evolution and asked to explain their choices | Field notes and end-of-activity discussion | 1 and 2 |
| | | Biscuit phylogeny | Building of a simple cladogram with biscuits (from DNA to Darwin, 2017) | Field notes and end-of-activity discussion | 1 and 2 |
| | | Cladogram | Using specimens in the museum (hand-on) to explore ideas about building a cladogram (modified from Sawyer Science, 2017) | Field notes and end-of-activity discussion | 1 and 2 |

(continued)

**Table 3** (continued)

| Workshop number | Activity group | Activity title | Details of activity | Data collection strategy | Research question data collection related to |
|---|---|---|---|---|---|
| 3 | B | Lesson planning | Exploring the museum specimens (hands-on) to develop a teaching sequence. Followed by presentation to the group and discussion of ideas | Field notes and end-of-activity discussion | 2 |
| | | Post-workshops questionnaire (as in Workshop 1) | Questionnaire completed individually | Mixture of open response and Likert-type questions | 1 |
| | | End-of-project Group Discussion | Group discussion with prompt questions | Field notes and transcription | 1 and 2 |

interesting points. These discussions were captured with field notes and the plans were collected for analysis.

The project ended with focus group discussions with the participants, facilitated by a number of prompt questions from the researchers which were designed to help us further understand how the use of objects in the museum had improved the participants' evolution SK and PK for teaching (e.g. What objects did you touch in the museum? What have you learnt from these workshops? Did holding the specimen play any importance? Have you used objects in your teaching of this topic before?). These were designed to encourage the participants to reflect on how their knowledge and understanding of evolution had developed and how the use of objects in their teaching may have changed. The researchers took an important role here, in guiding both the nature and direction of the discussion. The final workshop ended with the participants completing a copy of the questionnaire they had completed at the start.

## 2.2 *Data Analysis*

Each of the data collection tools—questionnaires, field notes, focus group discussion and plans—yielded a range of data. A quantitative analysis was carried out on the pre- and post-questionnaire for the closed-questions (Q5–14) to determine changes that took place in frequency of correct response. Having a sample of PSTs that was self-selecting, and presumably motivated by the project, presents a potential problem in terms of data analysis and precludes the use of parametric testing. Therefore, we decided to treat the PST sample as a paired (pre- and post-test) sample and applied Chi-squared analysis to the Likert responses of the level of agreement in their responses. Analysis of ordinal data of this type is a common approach allowing for direct comparison between data sets and is commonly seen in studies of this type (Cohen et al., 2013). Given the complex and extensive ways the rest of data (other questionnaire items, field notes, focus group discussion and lesson plans) was collected, and in order to provide a more comprehensive picture of the data, we drew on Qualitative Content Analysis (Elo & Kyngäs, 2008; Mayring, 2014). This analysis follows a protocol of various steps, including the development of concrete research questions and identification of the unit of analysis, formulation of inductive data categories and an iterative process where data categories are tested through the analysis procedure (further details can be found at Mayring 2002, 2014). Each Activity title listed in Table 3 was identified as a unit of analysis. The closed response questions in the questionnaire did not yield textual data but instead were Likert-style questions. Analysis of these questions involved calculations of mean scores and confidence scales for each answer.

# 3 Results

In the following section, we present the data as content-related categories that were derived from the quantitative data analysis (Table 4) and the qualitative inductive content analysis (Table 5). The qualitative data were collected, recorded and analysed using NVivo, whereby open coding was carried out. Both researchers coded the data separately and there was an agreement made on the formulation of the inductive data categories. The data related to exploring the development of SK are presented first, which then leads to the results related to pedagogies. In the *Discussion* section, we draw together key emerging ideas and suggest ways forward.

## 3.1 Evolution Subject Knowledge

*'Survival of the Fittest'*

Although students were good at recalling evidence for evolution (fossils, genetics and molecular evidence), they struggled more with defining evolution, where answers from Q1 of the pre-questionnaire ranged from specific answers referring to 'the process of change in living things through time … with changes in gene frequencies' to much less detailed answers relating to 'living things adapting'. The idea that evolution efficiently *designs* organisms for their environment was a theme that kept arising from the data. In the activity when students were asked to bring in an object related to evolution, one student brought in a mobile phone and argued that this reminded him of evolution, as 'it just gets better over time'. This misconception was also found on the questionnaire and other data, where students would discuss evolution as a way of perfectly adapting to the environment as only the strong survive. The common phrase 'survival of the fittest' (coined by Spencer, 1864) has come to symbolise natural selection. However, the term 'fittest' has come to mean 'best' in some circumstances whereas it should refer to the overall fitness in terms of reproductive success. In this sense, fittest means 'being fitter than those less fit'. This is less straightforward to talk about (and possibly to understand) and so is often shortened. An easier way to think about this is to consider a herd of prey animals running away from a predator. The fittest are those individuals that survive by not being caught. This could be the fastest, or any individual that is faster than the slowest—which does get caught and killed.

As a result of the project, the participants showed a considerable change in their understanding and confidence in using this term. This is supported by the quantitative analysis carried out on the questionnaire (Table 4). Question 14 (a Likert-style response to the statement 'Survival of the fittest means basically that only the strong survive') showed a significant improvement in answering this correctly and an improvement in how confident the participants were in their answer ($p = 0.0006$, see Table 4). Improvements were also witnessed during the various activities, with participants highlighting that they needed to be more careful with their language when

**Table 4** Data analysis of the pre- and post-questionnaire responses

| Question | Pre-workshops | | | Post-workshops | | | $X^2$ value | $p$ value | Significance |
|---|---|---|---|---|---|---|---|---|---|
| | Number of correct responses | Number of incorrect responses | Mean confidence | Number of correct responses | Number of incorrect responses | Mean confidence | | | |
| 4 | 4 | 10 | 2.8 | 12 | 2 | 4.6 | 9.33 | 0.002 | *** |
| 5 | 11 | 2 | 2.8 | 13 | 1 | 4.8 | 0.38 | 0.050 | * |
| 6 | 12 | 2 | 4.3 | 13 | 1 | 4.9 | 0.38 | 0.540 | ns |
| 7 | 12 | 2 | 4.5 | 13 | 1 | 4.9 | 0.38 | 0.540 | ns |
| 8 | 10 | 4 | 3.9 | 13 | 1 | 4.2 | 2.19 | 0.140 | ns |
| 9 | 12 | 2 | 4.1 | 13 | 1 | 4.8 | 0.38 | 0.540 | ns |
| 10 | 10 | 4 | 4.0 | 13 | 1 | 4.8 | 2.19 | 0.140 | ns |
| 11 | 8 | 6 | 3.2 | 11 | 3 | 4.1 | 1.47 | 0.220 | ns |
| 12 | 5 | 9 | 2.8 | 11 | 3 | 4.8 | 5.25 | 0.020 | ** |
| 13 | 12 | 2 | 4.5 | 13 | 1 | 4.8 | 0.38 | 0.540 | ns |
| 14 | 3 | 11 | 1.8 | 12 | 3 | 3.9 | 11.6 | 0.0006 | **** |

ns $P > 0.05$
*$P \leq 0.05$
**$P \leq 0.01$
***$P \leq 0.001$
****$P \leq 0.0001$

**Table 5** Themes that emerged from the different workshop activities after carrying out the Qualitative Content Analysis (QCA)

| Activity title | Subject knowledge emerging categories from QCA | Pedagogic knowledge emerging categories from the QCA |
|---|---|---|
| Skull adaptation | Ancestry<br>Survival of the fittest<br>The knowledgeable other<br>Affective links | Alternative uses of specimens<br>Use of evidence<br>Enquiry science |
| Homologous and analogous structures | Ancestry<br>The knowledgeable other<br>Affective links<br>Scientific theories | Alternative uses of specimens<br>Use of evidence |
| Personal object-related to evolution | Survival of the fittest<br>Ancestry | Use of evidence |
| Biscuit phylogeny | Survival of the fittest<br>Ancestry<br>Geological 'deep' time | Use of evidence<br>Enquiry science |
| Cladogram | Survival of the fittest<br>Ancestry<br>Geological 'deep' time<br>The knowledgeable other<br>Affective links | Enquiry science<br>Use of evidence |
| Lesson planning | Scientific theories | Alternative uses of specimens<br>Use of evidence<br>Enquiry science<br>Affective links |
| End-of-project group discussion | Survival of the fittest<br>Geological 'deep' time<br>The knowledgeable other<br>Affective links | Use of evidence<br>Enquiry science<br>Alternative use of specimens |

*Note* affective links are shown within subject knowledge but identified as a theme within both subject knowledge and pedagogic knowledge

discussing organisms effectively adapting to their environment and only the 'fittest' surviving. Objects in the museum were used to demonstrate this idea; for example, when students picked up the kiwi skeleton and presented the inefficient design of the kiwi bird, making reference to the tiny wings that could be seen. Students also discussed the advantage of having eyes on the side of your skull, such as in goats, in order to look out for predators and see all around (see Table 6). When handling the gorilla and other ape-like skulls, students debated about why these organisms did not have eyes on the side of their head if it meant they would have been able to see more.

*Ancestry: Homologous and Analogous Structures*

In the questionnaire, the interpretation of the cladogram was a challenge to most participants with only four participants correctly interpreting the image (Table 4,

**Table 6** Examples of data from the categories identified

| Categories identified | Examples of data |
|---|---|
| Ancestry: homologous and analogous structures | 'I never looked at animals' legs properly before, they really are different from the outside, but the skeletons are so similar. Such good evidence to show children that we all come from a common ancestor' |
| Affective links | Reference to learning through argumentation at a place 'like this' being powerful and special, especially being able to imagine naturalists actually arguing it out in this building years ago |
| Use of evidence | 'I disagree … With the coccyx, it shows there was some link to our evolutionary history. There is an element there that shows a hint of what we used to look like' |
| Geological 'deep' time | Realisation that the phylogenetic 'tree' tells a story of the history of time, and that it is not just a picture of how things 'link together' |
| Alternative uses of specimens | Reference to using the snake specimen in the lesson to introduce them to vertebrates and how the skeleton is used for protection and movement |
| Enquiry science | 'I want my students to use the specimens to generate lots of questions and discuss, I don't want it to be all about being told things' |
| Affective links | 'I have never really been exposed to animal biology, I have just been immersed in human biology. Give me an alveoli any day and there would be no problem! Outside of school has really been my only experience of this type of biology rather than in school, which I realise now is pretty bad' |
| The knowledgeable other | Question about the ancestry of a turtle if it has a fused vertebrate, but more directed at the nearby expert |
| Survival of the fittest | 'It doesn't matter whether eyes are better at the side, it matters what other organisms are out there. If they are slower and weaker, then it still gives the others an advantage' |
| Scientific theories | Reference made to the number of specimens available in the museum that help demonstrate the body of facts available for scientists to use as observational evidence to support evolution |

Q4), with mean confidence of all participants at 2.8 out of 5. It was clear that those who got the answer correct post-workshop (12 out of 14) were confident in their understanding, with a mean confidence score of 4.6. Incorrect answers showed a lack of understanding of how common ancestry can be determined from the root of the tree and how organisation of the tree tips can be read. The post-workshop questionnaire showed a significant shift in understanding and application of knowledge ($p = 0.002$). Discussions in the focus group (see Tables 5 and 6) revealed that the object-based task on the construction of cladograms developed participants' knowledge and confidence in common ancestry, in addition to helping them understand the evidence that homologous and analogous structures provide for evolution.

Seeing and handling homologous and analogous structures to understand the concept of ancestry was an effective way to improve students' knowledge of this difficult concept. Once students were familiar with the terms 'homologous' and 'analogous', and their link to evolution, they continued to refer to such structures in and around the museum as an explanation for evolution and a point of interest. The students did not necessarily refer to them as 'homologous' and 'analogous'. During the cladogram activity, students compared the anatomy of different structures and made reference to those with similar structures being 'related organisms', sometimes even mentioning the sharing of a 'common ancestor'.

The bird wing and the batwing were common structures readily discussed during the workshops, and the most popular example found in the post-questionnaire when asking students about analogous structures. When comparing the skulls of different organisms, the students were fascinated with the difference between the horns of goats, the horns of rhinos and the antlers of deer. These differences were also used in the students' lesson plans when designing their own lesson. Seeing and touching a human leg and sheep leg helped students apply the idea of homologous structures, where students compared how humans walked on their metatarsals whilst sheep walked on their phalanges. Likewise, the pentadactyl limb of a dugong and a gibbon seemed to help students' understanding of ancestry.

The objects in the museum provoked discussions surrounding ancestry that demonstrated a build-up of the students' understanding of evolution and the evidence that supports it. Below shows an example of such a conversation between three participants happening at the snake skeleton as they try to decide how vestigial limbs in Old World snakes developed:

> P1: That bone [vestigial limb bone in a python skeleton] is smaller because the animal doesn't use it.
>
> P2: But, that isn't how evolution works; it isn't that the animal 'makes a choice'. It could be that the limb has different functions to moving and that is why it is smaller.
>
> P1: If that were the case, then all snakes would have the little bone; these specimens [vipers] don't. I think we need to look at other bits of the skeleton to see if the bone is from the limb that that disappeared, or if it is a new structure.
>
> P3: OK, but what evidence? We would need to see the fossils between lizard things and snakes. Are there any?
>
> P1: That might work but I think you can use geography too. Look, the African snakes are older, more ancient, than those from America, so they are like the lizard ancestor – well more like the lizard ancestor.

*Geological 'Deep' Time*

Some students in the pre-questionnaire stressed the idea of changes happening over time in their written responses. However, looking at the questionnaire responses and coding the data, it is evident that not all students understood the importance of deep time. Question 12 was concerned with the idea of individuals within a species adapting to their environment. This is a common misconception among young people (Spindler & Doherty, 2009) and is suggestive of Lamarkian evolution (Kampourakis & Zogza, 2007, 2008, 2009). Whilst there is growing evidence that epigenetic modification takes places within individuals, some of which can also be inherited by subsequent generations, there is a consensus that in macro evolution this is not very important. As Table 4 shows, the participants changed in their ability to answer this question correctly ($p = 0.020$) and increased in confidence in their ability. We suggest that one way this may have been supported is through the students gaining a better understanding of the history of life on a much grander scale, where the profound idea of deep time is needed. Only later on in the workshops (see Table 5) did students begin to appreciate evolution occurring at the population level and over long periods of time. Even though these workshops did not explicitly address time as a key idea, it was a common theme when students were handling and observing the objects and it contributed to improvements in the participants' understanding of evolution. Reference to turtles being 'millions of years old' was discussed when looking at the turtles' fused rib cage. In addition, the time scale of when lizards and snakes shared a common ancestor was also questioned when handling these skeletons. Likewise, the prehistoric, jawless lamprey triggered discussions surrounding the rise of vertebrates and how the phylogenetic tree showed the history of the different organisms over time, rather than just linking them together (Table 6).

*The Knowledgeable Other*

It was evident from the qualitative data that the students' knowledge of evolution did improve. The quantitative data also showed a significant change to their understanding of four of the questions (Table 4). From the qualitative data, a considerable amount of the learning seemed to occur during the object-based activities whilst students conversed with those involved in the project (educators and museum curators). Students posed many questions that arose from the objects to the 'knowledgeable others' throughout the different activities (Table 5), where questions arose from the actual objects (Table 6). This highlights the need for someone with a secure level of knowledge to be present in a place where objects raise such curiosity and wonderment. The 'knowledgeable others' also corrected students' misconceptions and steered conversations to promote self-learning and inquiry. There was an explicit recognition from the students of wanting a secure knowledge of evolution before teaching their own students. The perceived confidence participants had in their own knowledge of evolution increased after the workshops, as shown in Table 4 of the questionnaire results. This is something that the post-workshop focus group discussion also revealed, where most of the participants agreed that the various activities they had carried out supported their developing knowledge and confidence in their understanding of evolution.

References were explicitly made to how helpful the knowledgeable others were: 'It has been really helpful having you guys that know so much about this here'.

When students were wandering around the museum, as opposed to being within a group with the lecturers present, they would often ask questions to themselves or their partner. However, it does seem from the field notes collected that even these personal questions were in hope that a lecturer would hear them. For example, when looking at the snake skeleton, one student asked 'I wonder what bone that is and if it is homologous to a human ribcage?'.

*Scientific Theories*

Question 5 of the questionnaire was concerned with the idea of scientific theories. In support of the inductive content analysis (see Tables 5 and 6), participants showed a significant shift in their confidence in interpreting what a scientific theory is and the reliability of theories. Albeit only two participants changed their mind, more significant is that the mean confidence level changed from 2.8 to 4.8 ($p = 0.05$, see Table 4), suggesting that the pre-service teachers' self-efficacy in teaching about biological evolution rose, something which has positive implications for their practice in the classroom. Furthermore, it suggests that participation in the workshops does help to strengthen understanding of what scientific theories are and how biological evolution can be explained as a theory.

## 3.2 Pedagogic Knowledge

The development of students' PK was also analysed through Qualitative Content Analysis. The categories that arose from the data related to this question are discussed below.

*The Use of Specimens*

From the beginning of the project, students were informed of the purpose of the project and also the outcomes from the project. They were told that by the end of the three workshops, they would have created a lesson plan that would model a lesson at the museum with a class of their choice. Analysing the students' lesson plans and the conversations that took place during the workshops, all students planned for activities that involved handling the specimens. Whilst some used new specimens, most used the same specimens that they had experienced. In some cases, the students developed lesson plans that went beyond the specific theme of evolution. A common extension of ideas moved into broader ideas about adaptation, such as those associated with movement, support and protection. For example, the skeletal systems of the dugong and human were compared for movement. The goat horns were used as an example of protection. The fused rib cage of the turtle and the pectoral girdle of the penguin and kiwi were also made available for students to explore these ideas (also see Tables 5 and 6). Emphasis was placed on *touching* specimens, with a lot of students recognising this as the key stimulating factor in the workshops, and referred to 'awe and wonder' when thinking about the size of a tiger skull or the shape of the beaver's teeth. There was a difference of opinion about whether the specimens needed to be real, but all agreed that the touching of the specimens was meaningful. For example: 'For me, it was about the physical more than it was about the physical being "real." I wouldn't have cared if it was plastic, it was just more about the fact that I could see it and touch it, and that really meant something'.

Participants also noted that museums sometimes loan out specimens and, encouragingly, there was a general consensus that non-museum quality specimens could be used in classrooms. These included specimens held in school or materials either the teacher or students might collect, objects once commonly found on a classroom 'nature table'. One student raised her frustrations as to why she had not thought about using objects in her classroom until now, considering one of the most memorable experiences for her as a child was when the teacher brought in rodent skeletons in her primary school.

*The Use of Evidence*

The specimens played a key role as evidence. This was initially seen in the first workshop in relation to evidence for evolution, but was then expanded to other scientific concepts, such as evidence for placing a particular organism into a certain phylum or class, or evidence for how teeth have different functions. From the pre-questionnaire, it was shown that students were already familiar with the different types of evidence to support evolution, which suggests they appreciate the importance of evidence to science. However, what this project showed was that students were keen to use the specimens that they saw and touched as evidence for the teaching of other scientific concepts (for example, see Tables 5 and 6). Furthermore, they made many references to the use of evidence in generating scientific theories, another theme previously identified, as shown in Table 6, and applying this to the theory of evolution.

*Enquiry Science*

Students valued the use of different types of discussions that occurred in the workshops and placed emphasis on presenting science to their students as a subject that cannot know all the answers but asks questions depending on the evidence available (see Tables 5 and 6). There was an appreciation for the tentative nature of science where participants designed enquiry activities where their students would not necessarily find out the 'correct answers' at the end. There was also recognition of the need for whole-group discussions (with lecturers), student-to-student explorative talk, presentations between students and the one-to-one discussions with lecturers to support the students' learning and curiosity. Although there was an appreciation for the nature of science, the need for precision was identified when talking about evolution in order to attempt to avoid misconceptions.

## 3.3 *Affective Links*

This short section is presented separately as we believe it impacts on both SK and pedagogy. The affective responses from students (see Tables 5 and 6) were evident throughout the workshops. Students were in awe of the place (e.g. 'It's amazing, you can imagine scientists arguing it out here hundreds of years ago'), the specimens (e.g. 'Is everything really real that we are touching?') and the subject of evolution itself (e.g. 'I feel bad that I have not spent my time getting to understand this more, I've always been involved in human biology.'). Whilst there were mixed feelings and discussions towards whether the objects needed to be real, and towards the logistics of taking students on a trip to the museum given the time constraints of covering the whole curriculum, there was a sense of real pleasure during the workshops. Most of the affective responses were seen either during the workshops, where the students handled the specimens, or in reference to the museum itself. These responses were mostly related to an appreciation that they had the opportunity to hold such species and be in such a historical place that held such treasured specimens.

## 4 Discussion

This project improved students' SK of evolution and developed their PK. It is difficult to identify the specific impact the *handling of the objects* had in comparison to the impact the *place* had on the student teachers, and it goes beyond the scope of this chapter. However, the data suggest that it is a combination of both. This suggests that it would be useful to transfer this work to a more formal learning environment (i.e. classroom), where work can focus of the impact of object-based learning, and also a similar setting (i.e. another museum), where work can focus on the impact of the space to investigate these key influences further.

Touching the specimens intrigued the students and triggered them to ask questions about their age, origin and link with other species. It engaged them in the topic of evolution in a novel way that suggests it would have a more lasting impact in their memory than by simply being 'told'. Students developed their understanding of natural selection by discovering how 'inefficient' evolution can be. The workshops provided students with a way to consider deep time through the use of specimens. The activities provided a scaffold to support students' curiosity about organisms that are closely related and those that are distant relatives from millions of years ago. Thinking about our ancestry in this way, and understanding cladograms, further developed the students' conceptual understanding of evolution. None of this should be recognised without the 'knowledgeable other' as identified in the data analysis. Playing a lead role in the fruitful and stimulating conversations, without these 'drivers' being part of the dialogue, improvements in understanding of evolution would be very different. A robust understanding of evolution is needed in order to be able to apply this workshop to a classroom setting without such experts available. Something that should be investigated further is the collaboration between such experts and the learner being immersed in such an awe-inspiring space, and what impact this could have on future teaching training development programmes.

All participants were keen to introduce more objects into their teaching and recognised the power of handling such objects. The importance of their own SK was recognised, especially in order to answer a variety of questions when talking to their own students. The activities they developed all had an enquiry nature about them, emphasising the tentative nature of science and how we do not know the answer to everything. The application of the specimens by the participants to other scientific subjects, such as classification and joints, suggests the importance of looking at object-based learning in different subject-specific areas in order to see if it would have a similar positive effect.

Students responded emotionally to the experience and built up an appreciation for evolution as a subject, as well as for the specimens themselves and the handling of such objects. Although this research is a case study and recognises the unique nature of the space at the Grant Museum and the rare opportunity to handle real specimens, it would be useful to transfer this work to other similar settings with similar opportunities to see if the outcomes are comparable to this research.

## References

Anderson, D. L., Fisher, K. M., & Norman, G. J. (2002). Development and evaluation of the conceptual inventory of natural selection. *Journal of research in science teaching, 39*(10), 952–978.
Bandura, A. (1997). *Self-efficacy: The exercise of control*. NJ: Prentice-Hall.
Baum, D. A., & Smith, S. D. (2013). *Tree thinking: An introduction to phylogenetic biology*. Roberts.
Baum, D. A., Smith, S. D., & Donovan, S. S. (2005). The tree-thinking challenge. *Science, 310*(5750), 979–980.
Braund, M., & Reiss, M. (2006). Towards a more authentic science curriculum: The contribution of out-of-school learning. *International Journal of Science Education, 28*(12), 1373–1388.
Burke da Silva, K. (2012). Evolution-centered teaching of biology. *Annual Review of Genomics and Human Genetics, 13,* 363–380.
Chatterjee, H. J. (2011). Object-based learning in higher education: The pedagogical power of museums. In *International Committee for University Museums and Collections (UMAC) Proceedings*, 3.
Chatterjee, H. J., & Hannan, L. (Eds.). (2015). Engaging the senses: Object-based learning in higher education. Ashgate Publishing, Ltd.
Cohen, L., Manion, L., & Morrison, K. (2013). *Research methods in education*. New York: Routledge.
Crawford, B. A., Zembal-Saul, C., Munford, D., & Friedrichsen, P. (2005). Confronting prospective teachers' ideas of evolution and scientific inquiry using technology and inquiry-based tasks. *Journal of research in science teaching, 42*(6), 613–637.
Critchley, H. (2008). Emotional touch: A neuroscientific overview. In H. Chatterjee (Ed.), *Touch in museums: Policy and practice in object handling* (pp. 61–71). Oxford: Berg.
Davies, P., & Lister, A. M. (2001). *Palaeoloxodon cypriotes*, the dwarf elephant of Cyprus: Size and scaling comparisons with *P. falconeri* (Sicily-Malta) and mainland *P. antiquus*. In *World of elephants conference proceedings* (pp. 479–480).
Davies, P., & Nicholl, J. (2017). Using object-based learning to understand animal evolution. In M. Mueller (Ed.), *Animals and science education: Ethics, curriculum and pedagogy* (Vol. 2, p 145).
DNA to Darwin. (2017). *DNA to Darwin: student guide*. http://www.dnadarwin.org/casestudies/INTRO/FILES/IntroSG1.0.pdf. Last accessed (DATE).
Diamond, J., & Evans, E. M. (2007). Museums teach evolution. *Evolution, 61*(6), 1500–1506.
Dominici, S., & Cioppi, E. (2012) Evolutionary theory and the florence paleontological collections. *Evolution: Education and Outreach, 5*(1), 9–13.
Elo, S., & Kyngäs, H. (2008). The qualitative content analysis process. *Journal of Advanced Nursing, 62*(1), 107–115.
Falchetti, E. M. (2012). Biological evolution on display: An approach to evolutionary issues through a museum. *Evolution: Education and Outreach, 5*(1), 104–122.

Gallace, A., & Spence, C. (2008). The cognitive and neural correlates of "tactile consciousness": A multisensory perspective. *Consciousness and Cognition, 17,* 370–407.
Gay, H. (2012). Talking about evolution in natural history museums. *Evolution: Education and Outreach, 5*(1), 101.
Giachritsis, C. (2008). The use of haptic interfaces in haptic research. In H. J. Chatterjee (Ed.), *Touch in museums: Policy and practice in object handling* (pp. 75–90). Oxford: Berg.
Hargreaves, A., & Fullan, M. (2012). *Professional capital: Transforming teaching in every school.* London: Routledge.
Hauf, P., & Paulus, M. (2011). Experience matters: 11-month-old infants can learn to use material information to predict the weight of novel objects. *Infant Behavior and Development, 34*(3), 467–471.
Jensen, M. S., & Finley, F. N. (1996). Changes in students' understanding of evolution resulting from different curricular and instructional strategies. *Journal of research in science teaching, 33*(8), 879–900.
Kampourakis, K. (2014). *Understanding evolution.* Cambridge: Cambridge University Press.
Kampourakis, K., & Zogza, V. (2007). Students' preconceptions about evolution: How accurate is the characterization as "Lamarckian" when considering the history of evolutionary thought? *Science & Education, 16*(3), 393–422.
Kampourakis, K., & Zogza, V. (2008). Students' intuitive explanations of the causes of homologies and adaptations. *Science & Education, 17*(1), 27–47.
Kampourakis, K., & Zogza, V. (2009). Preliminary evolutionary explanations: A basic framework for conceptual change and explanatory coherence in evolution. *Science & Education, 18*(10), 1313–1340.
Kavanagh, G., (2000) *Dream spaces: Memory and the museum.* London Continuum.
Martin, M., & Jones, G. V. (2009). Affect and alexithymia determine choice among valued objects. *Emotion, 9,* 340.
Mayring, P. (2002). Qualitative content analysis—Research instrument or mode of interpretation? In M. Kiegelmann (Ed.), *The role of the researcher in qualitative psychology* (pp. 139–148). Tübingen: Verlag Ingeborg Huber.
Mayring, P. (2014). *Qualitative content analysis: Theoretical foundation, basic procedures and software solution.* Open Access Repository. http://www.ssoar.info/ssoar/bitstream/handle/document/39517/ssoar-2014-mayring-Qualitative_content_analysis_theoretical_foundation.pdf.
McAlpine, J. (2002). Loan star. *Museums Journal, 102*(1), 26.
Nehm, R. H., & Reilly, L. (2007). Biology majors' knowledge and misconceptions of natural selection. *BioScience, 57*(3), 263–272.
Pajares, M. F. (1992). Teachers' beliefs and educational research: Cleaning up a messy construct. *Review of educational research, 62*(3), 307–332.
Pye, E. (Ed.). (2016). *The power of touch: Handling objects in museum and heritage context.* Routledge.
Sa'adah, S., Hidayat, T., & Sudargo, F. (2017). Undergraduate students' initial ability in understanding phylogenetic tree. *Journal of Physics: Conference Series, 824*(1), 012040 (IOP Publishing).
Sanders, M., & Ngxola, N. (2009). Addressing teachers' concerns about teaching evolution. *Journal of biological education, 43*(3), 121–128.
Sawyer Science (2017). *How to build a cladogram.* http://sawyerscience.com/pdfs/5-Unit3Evolution/building_a_cladogram_practice.pdf. Last accessed (DATE).
Shtulman, A. (2006). Qualitative differences between naïve and scientific theories of evolution. *Cognitive Psychology, 52*(2), 170–194.
Silva, H. M., Araújo, E. S. N. N. D., Gibram, D. E., & Carvalho, G. S. D. (2014). Conceptual change about evolution and origins of life throughout an undergraduate course of Biological Sciences. In *Proceedings of INTCESS 14-international conference on education and social science proceedings* (pp. 1249–1258). INTCESS.
Smith, M. U. (2010). Current status of research in teaching and learning evolution: II. Pedagogical issues. *Science & Education, 19*(6–8), 539–571.

Solway, R., Camic, P. M., Thomson, L. J., & Chatterjee, H. J. (2015). Material objects and psychological theory: A conceptual literature review. *Arts & Health, 8,* 1–20.
Spencer, H. (1864). *The principles of biology*. Boston: Appleton.
Spindler, L. H., & Doherty, J. H. (2009). Assessment of the teaching of evolution by natural selection through a hands-on simulation. *Teaching Issues and Experiments in Ecology, 6,* 1–20.
STEM. (2016). https://www.stem.org.uk/cpd/ondemand/226530/developing-subject-understanding-evolution-year-6. Last accessed (DATE).
Taber, K. S. (2017). Representing evolution in science education: The challenge of teaching about natural selection. In *Science education: A global perspective* (pp. 71–96). Springer International Publishing.
Thomas, G. (2011). A typology for the case study in social science following a review of definition, discourse, and structure. *Qualitative inquiry, 17*(6), 511–521.
Were, G. (2008). Out of touch? Digital technologies, ethnographic objects and sensory orders. In H. J. Chatterjee (Ed.), *Touch in museums: Policy and practice in object handling* (pp. 127–131). Oxford: Berg.
Yates, T. B., & Marek, E. A. (2015). A study identifying biological evolution-related misconceptions held by prebiology high school students. *Creative Education, 6*(08), 811.

**Jo Nicholl** is a Lecturer in Science Education at UCL Institute of Education. She has a B.Sc. in biology and was formerly the science advisor for a London High School before moving into Initial Teacher Education. She is currently studying for her Ph.D. looking at the role of zoos in science education. Jo teaches on both the Secondary Science PGCE and is the programme leader for the Science Education Masters programme. Her research interests include non-formal learning in biology and curriculum development.

**Paul Davies** is Head of Science and Director of Teaching and Learning at Queen's College, London. Before returning to teaching, Paul was a Senior Lecturer in Science Education at UCL Institute of Education, where he remains an Associate Lecturer. He began his career as a palaeontologist before training as a science teacher, with a specialism in biology.

Paul's interests include the philosophy of biology and, in particular evolution, the role that fieldwork plays in supporting biology education. He co-leads the science education the Biology Education Research Group.

# What Now for Evolution Education?

Michael J. Reiss and Ute Harms

## 1 The Present State of Evolution Education

While evolution is universally regarded by biologists as a core, possibly *the* key, aspect of biology, the understanding of it as a concept is poor among school students, other learners in full-time education and the general population alike. There are two main classes of reasons for this. One is to do with the cognitive difficulties of the theory. Even if we simply focus on natural selection, this is a challenging concept for learners. It requires powers of abstract reasoning and there are a number of steps in the argument, each of which needs to be comprehended if the overarching concept of natural selection itself is to be understood. And then, of course, there is so much more to evolution than the theory of natural section. For a start, an appreciation of 'Deep Time' is needed—and this is itself a difficult concept for many students, one that can literally be unimaginable. In addition, a learner needs to have an appreciation of the different sorts of competition between organisms, whereas many learners think only of predator–prey relationships. These various cognitive challenges are extensively referenced in the Chapter 'The Present Status of Evolution Education'.

The second class of reasons why the understanding of evolution is poor is to do with cultural issues that the theory raises among some learners. The most straightforward reading of the scriptures of a number of the world's religions seems to argue against the core evolutionary notion that all organisms are related through descent (vertical transmission) as well as, as we now know, through the horizontal transmission of genetic material. In the case of the Judaeo-Christian scriptures, but not in the Qur'an, there is even an apparent timescale for the early history of life

M. J. Reiss (✉)
Institute of Education, University College London, 20 Bedford Way, London WC1H 0AL, UK
e-mail: m.reiss@ucl.ac.uk

U. Harms
IPN - Leibniz Institute for Science and Mathematics Education, Biology Education, Olshausenstr. 62, 24118 Kiel, Germany
e-mail: harms@ipn.uni-kiel.de

 
U. Harms and M. J. Reiss (eds.), *Evolution Education Re-considered*,
https://doi.org/10.1007/978-3-030-14698-6_18

(some of the 'six days' in *Genesis*) which runs completely counter to the aeons that evolution entails—life usually being thought to have evolved on our planet over some 3.8 billion years, timescales that differ by a factor of more than $10^{11}$. Most theologians have long argued that the scriptures should not be read in this literal way. Nevertheless, many religious believers do so read them. Unsurprisingly, faced with a choice between believing what they are told at home and sometimes in their places of worship versus what they are told in their school science lessons, many youngsters ignore or actively reject what they are told at school, thus hampering their learning about evolution (Reiss, 2009).

There is an additional cultural reason why evolution may be rejected, though the literature about this is much sparser. That is because some of the key notions of evolution—that the universe may not have some pre-determined aim, that chance has play a major role in our being here, and so on–can cause existential anxieties (Tracy, Hart, & Martens, 2011; Newall, 2017). Faced with these, a learner may feel safer consciously or unconsciously pushing evolutionary ideas to the back of their mind.

So, evolution is a particularly difficult concept for learners, for a number of reasons. In addition, as we reference in the Chapter 'The Present Status of Evolution Education', school students often have teachers who themselves are not especially well-informed about the subject (and see Nehm, 2018). In the rest of this chapter, we therefore do three things. First, we look at what the various chapters in this book have to say about successful teaching for understanding about evolution; in fact, most of the interventions reported in this book attempt to enhance learners' understanding of evolution. Secondly, we look at what the two chapters in this book that substantively examine attitudes have to say about successful teaching for attitude change in relation to evolution. Finally, we pull together what we feel both this book and the existing literature tell us about how evolution education can be undertaken successfully.

## 2 Successful Teaching for Understanding About Evolution

In the Chapter 'Evidence for the Success of a Quantitative Assessment Instrument for Teaching Evolution in Primary Schools in England', Loredana Buchan, Momna Hejmadi and Laurence Hurst note that the dearth of experimental evidence for what 'works' in evolution education is particularly acute for primary-aged children. Related to this is the question of whether abstract concepts of genetics and evolution can even be taught to primary children, given the complexities and abstract nature of these topics. At present, children are rarely taught about genetics and DNA until they are around 15 years of age. However, they are exposed to these concepts through the media, comics, games and films and children as young as five are able to grasp some ideas about genetics and natural selection given the correct type of instruction (Legare, Lane, & Evans, 2013). In the intervention, upper primary pupils in England were taught four schemes of work—variation, natural selection/microevolution, geological time and macroevolution—with each scheme of work being taught for four lessons. Enquiry-based learning was employed.

Buchan et al. used a multiple choice assessment instrument, developed by selecting items from the AAAS science assessment website based on the research of Flanagan and Roseman (2011) and then adapting thee to reduce reading difficulty and cognitive load. Pupils were assessed at three different time points: pre-teaching to establish a priori knowledge, immediately after teaching to establish changes in understanding due to the teaching programme and three to six months later to evaluate retention. Gains in pupil knowledge were found both immediately after teaching and, though to a lesser degree, three to six months later. Interestingly, while teacher acceptance of evolution had a significant effect on class performance, none of teacher understanding of evolution, religiousness, highest biology qualification, formal evolution education, gender or years of experience did.

Another study that employed inquiry-based science education (IBSE) learning was that presented by Alexandra Buck, Sofoklis Sotiriou and Franz Bogner in the Chapter 'Bridging the Gap Towards Flying: *Archaeopteryx* as a Unique Evolutionary Tool to Inquiry-Based Learning'. Their study applied an educational module to sixth graders (11 years old) with no pre-knowledge of evolution who were given the opportunity to explore evolution in an unconstrained, playful and creative learning-by-doing-way based on IBSE principles. Students rotated round workstations that were structured into three sections: (i) arts in science with a high-quality, life-sized *Archaeopteryx* fossil replica; (ii) multimedia; and (iii) hands-on experiments. Multimedia tools employed wildlife documentary videos and virtual flight simulation; hands-on stations included feathers, dinosaur bones and stuffed birds (seagull, hawk and blackbird) as demonstration objects; for arts in science, collaborative handicraft artwork with natural fossils and paper flight modelling was applied. Students learned about the evolutionary link between birds and dinosaurs by discovering distinct skeleton features through direct hands-on experience with the *Archaeopteryx* fossil replica. While experimenting with phenomena such as fossilisation or gliding with thermal uplift, students slipped into the role of a 'science researcher' following the creative and multisensory approach of scientific thinking. Considerable knowledge gain was found after participation in the module.

In the Chapter 'Developmental Progression in Learning About Evolution in the 5–14 Age Range in England', Terry Russell and Linda McGuigan provide an overview of their research into the teaching and learning of evolution across the 5–14 age range in England. Their work looks at the five interrelated sub-domains of 'Deep time', 'Fossils', 'Variation', 'Inheritance' and 'Macroevolution'. One finding was that selective breeding, the deliberate management of heritable features for transmission to offspring, proved to be more accessible to younger children than the process of natural selection. As Russell and McGuigan point out, this was perhaps because the outcomes of selective breeding are observable over much shorter timescales; selective breeding is also controlled rather than the result of trial and error. A second finding was that introducing cladograms—particularly when these were accompanied by timelines and modelled by actual bits of a tree showing branches—helped the pupils to understand macroevolution.

More generally, the work by Russell and McGuigan shows the value of (i) a metacognitive approach (e.g. Schraw, Crippen, and Hartley, 2006)—pupils were explicitly invited to think about their own and others' thinking, (ii) multimodality (e.g. Tang, Delgardo, and Moje, 2014)—manifest in the use of alternative formats to encapsulate ideas and (iii) argumentation (e.g. Driver, Newton, and Osborne 2000). Of course, the literature about their benefits of all three of these for learning is extensive—but it is good to see these approaches used with quite young learners and shown to be of value in promoting learning about evolution.

Recent years have shown an increase in the attention paid within science education to learning progression (e.g. Duncan & Rivet, 2013). In the Chapter 'Teaching Evolution Along a Learning Progression: An Austrian Attempt with a Focus on Selection', Martin Scheuch, Jaqueline Scheibstock, Heidemarie Amon and Helene Bauer start from the observation that in the Austrian state curriculum, the topic of evolution is only mentioned in grade 7 and in grade 12. Accordingly, they developed a learning progression for grades 8, 9 and 10. Unsurprisingly, interviews with students who had received these grade 8, 9 and 10 lessons showed that they had benefitted from them in terms of their understanding of a number of key evolutionary concepts. Scheibstock et al. argue that a curriculum which addresses topics within evolution each year provides stepping stones for conceptual learning for change from everyday conceptions to more scientific ones; in other words, conceptual reconstruction is facilitated.

In the Chapter 'Examining Teaching Assistants' (TA) Experiences Facilitating Traditional Versus Active Learning-Based Tree-Thinking Curricula: TA Perceptions, Student Outcomes, and Implications for Teaching and Learning About Evolution', Yi Kong, Nancy Pelaez, Trevor Anderson and Jeffrey Olimpo examine the benefits of an innovative curriculum based on the use of evolutionary trees for USA undergraduates. This curriculum explicitly engaged students in exercises to assist them in developing an understanding of the chronology depicted with evolutionary trees; it proved to enhance student learning. Interviews with the students' graduate teaching assistants helped to uncover the reasons for this. First, the teaching assistants felt that the new curriculum succeeded in terms of the specific activities it required of students (e.g. students constructed evolutionary trees using data sources collected from organisms to understand different methods used to build trees). Secondly, the new curriculum resulted in students showing more enthusiasm and being more engaged with classroom activities. Thirdly, some of the teaching assistants maintained that their own understanding had improved as a result of the new curriculum.

Context-based learning in science education has long had its advocates (e.g. Campbell et al., 1994) and in the Chapter 'Utility of Context-Based Learning to Influence Teacher Understanding of Evolution and Genetics Concepts Related to Food Security Issues in East Africa', Tim Goodale discusses the use of the agricultural crop *Cassava* and issues of food access and security in East Africa as the context for teaching evolution and genetics. Working with beginning science teachers in the USA, Goodale produced a voluntary six-hour professional development programme that consisted of six activities: (i) learning how to mimic the genetic mapping of the cassava plant by utilising classroom exercises that demonstrate how scientists isolate DNA, determine sequencing of codons and then determine the eventual mapping of genes and their expression of traits; (ii) identifying mutation types associated with vector transmission and subsequent evolutionary changes that lead to cassava diseases; (iii) utilising gel electrophoresis to identify infected plants; and (iv) utilising argumentation to propose long-term solutions involving gene therapy, GMOs and large-scale vector control. Participants also utilised a gel electrophoresis lab and identified an infected/sick plant sample compared to a healthy sample. Lastly, using guiding principles of scientific argumentation, participants had to propose a one-page solution to the food security issue surrounding cassava mosaic disease and discuss the relevant pros and cons.

Evaluation of the programme showed that, while there were successes, teachers had deficiencies in content knowledge related to evolution and, for religious reasons, not infrequently had mixed feelings as to their overall acceptance of key concepts in evolution. In contrast, the same participants exhibited strong content knowledge related to genetics and had little conflict with accepting the science related to the topic.

In the Chapter 'Using Human Examples to Teach Evolution to High School Students: Increasing Understanding and Decreasing Cognitive Biases and Misconceptions', Briana Pobiner, William Watson, Paul Beardsley and Connie Bertka examine the impact of using a constructivist, guided-inquiry pedagogical approach using human evolution case studies to teach high school students in the USA about natural selection in an attempt to better understand the most promising approaches to support teachers in helping students learn about evolution in general, and natural selection in particular, and overcome common cognitive biases and misconceptions. They also investigated the effect of these curricular materials when used in tandem—or not—with teaching strategies that explicitly acknowledge the cultural controversy around evolution.

Pobiner et al. had three mini-units that focused on natural selection in modern humans: one on adaptation to altitude; one on the evolution of human skin colour; and one to do with malarial parasite resistance to antimalarial drugs. They also produced a 'Cultural and Religious Sensitivity (CRS) Teaching Strategies Resource'. The purpose of this was to encourage and equip high school teachers to help students manage any tension they may experience between a scientific study of evolution and their religious and cultural beliefs, and to create a classroom environment that supports both an increased understanding of the nature of science and a scientific understanding of evolution. The resource was not meant to resolve any conflict some students see between their personal worldviews and the scientific account of human evolution, but to help create a non-threatening classroom environment.

Pobiner et al. found only limited evidence that using humans as opposed to mouse contexts helped students learn more. Intriguingly, the CRS activities seemed to pave the way for greater increases in understanding and decreases in cognitive biases and misconceptions in the mouse context, but not the human context. Perhaps, for some students, paying attention to cultural and religious sensitivities can help them learn about evolution in the relatively non-threatening context of mice, whereas human evolution is still too threatening.

There is a growing emphasis on the use of models in science education as these have been shown to lead to cognitive gains in several science disciplines (e.g. Malone, Schuchardt, and Schunn, 2018). In the Chapter 'Models and Modeling in Evolution', Kathy Malone, Anita Schuchardt and Zakee Sabree discuss an evolution unit grounded in the use of modelling and its effects on learning in evolution and attitudes towards science in general. Models can be understood as explicit representations (e.g. graphs, mathematical equations, pictures and physical models) that students use to make science phenomena more understandable and predictable.

Malone et al. developed an activity in which students simulate the life and reproduction cycles of a lizard population with two traits of interest: mouthparts and skin colour. The lizards undertake 16 rounds (generations) that model feeding and reproducing, under conditions of (i) no selective pressure and (ii) selective pressure. The data collected over these multiple rounds are analysed to permit students to develop a model of what happens to a population's traits with and without selective pressure. Participants were 15- and 16-year-old students from the USA. Compared to control students (taught conventionally), on average, students who were taught using the modelling approach manifested greater conceptual understanding of natural selection, had fewer alternative conceptions and made greater use of multiple representations (e.g. graphs).

In the Chapter 'Cultural Diversity and Evolution: Looking for a Dialogical Teaching Perspective', Alma Adrianna Gómez Galindo, Alejandra García Franco, Leonardo Gonzáles Galli and José de la Cruz Torres Frías argue that evolution education has not sufficiently explored the cultural and contextual aspects related to learning. Gómez Galindo et al. point out that culturally relevant educational practices look to facilitate students' dialogue between their own ways of knowing and scientific ways of knowing (cf. Aikenhead, 2001). Such practices provide opportunities for greater focus on real-world issues that are relevant for students' lives, thus making science education a space where there are opportunities for personally significant experiences.

Gómez Galindo et al. worked with indigenous secondary students in the Mayan Highlands in Mexico. Given the importance of maize in Mexico, maize was chosen as the focal point for the teaching. In an initial, exploratory study, it soon became clear that the students were unable to explain the existence of different varieties of maize; they lacked the theoretical elements that would allow them to explain the origin of the diversity of maize from a biological perspective. In a subsequent study, students were therefore encouraged to make a comparative table of maize races, bring information from their communities around the selection of kernels, share experiences around sowing and establish any connections between specific varieties and meteorological conditions. It was clear that there was a strong emotional relationship of students to maize. Gómez Galindo et al. see their work as embodying an intercultural dialogic approach to evolution education that includes context and culture as central aspects in the analysis of learning difficulties and in the generation of teaching proposals.

In the Chapter 'Transforming a College Biology Course to Engage Students: Exploring Shifts in Evolution Knowledge and Mechanistic Reasoning', Lisa Kenyon, Emily Walter and William Romine begin by noting that active, evidence-based learning has significant advantages over traditional, lecture-based approaches (Freeman et al., 2014). Accordingly, they transformed an introductory biology course for undergraduates in the USA to a more practice-based learning environment, in which students constructed knowledge about evolution through explanation and argumentation. Kenyon et al. were particularly interested in enabling students to think about causal mechanisms (i.e. to engage in mechanistic reasoning). The reason for this was a presumption that engaging in these practices may promote a cognitive and social approach to evolution, thereby engaging affective and logic-driven pathways to students' evolution acceptance.

Kenyon et al. therefore designed a course in which multiple activities were integrated to emphasise empirical reasoning skills and experience with evolutionary phenomena. Students explored questions about the evolutionary origin of animals and plants, their morphology and physiology, and the ecological interactions between organisms and the ecosystems they inhabit. Something of the intention of the course is indicated by citing the one assessment item they used to solicit mechanistic reasoning:

> African elephants are known for their large tusks, which the animals use for digging and defense. These tusks are valuable to people because of their ivory, which can be used in jewelry and decorations. Poachers hunt and kill elephants for their tusks, often before elephants are able to reproduce. Some elephants never grow tusks. In 1930, 1% of adult elephants didn't have tusks. In some areas today, up to 38% of adult elephants don't have tusks.
>
> How and why is the percentage of elephants without tusks higher today than it was in 1930?

Some students manifested improvements in mechanistic reasoning and gains in knowledge about natural selection (as measured by another assessment instrument). Overall, Kenyon et al. concluded that the intervention helped to promote sense making, evaluating, argumentation and consensus building while providing meaningful learning about natural selection.

In the Chapter 'Improving Student Understanding of Randomness and Probability to Support Learning About Evolution', Ute Harms and Daniela Fiedler begin with the premise that the conceptual difficulties that students face in learning about evolution are strongly related to the difficulties that students face in understanding abstract concepts like randomness and probability (Tibell & Harms, 2017). They worked with German university students studying for a Master of Education degree. Students were divided into three different treatment groups with one group engaging in a simple computer activity to simulate mutations, another engaging in a text-based study to help them better understand what is meant by 'randomness' by reference to both everyday and scientific events, and the third group undertaking mathematical tasks in the field of probability. Analysis revealed that understanding of randomness and probability was significantly positively related to student understanding of evolution, quantitative abilities and the student's last grade in mathematics. It was also found that of the three treatment groups, the mathematical tasks proved most helpful in enhancing learning about evolution.

Most science education research takes place in the formal setting of schools yet much, possibly most, learning about science takes place in informal settings (Braund & Reiss, 2006; Bell, Lewenstein, Shouse, and Feder, 2009). In the Chapter 'Evolution

Learning and Creationism: Thinking in Informal Learning Environments', Jorge Groß, Kerstin Kremer and Julia Arnold present two case studies in Germany that research the interplay between creationist conceptions and evolution understanding in informal learning environments. In the first case study, an exhibition was designed, in collaboration with Ulrich Kattmann and Annette Scheersoi, to show visitors to an IKEA store evolution in the context of daily life: everyday conceptions regarding evolution were used as starting points for the presentation of the biological topics, and assorted media (interactive media, short texts, original artefacts, etc.) offered numerous entry points to the exhibition's theme.

In the study by Groß et al., semi-structured interviews were undertaken with students in the 10–18 year age range. The results showed that the intervention succeeded in awakening the visitors' interest and clarifying misunderstandings about the theory of evolution. However, there was no conceptual change in the argumentation of students with creationist beliefs. Accordingly, the second case study focused on a guided tour for fifth grade pupils through a natural history museum's permanent exhibition with an associated workshop on fossils to see how far knowledge gain can affect creationist beliefs in novice learners. It transpired that there were both knowledge gains on evolution and natural history initiated by the education programme and a reduction in the extent of creationist beliefs.

In the final chapter that reports the results of an intervention, Jo Nicholl and Paul Davies in the Chapter 'Participating in an Object-Based Learning Project to Support the Teaching and Learning of Biological Evolution: A Case Study at the Grant Museum of Zoology' look at the use of object-based learning (Chatterjee, 2011) in a small natural history museum to support teaching and learning about biological evolution. A series of workshops were conducted at the Grant Museum of Zoology in London where pre-service teachers (graduate students) were given the opportunity to handle and touch real specimens. The workshops were found to increase pre-service teachers' knowledge of biological evolution as well as improve their confidence about what they already knew. The use of the objects encouraged them to make appropriate observations, ask questions and engage in discussions that questioned their understanding of biological evolution. In addition, the pre-service teachers identified and valued a range of pedagogies associated with object-based learning that could be applied either within an informal museum setting or a more formal classroom setting.

## 3 Successful Teaching for Attitude Change in Relation to Evolution

In the Chapter 'Learning About Evolution and Teaching in a Cross-Curricular Teacher Education Session: Findings from a Small-Scale Study with Pre-service Primary School Teachers', Berry Billingsley, Manzoorul Abedin, Keith Chappell and Chris Hatcher explore pre-service primary teachers' perceptions in England of a cross-curricular teaching session. The intervention in their study was a cross-curricular session, designed to provide pre-service primary teachers with a space in which they could explore the relationships between science and religion prior to a follow-up session that focused on developing pedagogies and subject knowledge relating to science. The pre-service teachers were initially asked to give their perceptions of how the media typically describe the relationship between science and religion. The discussion turned then to the notion that a school teacher can resist and critique perspectives that appear in the media; participants then examined and shared examples of ways that the relationship is described in scholarship. The session then drew participants' attention to particular areas of confusion or gaps that are common in survey responses and sought to address these. Billingsley et al. found that the session moved many of the pre-service teachers away from a perception of necessary conflict between science and religion.

In the Chapter 'Overcoming Motivational Barriers to Understanding and Accepting Evolution Through Gameful Learning', David Owens tested gameful, inquiry-based learning intervention with the intention of enhancing motivation to learn in the context of plant evolutionary life history. Owens points out that learning through inquiry typifies scientific thinking; a classroom environment rooted in inquiry-based learning allots time for learners to make discoveries, understand new ideas, develop questions that require significant cognitive engagement to be answered, and critique those questions and answers using evidence—all of which should be conducive to conceptual change (Pugh, Linnenbrink-Garcia, Koskey, Stewart and Manzey, 2010). Given the popularity of video games, it makes sense to research their efficacy in science education. Owens points out that two common elements of gamification are the leaderboard and repeat-testing. *Leaderboards* advertise each player's points and rank, including badges that highlight achievements or accomplishments, and promote the demonstration of competence and enable comparisons among students. The *repeat-testing* element of games enables individuals to repeat levels with minimal risk until satisfied with the competence they have developed.

Owens worked with USA undergraduates. He found that some individuals did not like the competitive nature brought on by the leaderboard. Nevertheless, both the leaderboard and the repeat-testing proved overall to enhance student motivation to learn about biology in the context of plant evolutionary life history. Participants indicated that the leaderboard increased interest, engagement and motivation to prepare, while repeat-testing reduced test anxiety and made the material easier to learn.

## 4 Lessons Learnt

The chapters in this book demonstrate that understanding of evolution is aided by the sorts of pedagogical approaches that are known to work well in other areas of science education and beyond. In particular, there is evidence that the appropriate use of teaching for metacognition, multimodal approaches, argumentation, inquiry-based science education, reinforcement of learning, context-based learning, models, intercultural dialogic approach and object-based learning can all help promote learning about evolution. To this list, we can add that teacher expertise is of great importance. At the same, it is notable that almost all studies use only one of these pedagogical approaches. We know almost nothing about the extent to which and how the use of more than one of these approaches might benefit learning.

Then there is much in the chapters in this book about issues that are particular to evolution education. In the Chapter 'Inequitable Foundations? Educational Equality in Evolution', Jaimie Miller-Friedmann, Susan Sunbury and Philip Sadler found that the best predictor for competence in evolution understanding among USA middle and high school students was a general comprehension of life science. In a number of other chapters (e.g. Terry Russell and Linda McGuigan in the Chapter 'Developmental Progression in Learning About Evolution in the 5–14 Age Range in England' and Ute Harms and Daniela Fiedler in the Chapter 'Improving Student Understanding of Randomness and Probability to Support Learning About Evolution'), specific approaches to teaching particular aspects of evolution are trialled and shown to help promote learning. Of course, there is much that still remains to be done but there are reasons for the beginnings of optimism as the biology education community is starting to build up a corpus of knowledge about what works well when teaching specific aspects of evolution.

One feature that is distinctive to evolution education within biology education is that a not inconsiderable number of learners come from backgrounds where at least some aspects of the theory of evolution (aspects of macroevolution rather than microevolution) are actively rejected on the basis of a supposed clash with religion. As is widely acknowledged, it is not easy for teaching to change the views of those who old creationist beliefs (e.g. Long, 2011). However, research suggests that careful and respectful teaching about evolution can lead to students who initially reject the theory of evolution becoming more likely to accept at least some aspects of it. Winslow, Staver, and Scharmann (2011) found that undergraduates who had been raised by their families to believe in creationism could come to accept evolution by evaluating the evidence for it, negotiating the meanings of *Genesis*, recognising evolution as a non-salvation issue and observing their teachers as Christian role models who accepted evolution. In the Chapter 'Evolution Learning and Creationism: Thinking in Informal Learning Environments', Groß et al. showed that knowledge gains about evolution led to a reduction in the extent of creationist beliefs.

This is an appropriate point to mention that there is something of a controversy in the literature about what should be the precise aim of teaching evolution to creationists and others who do not accept it (Hermann, 2008; Reiss, 2011; Williams, 2015). A common view is that evolution educators should aim to get such learners to come to *accept* the theory of evolution; another possibility is that the aim should be to get such learners to *understand* the theory of evolution—leaving it up them whether or not they accept it. Fortunately for those who accept the theory of evolution, as the evidence cited above indicates, the difference between these two views, while philosophically important, may be smaller than is sometimes supposed as far as classroom practice goes. Educators (whether in a school, a museum or elsewhere) can simply do their best to convey the evidence for evolution and ensure an understanding of evolution, while being respectful of their learners.

It should not be surprising that successful evolution education is not straightforward for teachers and others to teach or students and others to learn. One way of interpreting the move from a lack of understanding or acceptance of evolutionary theory to an understanding or acceptance of it is to see this move as an instance of conceptual change (Reiss, 2017). In his book, *The Examined Life*, the psychoanalyst Grosz (2014) relates the story of Marissa Panigrosso who was on the 98th floor of the World Trade Centre South Tower on 11 September 2001, talking with two of her co-workers, when the first plane hit the North Tower. The fire alarm went off and a wave of anxiety swept through the office. Marissa Panigrosso did not stop to turn her computer off or even to pick up her purse. She walked to the nearest emergency exit and left the building. The two women with whom she was talking did not leave. In fact, many people in her office ignored the firm alarm—despite what they could see happening in the North Tower. Some of her colleagues went into a meeting. A friend of Marissa turned back after walking down several flights of stairs saying 'I have to go back for my baby pictures' (Grosz, 2014, p. 122). This friend lost her life, as did the two women with whom Marissa Panigrosso was talking and the colleagues who went into a meeting. Marisso survived. The conclusion that Grosz draws from this is that change can be difficult: 'Committing ourselves to a small change, even one that is unmistakably in our best interests, is often more frightening than ignoring a dangerous situation' (p. 123).

We conclude by noting that there is much that still remains to be done in researching evolution education. While the chapters in this book show that there are an increasing number of research-based interventions in evolution education that are leading to greater conceptual understanding, the number of such interventions is small and the curriculum and pedagogical approaches that these interventions draw on are rarely ones that are robustly established. Furthermore, we still lack consensus about the best instruments for measuring cognitive gains and there is a paucity of longitudinal studies. Overall, the field of evolution education, while perhaps no longer in its infancy, still has a long way to go before it matures.

## References

Aikenhead, G. (2001). Integrating western and aboriginal sciences: Cross-cultural science teaching. *Research in Science Education, 31,* 337–355.

Bell, P., Lewenstein, B., Shouse, A., & Feder, M. A. (2009). *Learning science in informal environments: People, places and pursuits.* Washington, DC: National Academies Press.

Braund, M., & Reiss, M. J. (2006). Towards a more authentic science curriculum: The contribution of out-of-school learning. *International Journal of Science Education, 28*(12), 1373–1388.

Campbell, B., Lazonby, J., Millar, R., Nicolson, P., Ramsden, J., & Waddington, D. (1994). Science: The Salters' approach—A case study of the process of large scale curriculum development. *Science Education, 78*(5), 415–447.

Chatterjee, H. J. (2011). Object-based learning in higher education: The pedagogical power of museums. In *International Committee for University Museums and Collections (UMAC) Proceedings* (Vol. 3).

Driver, R., Newton, P., & Osborne, J. (2000). Establishing the norms of scientific argumentation in classrooms. *Science Education, 84*(3), 287–312.

Duncan, R. G., & Rivet, A. E. (2013). Science learning progressions. *Science, 339*(6118), 396–397.

Flanagan, J. C., & Roseman, J. E. (2011). *Assessing middle and high school students' understanding of evolution with standards-based items*. Paper presented at the National Association of Research in Science Teaching Annual Meeting, Orlando, FL.

Freeman, S., Eddy, S. L., McDonough, M., Smith, M. K., Okoroafor, N., Jordt, H., & Wenderoth, M. P. (2014). Active learning increases student performance in science, engineering, and mathematics. *Proceedings of the National Academy of Sciences, 111*, 8410–8415.

Grosz, S. (2014). *The examined life: How we lose and find ourselves*. London: Vintage.

Hermann, R. S. (2008). Evolution as a controversial issue: A review of instructional approaches. *Science & Education, 17*, 1011–1032.

Legare, C. H., Lane, J. D., & Evans, E. M. (2013). Anthropomorphizing science: How does it affect the development of evolutionary concepts? *Merrill-Palmer Quarterly, 59*(2), 168–197.

Long, D. E. (2011). *Evolution and religion in American education: An ethnography*. Dordrecht: Springer.

Malone, K. L., Schuchardt, A. M., & Schunn, C. D. (2018). Improving conceptual understanding and representation skills through Excel-Based modeling. *Journal of Science Education and Technology, 27*(1), 30–44.

Nehm, R. H. (2018). Evolution. In K. Kampourakis & M. J. Reiss (Eds.), *Teaching biology in schools: Global research, issues and trends* (pp. 164–177). New York: Routledge.

Newall, E. (2017). Evolution, insight and truth? *School Science Review, 99*(367), 61–66.

Pugh, K. J., Linnenbrink-Garcia, L., Koskey, K. L., Stewart, V. C., & Manzey, C. (2010). Motivation, learning, and transformative experience: A study of deep engagement in science. *Science Education, 94*, 1–28.

Reiss, M. J. (2009). The relationship between evolutionary biology and religion. *Evolution, 63*, 1934–1941.

Reiss, M. J. (2011). How should creationism and intelligent design be dealt with in the classroom? *Journal of Philosophy of Education, 45*, 399–415.

Reiss, M. J. (2017). Teaching the theory of evolution in informal settings to those who are uncomfortable with it. In P. G. Patrick (Ed.), *Preparing informal science educators: Perspectives from science communication and education* (pp. 495–507). Cham: Springer.

Schraw, G., Crippen, K. J., & Hartley, K. (2006). Promoting self-regulation in science education: Metacognition as part of a broader perspective on learning. *Research in Science Education, 36*(1–2), 111–139.

Tang, K.-S., Delgardo, C., & Moje, E. B. (2014). An integrative framework for the analysis of multiple and multimodal representations for meaning-making in science education. *Science Education, 98*(2), 305–326.

Tibell, L. A. E., & Harms, U. (2017). Biological principles and threshold concepts for understanding natural selection—Implications for developing visualizations as a pedagogic tool. *Science & Education (SCED), 26*(7), 953–973.

Tracy, J. L., Hart, J., & Martens, J. P. (2011). Death and science: The existential underpinnings of belief in intelligent design and discomfort with evolution. *PLoS ONE, 6*(3), e17349. https://doi.org/10.1371/journal.pone.0017349.

Williams, J. D. (2015). Evolution versus creationism: A matter of acceptance versus belief. *Journal of Biological Education, 49*, 322–333.

Winslow, M. W., Staver, J. R., & Scharmann, L. C. (2011). Evolution and personal religious belief: Christian university biology-related majors' search for reconciliation. *Journal of Research in Science Teaching, 48*, 1026–1049.

**Michael J. Reiss** is Professor of Science Education at UCL Institute of Education, University College London, Visiting Professor at the Universities of York and Kiel and the Royal Veterinary College, Honorary Fellow of the British Science Association, Docent at the University of Helsinki and a Fellow of the Academy of Social Sciences. After undertaking a Ph.D. and post-doctoral research in evolutionary biology and population genetics, he trained to be a science teacher and taught in schools for five years before returning to higher education. The former Director of Education at the Royal Society, his academic interests are in science education, bioethics and sex education and he has published widely on issues to do with creationism in schools.

**Ute Harms** is Director at the IPN - Leibniz Institute for science and mathematics education, Full Professor for Biology Education at the University of Kiel (Germany) since 2007, and Fellow of the Royal Society of Biology (Great Britain). She has a Ph.D. in cell biology and has worked as a high school teacher for several years. In 2000, she got her first Professorship for Biology Education at the Ludwig-Maximilians-University in Munich (Germany). From 2006 to 2007, she held a chair in Biology Education at the University of Bremen. Her main research interests are conceptual learning in biology and in science, focusing on evolution and energy, biology teacher education, biology-related competitions and transfer of contemporary topics in the life sciences to the public.

# Index




U. Harms and M. J. Reiss (eds.), *Evolution Education Re-considered*,
https://doi.org/10.1007/978-3-030-14698-6

Printed by Printforce, the Netherlands